Pragmatic Microservices with C# and Azure

Build, deploy, and scale microservices efficiently
to meet modern software demands

Christian Nagel

Pragmatic Microservices with C# and Azure

Publishing Product Manager: Kunal Sawant

Book Project Manager: Prajakta Naik

Senior Editor: Rounak Kulkarni

Technical Editor: Jubit Pincy

Copy Editor: Safis Editing

Proofreader: Rounak Kulkarni

Indexer: Hemangini Bari

Production Designer: Alishon Mendonca

Senior Developer Relations Marketing Executive: Shrinidhi Monaharan

Business Development Executive: Samriddhi Murarka

First published: May 2024

Production reference: 1240524

Published by Packt Publishing Ltd.

Grosvenor House

11 St Paul's Square

Birmingham

B3 1RB, UK

ISBN 978-1-83508-829-6

www.packtpub.com

To my mother, Margarete Nagel, and the memory of my father, Ernst, for always supporting me when help was needed. To my wife, Angela, and my children, Stephanie, Matthias, and Katharina – I love you all! Thank you for being here!

– Christian Nagel

Contributors

About the author

Christian Nagel is a Microsoft MVP for Microsoft Azure and Development Technologies, software architect, and veteran developer who has been building solutions with .NET technologies since 2000 and working with UNIX and OpenVMS since before that. He has authored many acclaimed .NET books, and speaks at international conferences such as TechEd, TechDays, Thrive, and BASTA! Christian is a Microsoft Certified Trainer, DevOps Engineer Expert, and Azure Developer Associate. He's the founder of CN innovation, a company offering training, consulting, and development. Even after many years in software development, Christian still loves learning, using new technologies, and teaching others how to use them. Using his profound knowledge of Microsoft technologies, he has written various books. Contact him via his website at https://www.cninnovation.com, and follow his posts on **X** (formerly Twitter) at @christiannagel.

This book was a great journey that started before the first public preview of .NET Aspire. I would like to thank the .NET Aspire team for their fast fixes and their support; special thanks to David Fowler with whom I had interesting discussions. I would like to thank my reviewers: István Novák, who has already reviewed several of my books and always helps remind me when I miss something that helps you – the reader – to better follow the content, and Brady Gaster, for his great inspirations and insights on Microsoft technologies, and, if he can't help, he always has contacts who can. I also thank Sebastian Szvetecz, who is now part of CN innovation and has great inputs to the Codebreaker repository. Last but not least, I thank the team at Packt for making this book a reality.

About the reviewers

Brady Gaster works on the .NET tools team, where he collaborates with product teams such as ASP.NET, Azure Container Apps, Functions, App Service, and the Azure SDK to help .NET developers using ASP.NET, SignalR, Orleans, and .NET Aspire have a great experience building apps with .NET and Azure. When Brady's not learning with (or from) his two sons or hacking on demos and samples, he's probably in his home music studio making music with various analog synthesizers and drum machines, burning calories by bouncing up and down as he spins drum and bass or dubstep, or finding ways to combine some or all of the above into a new mashup with a cacophony of cables.

Istvan Novak is a freelance technology consultant and commonsense coach. In the last 30 years, he participated in over 50 enterprise software development projects.

In 2002, he co-authored the first Hungarian book on .NET development. In 2007, he was awarded the Microsoft MVP title, holding it for 16 years. In 2011, he became a Microsoft Regional Director for six years.

Istvan holds a master's degree from the Technical University of Budapest, Hungary, and a doctoral degree in software technology. He is a passionate scuba diver. You may have a good chance of meeting him underwater in the Red Sea in any season of the year.

Table of Contents

3

Writing Data to Relational and NoSQL Databases   61

4

Creating Libraries for Client Applications   99

Part 2: Hosting and Deploying

5

Containerization of Microservices   123

6

Microsoft Azure for Hosting Applications   151

7

Flexible Configurations   183

8

CI/CD – Publishing with GitHub Actions   211

9

Authentication and Authorization with Services and Clients   247

Part 3: Troubleshooting and Scaling

10

All About Testing the Solution   277

11

Logging and Monitoring   303

12

Scaling Services   343

Part 4: More communication options

13

14

15

16

Running Applications On-Premises and in the Cloud   453

Preface

.NET Aspire is a new, opinionated framework offering tools and libraries for creating microservices with .NET, no matter whether they should run on-premises, on Microsoft Azure, or in any other cloud environment. In this book, you will learn how to get the most out of .NET Aspire when you build your solutions.

Creating ASP.NET Core minimal APIs (an easy and fast option for creating REST services) is just a small part of creating applications using a microservices-based architecture. This book covers all the different aspects that are needed for building a successful solution. Accessing databases, both relational as well as NoSQL; using Docker and deploying Docker images; automatic deployment with GitHub actions; monitoring the solution with logging, metrics data, and distributed tracing; creating unit tests, integration tests, and load tests; automatically publishing the solution to different environments; and using binary, real-time, and asynchronous communication – all of this is covered in the book.

Through the code provided in this book, you'll work on a backend solution to run a cool game. From *Chapter 2*, you'll already have usable and testable functionality, and it will be enhanced chapter by chapter to cover all the important aspects relating to microservices. If you don't want to work through each chapter in sequence, we have provided code that you can start with for each chapter.

The application can run deployed to Microsoft Azure, using several Azure services such as Azure Container Apps, Container Registry, Cosmos DB, App Configuration, Key Vault, Redis, and SignalR Service. It can also run in an on-premises environment on a Kubernetes cluster using Kafka, Redis, and other resources.

By the end of this book, you'll have become confident in implementing a stable, performant, and scalable solution and using various Azure services that are great for hosting such a service-based solution. While the solution of the book is a game, the knowledge learned will help you to create any business-related service architecture.

Who this book is for

The book is for developers and software architects who are familiar with C# and .NET, have basic knowledge of Microsoft Azure, and want to know about all the aspects required for creating modern microservices with .NET and Microsoft Azure.

What this book covers

Chapter 1, Introduction to .NET Aspire and Microservices, gives you an introduction to **.NET Aspire**, as well as tools and libraries that can be of great help when creating microservices. You'll start an initial .NET Aspire project, and we will look at what it consists of and the parts of .NET Aspire you can take advantage of. You'll see the services that the Codebreaker application (this is the solution we are going to build in all the chapters of the book) is made up of and learn about the services that are used with Microsoft Azure.

Chapter 2, Minimal APIs – Creating REST Services, acts as the kick-off for creating the Codebreaker application. You will learn how to use the **ASP.NET Core minimal API** technology to efficiently create REST services, use OpenAPI to describe the services, and test the services with HTTP files.

Chapter 3, Writing Data to Relational and NoSQL Databases, uses just the in-memory storage in *Chapter 2* and adds database storage using **Azure SQL** and **Azure Cosmos DB**, comparing relational to non-relational databases and using **EF Core** with both variants.

Chapter 4, Creating Libraries for Client Applications, adds client libraries to access the services, with one variant using an HTTP client factory and another making use of **Kiota** to automatically generate code for the client.

Chapter 5, Containerization of Microservices, dives into all the important concepts of **Docker** and how to create Docker images from the services created so far. You will learn the concepts of Docker before using the .NET CLI to create Docker images. A **.NET native AOT** version of a service will be created that allows for creating Docker images without a .NET runtime, just native code.

Chapter 6, Microsoft Azure for Hosting Applications, as Docker images have been created in the previous chapter, will now cover how to publish the application to an **Azure Container Apps** environment. Before that, important concepts of Azure are covered. Then, Azure resources will be created with the help of the Azure Developer CLI and .NET Aspire.

Chapter 7, Flexible Configurations, dives into the .NET configuration. You'll understand configuration providers with .NET, learn how configuration applies to the app model of .NET Aspire, add configurations and secrets with Azure Container Apps, and integrate **Azure App Configuration** and **Azure Key Vault**. For easier access without the need to store secrets, **Azure managed identities** are covered here as well.

Chapter 8, CI/CD – Publishing with GitHub Actions, builds on the fact that continuous integration and continuous delivery are important aspects of a microservices solution. This chapter covers how to automatically build and test applications with **GitHub actions** and how to automatically update solutions running on Microsoft Azure. To support modern deployment patterns, feature flags available with Azure App Configuration are integrated in this chapter.

Chapter 9, Authentication and Authorization with Service and Clients, covers two versions to authenticate and authorize applications and users: integration of **Azure Active Directory B2C** for the cloud version of the Codebreaker solution and **ASP.NET Core identities** for the on-premises solution. To avoid dealing with authentication with every service, a gateway using **YARP** is created.

Chapter 10, All About Testing the Solution, states that no change should break the application and errors should be detected as early as possible. In this chapter, you will learn about creating unit tests, integration tests with .NET Aspire (which makes testing a lot simpler), and using **Playwright** for end-to-end tests.

Chapter 11, Logging and Monitoring, delves into what is going on in the Codebreaker application. Memory leaks should be detected early. During development, we should look into the details of how the application communicates. This chapter covers efficient high-performance logging, writing custom metrics data, and distributed tracing – including the coverage of **OpenTelemetry** and how this integrates with .NET Aspire. In this chapter, we use **Prometheus** and **Grafana** for the on-premises solution and **Azure Application Insights** and **Azure Log Analytics** with the cloud.

Chapter 12, Scaling Services, delves into scaling services, which is one of the important reasons for using a microservices architecture. Using **Azure Load Testing**, we create a huge load on the main service of the application and find out what the bottlenecks are, decide between scaling up and scaling out, and add caching to increase performance using Redis.

Chapter 13, Real-Time Messaging with SignalR, covers informing clients in real time using **SignalR**. Using a REST API, a SignalR hub is invoked, passing real-time information about completed games, and the SignalR hub passes on this information to a group of clients. **Azure SignalR Service** is used to reduce the load from the services.

Chapter 14, gRPC for Binary Communication, increases the performance by changing communication to **gRPC** with service-to-service communication. You will learn how to create a Protocol Buffers definition, implement services and clients using this binary platform-independent communication, and how .NET service discovery and .NET Aspire can be used with gRPC.

Chapter 15, Asynchronous Communication with Messages and Events, deals with the fact that often answers are not required immediately after the request is sent – message queues and events come into play. Here, **Azure message queues**, **Azure Event Hubs**, and **Kafka** for an on-premises environment come into play to be used.

Chapter 16, Running Applications On-Premises and in the Cloud, discusses what's needed in a production environment when running on Azure, what the differences between production and development environments are, and how the Codebreaker application fulfills requirements relating to scalability, reliability, and security. Up to this chapter, the application was running either on the local development system or within an Azure Container Apps environment. In this chapter, the application is deployed to **Azure Kubernetes Service** and can be deployed in a similar way to an on-premises Kubernetes cluster using the **Aspir8** tool.

To get the most out of this book

You will need to know C# and .NET and have some fundamental knowledge about Microsoft Azure. The following tools and applications need to be installed:

Tool	Installation
Visual Studio 2022 (optional)	`https://visualstudio.microsoft.com/downloads/`
Visual Studio Code	`https://code.visualstudio.com/download`
Docker Desktop	`https://docs.docker.com/desktop/install/windows-install/`
	`https://docs.docker.com/desktop/install/mac-install/`
	`https://docs.docker.com/desktop/install/linux-install/`
Microsoft Azure subscription	`https://azure.microsoft.com/en-us/free/`
Azure CLI	`https://learn.microsoft.com/en-us/cli/azure/install-azure-cli`
Azure Developer CLI	`https://learn.microsoft.com/en-us/azure/developer/azure-developer-cli/install-azd`
Azure Cosmos DB emulator	`https://learn.microsoft.com/en-us/azure/cosmos-db/how-to-develop-emulator`
.NET Aspire	`https://learn.microsoft.com/en-us/dotnet/aspire/fundamentals/setup-tooling`

Visual Studio 2022 is the only software on this list that requires Windows. All other tools can be used on Windows, macOS, or Linux. On platforms other than Windows, you can use Visual Studio Code or other tools to work with .NET and C#.

To use .NET Aspire with Visual Studio, at least version 17.10.0 needs to be installed. You can use Visual Studio 2022 Community Edition.

Starting with *Chapter 6*, you need a Microsoft Azure subscription. You can activate Microsoft Azure for free at `https://azure.microsoft.com/free`, which gives you about USD 200 in Azure credit that's available for the first 30 days and several services that can be used for free for the time after. If you have a Visual Studio Professional or Enterprise subscription, you also get a free amount of Azure resources every month. You just need to activate this with your Visual Studio subscription: `https://visualstudio.microsoft.com/subscriptions/`.

If you are using the digital version of this book, we advise you to type the code yourself or access the code from the book's GitHub repository (a link is available in the next section). Doing so will help you avoid any potential errors related to the copying and pasting of code.

Download the example code files

You can download the example code files for this book from GitHub at `https://github.com/PacktPublishing/Pragmatic-Microservices-with-CSharp-and-Azure`.

The source code is using .NET 8. After .NET 9 is released, the source code in the main branch will be using .NET 9 with changes described in readme files. At that time, the .NET 8 source code will be available in the dotnet8 branch. During the time the code is updated to newer versions, you can check extra branches which contain changes.

We also have other code bundles from our rich catalog of books and videos available at `https://github.com/PacktPublishing/`. Check them out!

Conventions used

There are a number of text conventions used throughout this book.

`Code in text`: Indicates code words in text, database table names, folder names, filenames, file extensions, pathnames, dummy URLs, user input, and Twitter handles. Here is an example: "Mount the downloaded `WebStorm-10*.dmg` disk image file as another disk in your system."

A block of code is set as follows:

```
public static class LiveGamesEndpoints
{
  public static void MapLiveGamesEndpoints(this
    IEndpointRouteBuilder routes, ILogger logger)
  {
    var group = routes.MapGroup("/live")
```

When we wish to draw your attention to a particular part of a code block, the relevant lines or items are set in bold:

```
var builder = WebApplication.CreateBuilder(args);

builder.AddServiceDefaults();
builder.AddApplicationServices();
```

Any command-line input or output is written as follows:

```
dotnet new console -o LiveTestClient
```

Bold: Indicates a new term, an important word, or words that you see onscreen. For instance, words in menus or dialog boxes appear in **bold**. Here is an example: "Click the **Enable Live Trace** checkbox, and click to collect information on **ConnectivityLogs**, **MessagingLogs**, and **HttpRequestLogs**."

> **Tips or important notes**
> Appear like this.

Get in touch

Feedback from our readers is always welcome.

General feedback: If you have questions about any aspect of this book, email us at customercare@packtpub.com and mention the book title in the subject of your message. For questions and issues, you have with the source code, you can use the **Discussions** forum with the GitHub repository of this book: https://github.com/PacktPublishing/Pragmatic-Microservices-with-CSharp-and-Azure/discussions

Errata: Although we have taken every care to ensure the accuracy of our content, mistakes do happen. If you have found a mistake in this book, we would be grateful if you would report this to us. Please visit www.packtpub.com/support/errata and fill in the form.

Piracy: If you come across any illegal copies of our works in any form on the internet, we would be grateful if you would provide us with the location address or website name. Please contact us at copyright@packt.com with a link to the material.

If you are interested in becoming an author: If there is a topic that you have expertise in and you are interested in either writing or contributing to a book, please visit authors.packtpub.com.

Share Your Thoughts

Once you've read *Pragmatic Microservices with C# and Azure*, we'd love to hear your thoughts! Scan the QR code below to go straight to the Amazon review page for this book and share your feedback.

https://packt.link/r/1835088295

Your review is important to us and the tech community and will help us make sure we're delivering excellent quality content.

Download a free PDF copy of this book

Thanks for purchasing this book!

Do you like to read on the go but are unable to carry your print books everywhere?

Is your eBook purchase not compatible with the device of your choice?

Don't worry, now with every Packt book you get a DRM-free PDF version of that book at no cost.

Read anywhere, any place, on any device. Search, copy, and paste code from your favorite technical books directly into your application.

The perks don't stop there, you can get exclusive access to discounts, newsletters, and great free content in your inbox daily

Follow these simple steps to get the benefits:

1. Scan the QR code or visit the link below

https://packt.link/free-ebook/9781835088296

2. Submit your proof of purchase

3. That's it! We'll send your free PDF and other benefits to your email directly

Part 1:
Creating Microservices
with .NET

Part 1 introduces the fundamental functionality of a microservices application. Before delving into the development of the Codebreaker application, you will explore .NET Aspire - a new cloud-ready stack for service construction. This section covers the technology's offerings, essential features, an introduction to Microsoft Azure, and an overview of the components comprising the Codebreaker application. Subsequently, you will engage in coding using ASP.NET Core minimal APIs, crafting code for data interaction with both relational and NoSQL databases through **Entity Framework** (**EF**) Core, utilizing Azure Cosmos DB and SQL Server, and generating client libraries to access the REST service. One approach involves leveraging the HTTP client factory, while the other employs Microsoft Kioata.

Each chapter in this section provides a functional application that evolves with each subsequent chapter, enhancing the learning experience.

This part has the following chapters:

- *Chapter 1, Introduction to .NET Aspire and Microservices*
- *Chapter 2, Minimal APIs – Creating REST Services*
- *Chapter 3, Writing Data to Relational and NoSQL Databases*
- *Chapter 4, Creating Libraries for Client Applications*

1

Introduction to .NET Aspire and Microservices

Welcome to creating a solution consisting of Microservices. The first chapter provides the foundations for the microservices solution that will be developed in this book.

Here, you will learn which features .NET Aspire offers for microservices. In this book, we create the **Codebreaker** solution. You will learn what Codebreaker is and the parts it consists of. In the last section of this chapter, you'll learn which Azure services are used while we create the application on the tour up to the last chapter.

The first chapter lays the foundation.

In this chapter, you will learn about the advantages that are offered by .NET Aspire on creating microservices and you will gain the foundational knowledge needed to work with this technology, including how to define the app model, what it means for development and deployment, how service discovery is used, and how Azure resources are deployed while debugging the solution locally.

You will get an overview of the application we built in this book, the parts of the solution, and how the different services are connected.

In this chapter, you will learn about the following:

- Creating .NET Aspire projects
- The parts of the Codebreaker solution
- Using Microsoft Azure with .NET Aspire
- Azure services used by the Codebreaker solution

Technical requirements

In this chapter, you need .NET 8 with the .NET Aspire workload, either Visual Studio or Visual Studio Code, Docker Desktop, and a Microsoft Azure subscription. Information about the installation is explained in this chapter and the readme file of the source code repository.

The code for this chapter can be found in this GitHub repository: https://github.com/ PacktPublishing/Pragmatic-Microservices-with-CSharp-and-Azure.

In the ch01 folder, you'll see the projects with the results of this chapter. You'll see these folders:

- Aspire: This folder contains four projects created with a .NET Aspire template to run a .NET Aspire project including one service and a web application
- Azure: This folder contains the same four projects from the previous folder, enhanced by using an Azure resource

Starting with .NET Aspire

.NET Aspire is a new .NET technology offering tools and libraries that help create, debug, and deploy .NET solutions built using microservices. With all the chapters of this book, we'll take advantage of .NET Aspire.

> **Note**
> In this chapter, you'll get a core understanding of how .NET Aspire works. In all the other chapters, we'll make use of .NET Aspire and get into the details.

You can install it using the .NET **Command Line Interface (CLI)** or using Visual Studio 2022. The first version of .NET Aspire is based on .NET 8, thus at least .NET 8 is required to use .NET Aspire.

.NET Aspire requires .NET 8, and can be installed by installing a .NET workload:

```
dotnet workload install aspire
```

To see the workloads installed, and the version of .NET Aspire, use the following:

```
dotnet workload list
```

If you use Visual Studio, use the Visual Studio Installer, and select the **.NET Aspire SDK** component to install .NET Aspire.

.NET Aspire apps are designed to run in containers. Running the application locally, projects run directly on the system without the need for a Docker engine. Docker containers are used when deploying the solution. We can (and will) use available Docker images as part of the application. Here, the container runtime is required to run locally. In this book, we use the most used container runtime – **Docker Desktop**. Docker Desktop is free for personal use and for small companies. .NET Aspire also supports running containers with **Podman**.

After the installation of .NET Aspire, create a new project.

Creating a .NET Aspire project

When .NET Aspire is installed, you can create a new project containing an API service and a Blazor client application using the following:

```
dotnet new aspire-starter -o AspireSample
```

With this template, four projects are created:

- `AspireSample.ApiService`: This project contains a REST service that uses ASP.NET Core minimal APIs

- `AspireSample.Web`: An ASP.NET Core Blazor application that sends requests to the API service

- `AspireSample.ServiceDefaults`: A library project with shared initialization code for all services of the solution

- `AspireSample.AppHost`: The app host project defines the app model of the solution, and how all the resources are connected

Let's build and start the solution next.

The .NET Aspire dashboard

When you start the newly created project (the AppHost project needs to be the starting project), a console opens, showing the logs of the AppHost, and the browser opens a dashboard that shows the resources of the project, as you can see in *Figure 1.1*.

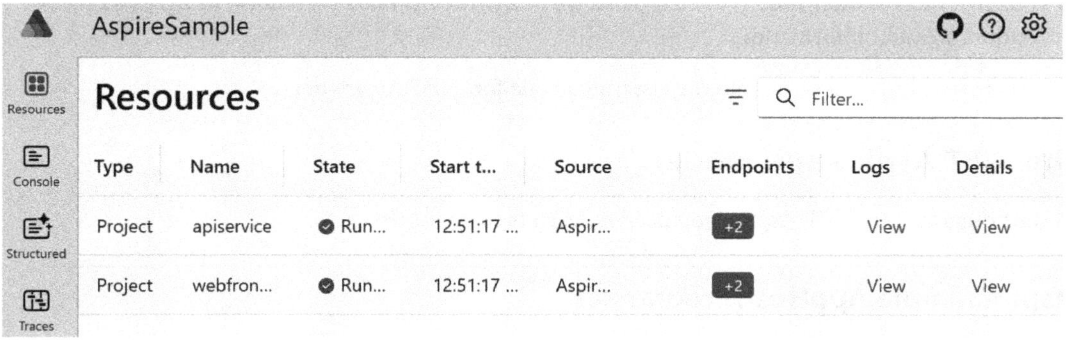

Type	Name	State	Start t...	Source	Endpoints	Logs	Details
Project	apiservice	● Run...	12:51:17 ...	Aspir...	+2	View	View
Project	webfron...	● Run...	12:51:17 ...	Aspir...	+2	View	View

Figure 1.1 – Aspire dashboard

With the .NET Aspire dashboard, you can see the resources running (`apiservice` and `webfrontend` in this image), the state of the resources, and the endpoints, and can access details and logs. In the left pane, you have access to logs, traces, and metrics data. While the dashboard is typically not used in production environments (we have **Prometheus**, **Grafana**, **Azure Application Insights**, and other environments), it's great to know all this information during development time. Are there memory leaks with services? How does the interaction with services happen? Where are the bottlenecks? You can find this information using the dashboard. This is discussed in detail in *Chapter 11*.

> **Note**
>
> Because the .NET Aspire dashboard is that great, it's available as a Docker image and can be used in small scenarios in production as well, but it has limitations outside of the development environment.

When you click on the link of the webfrontend, the application opens. In case you already created Blazor applications, you already know the links from the application, as shown in *Figure 1.2*.

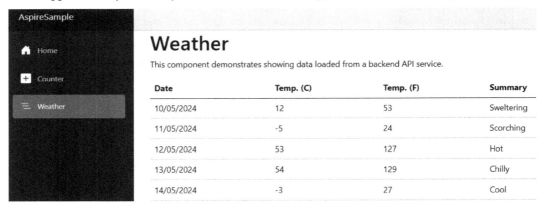

Figure 1.2 – webfrontend

As you click the **Weather** link in the left pane, webfrontend makes a request to apiservice for random weather information.

The app is running, so let's get into the generated code for .NET Aspire next.

The .NET Aspire app model

To start digging into .NET Aspire, you need to learn the app model:

AspireSample.AppHost/Program.cs

```
var builder = DistributedApplication.CreateBuilder(args);

var apiService = builder.AddProject<Projects.AspireSample_
ApiService>("apiservice");
// code removed for brevity
```

If you are used to the app builder pattern with .NET applications and the `Host` class to configure the DI container, app configuration, and logging, you can see some similarities. Here, a `DistributedApplication` class is used to create `IDistributedApplicationBuilder` with the `CreateBuilder` method. The returned builder is used to define all the resources needed by the solution. With the generated code, two projects are mapped using the `AddProject` method. The projects are referenced with a generic type, for example, `Projects.AspireSample_ApiService`. This type was created by adding a project reference to the `AspireSample.ApiService` project. You can see the reference when you open the `AspireSample.AppHost.csproj` project file.

Using project types with `AddProject` is convenient, but it's not a requirement. You can also pass a string of a directory where the project resides.

Other than adding projects, it's possible to add executables (`AddExecutable`) or Docker images (`AddContainer`).

.NET Aspire also offers a huge list of predefined resources, for example, RabbitMQ, Kafka, Redis, and SQL Server, and resources running within Microsoft Azure, such as Azure Cosmos DB, Azure Key Vault, and Azure Event Hub. To add resources to the app model, NuGet packages are prefixed with `Aspire.Hosting` and `Aspire.Hosting.Azure` needs to be added.

> **Note**
>
> In this book, many new resources are added to the Codebreaker solution. *Chapter 3* adds SQL Server and Azure Cosmos DB, *Chapter 5* adds Docker containers, *Chapter 7* adds Azure App Configuration and Azure Key Vault, *Chapter 11* adds Azure Log Analytics, Prometheus, and Grafana, *Chapter 13* adds Azure SignalR Services, and so on.

The name `"apiservice"` that's passed as a parameter passed to the `AddProject` method defines the name of the resource. We'll use this name later in the *Using service discovery* section.

`AddProject` returns an object of the `IResourceBuilder<ProjectResource>` type. The `IResourceBuilder` objects can be used to connect multiple resources within the app model. The `ProjectResource` type derives from the `Aspire.Hosting.ApplicationModel.Resource` base class and implements several resource interface types, such as `IResourceWithEnvironment` and `IResourceWithServiceDiscovery`.

Let's use this resource object to connect another resource:

Aspire/AspireSample.AppHost/Program.cs

```
// code removed for brevity
builder.AddProject<Projects.AspireSample_Web>("webfrontend")
   .WithExternalHttpEndpoints()
   .WithReference(apiService);

builder.Build().Run();
```

The `apiService` variable returned from the first `AddProject` method is referenced with the second project – a web frontend – using the `WithReference` method. This allows accessing the web frontend to access the API service. The URL of the API service is assigned as an environment variable to the web frontend – this is what the `IResourceWithServiceDiscovery` interface is used for. While the API service does not need to be accessed externally (only the web frontend needs access), the web frontend should be accessible from the outside. That's why the `WithExternalHttpEndpoints` method is used with the web frontend project. This configuration information is used to specify how the Ingress controller added as a proxy to the resource is configured.

Before looking into the projects that are referenced by the AppHost, let's get into the shared `AspireSample.ServiceDefaults` project.

The shared project for common configuration

The `AspireSample.ServiceDefaults` project is a library with a common configuration that can be used by all the resource projects:

Aspire/AspireSample.ServiceDefaults/Extensions.cs

```
public static class Extensions
{
    public static IHostApplicationBuilder AddServiceDefaults(this
      IHostApplicationBuilder builder)
    {
      builder.ConfigureOpenTelemetry();
      builder.AddDefaultHealthChecks();
      builder.Services.AddServiceDiscovery();
      builder.Services.ConfigureHttpClientDefaults(http =>
      {
        http.AddStandardResilienceHandler();
        http.AddServiceDiscovery();
      });
      return builder;
    }
    // code removed for brevity
```

This shared project contains the `AddServiceDefaults` extension method that implements a common configuration for the resource applications. With the implementation, `ConfigureOpenTelemetry` is invoked, which is another extension method defined by the `Extensions` class. The parts that are common for logging, metrics, and distributed tracing are implemented here. This is covered in *Chapter 11*. `AddDefaultHealthChecks` configures health checks for the services, which can include health checks for the .NET Aspire components that are used.

AddServiceDiscovery makes use of the Microsoft.Extensions.ServiceDiscovery library, which is also new since the first release of .NET Aspire, but can also be used independently of .NET Aspire. The AddServiceDiscovery method registers default service endpoint resolvers. Service discovery is not only configured with the DI container but also with the configuration of the HTTP client, with the lambda parameter of the ConfigureHttpClientDefaults method. Service discovery is discussed in the next section. ConfigureHttpClientDefaults is part of the Microsoft.Extensions.Http library, the HTTP client factory. The package that's referenced from the ServiceDefaults library is Microsoft.Extensions.Http.Resiliency. This library is new since .NET 8 and offers extensions to the Polly library. With a distributed application, invocations sometimes fail on transient issues. Retrying invocations to these resources can succeed when invoked another time. This functionality is built into .NET Aspire with default resiliency configuration in AddStandardResilienceHandler.

But now, let's get into service discovery.

Using service discovery

webfrontend needs to know about the link of apiservice to get the weather information. This link is different depending on the environment the solution is running with. Running the application locally on the development system, we use localhost links with different port numbers, and depending on the environments where the solution is running (for example, Azure Container App environments, Kubernetes, etc.), different configurations are required.

With the new service discovery, logical names can be used for the services, which are resolved using different providers. Thus, the same functionality works in different environments.

The Blazor client application configures HttpClient:

Aspire/AspireSample.Web/Program.cs

```
builder.Services.AddHttpClient<WeatherApiClient>(client =>
{
  client.BaseAddress = new("https+http://apiservice");
});
// code removed for brevity
```

The apiservice name comes from the app model definition – the name that has been passed to the AddProject method. Before the colon, the schema, for example, http or https can be specified. Separating schemas with + allows the use of multiple schemas, and the first one is preferred.

The `AddServiceDiscovery` method that was added to the DI container earlier adds a configuration-based endpoint resolver by default. With this, the configuration can be added to a JSON configuration file, for example, as follows:

```
{
  "Services": {
    "apiservice": {
      "https": [
        "localhost:8087",
        "10.466.24.90:80"
      ]
    }
  }
}
```

With the configuration, the section needs to be named `Services`. Within the `Services` section, the named service is looked for (`apiservice`), and there, the values below the schema name (`https`) are resolved. The port numbers are randomly created and will differ with your environment.

With the AppHost, as `apiservice` is referenced from the web frontend, the URIs to the API service are added as environment variables. Open the .NET Aspire dashboard, and in the **Details** column, click **View** with `webfrontend`. There, you can see the `services__apiservice_http__0` and `services__apiservice_https_0` environment variables, and the `http://localhost:5395`, and `https://localhost:7313` values. The URIs are specified within `Properties/launchsettings.json`:

Aspire/AspireSample.ApiService/Properties/launchSettings.json

```
"profiles": {
  "http": {
    "commandName": "Project",
    "dotnetRunMessages": true,
    "launchBrowser": true,
    "launchUrl": "weatherforecast",
    "applicationUrl": "http://localhost:5395",
    "environmentVariables": {
      "ASPNETCORE_ENVIRONMENT": "Development"
    }
  },
  "https": {
    "commandName": "Project",
    "dotnetRunMessages": true,
    "launchBrowser": true,
    "launchUrl": "weatherforecast",
```

```
    "applicationUrl": "https://localhost:7313;http://localhost:5395",
    "environmentVariables": {
      "ASPNETCORE_ENVIRONMENT": "Development"
    }
  }
}
```

The `applicationUrl` setting defines the URLs used on starting the application, and this is the link that is used to add it to the environment variable. Because environment variables are part of the .NET configuration, these values are retrieved by the service discovery configuration provider.

Azure Container Apps and Kubernetes offer service discovery features without using a service discovery library. With applications deployed there, a pass-through provider is configured using `DnsEndPoint`.

Running the .NET Aspire solution locally, the process of `webfrontend` and `apiservice` use random ports. A reverse proxy is automatically added before these processes, and the reverse proxy is accessible via the configured launch settings.

This allows the changing of the number of replicas with the app model:

Aspire/AspireSample.AppHost/Program.cs

```
var apiService = builder.AddProject<Projects.AspireSample_
ApiService>("apiservice")
    .WithReplicas(3);
```

With the app model configuration in the AppHost, using `WithReplicas(3)` starts three instances of the service using three random ports, and the same port number from the reverse proxy as shown in *Figure 1.3*.

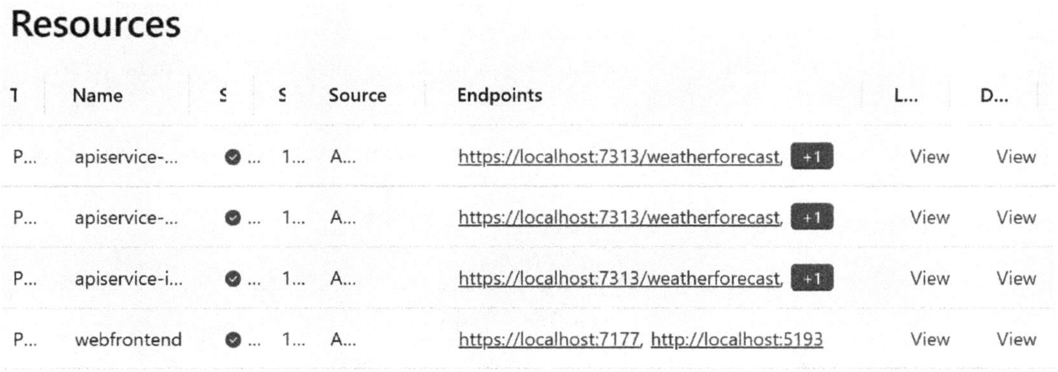

Resources

T	Name	S	S	Source	Endpoints	L...	D...
P...	apiservice-...	✓	1...	A...	https://localhost:7313/weatherforecast, +1	View	View
P...	apiservice-...	✓	1...	A...	https://localhost:7313/weatherforecast, +1	View	View
P...	apiservice-i...	✓	1...	A...	https://localhost:7313/weatherforecast, +1	View	View
P...	webfrontend	✓	1...	A...	https://localhost:7177, http://localhost:5193	View	View

Figure 1.3 – Multiple replicas

You can see three `apiservice-` services running with different postfixes, and three processes with the same port number, as shown with the endpoints. The endpoint defined from the launch settings is the endpoint of the reverse proxy. When you open **Details**, you can see different target ports with every service. The reverse proxy acts as a load balancer to choose one of the replicas.

> **Note**
>
> To start the solution with the `http` launch profile, you need to add the `ASPIRE_ALLOW_UNSECURED_TRANSPORT` environment variable to the launch settings of the AppHost project and set it to `true`.

This was an important core functionality from .NET Aspire. However, there's more.

.NET Aspire components

.NET Aspire components make it easy to use Microsoft and third-party features and services from within the applications that are configured. Azure Cosmos DB, Pomelo MySQL Entity Framework Core, and SQL Server are components available to access databases, and RabbitMQ, Apache Kafka, and Azure Service Bus are components for messaging. There's a list available at `https://learn.microsoft.com/en-us/dotnet/aspire/fundamentals/components-overview`.

To use a component, typically with the AppHost, a resource needs to be configured by adding a host NuGet package, for example, for the Azure Cosmos DB EF Core component, you would add the `Aspire.Hosting.Azure.CosmosDB` package. The component itself is then used by adding the `Aspire.Microsoft.EntityFrameworkCore.Cosmos` package to the service that accesses the database, for example, the API service.

What does a component have to offer? Do you know what names are used by a technology to turn on logging metrics data? Aspire components know this, and it's easy to configure it. When an Azure Cosmos DB resource is added to the app model, and is referenced by a service project, the connection string is configured as an environment variable (or stored within a secret store) and can be accessed by the project that needs the connection.

In many of the book chapters, we'll add some new components, thus we don't get into more details here.

Creating the app model manifest

With the app model defined in the `AppHost` project, we can create a JSON manifest file that describes the resources. You need to stop the project to allow a rebuild if it's still running:

```
cd ApireSample.AppHost
dotnet run --publisher manifest --output-path aspire-manifest.json
```

An extract of this manifest file is shown in the following snippet:

Aspire/AspireSample.AppHost/aspire-manifest.json

```
"webfrontend": {
  "type": "project.v0",
  "path": "../AspireSample.Web/AspireSample.Web.csproj",
  "env": {
    "services__apiservice__http__0": "{apiservice.bindings.http.url}",
    "services__apiservice__https__0": "{apiservice.bindings.https.
url}"
  },
  "bindings": {
    "https": {
      "scheme": "https",
      "protocol": "tcp",
      "transport": "http",
      "external": true
    }
  }
}
```

The manifest contains information about the resource type, environment variables, binding, and more. With the app model, we can also specify the use of Azure resources. This manifest file can now be used by tools to deploy the solution, (e.g., by using the Azure Developer CLI to deploy it to Microsoft Azure). Creating Azure resources is covered in *Chapter 6* and continued from there in other chapters.

Using Aspir8 (an open-source project, see https://github.com/prom3theu5/aspirational-manifests/), it's possible to deploy the solution to a Kubernetes cluster. This is used in *Chapter 16*.

The app model can be customized based on different launch profiles. With this, different manifest files can be created to deploy to (e.g., Azure and use specific Azure resources and to an on-premises Kubernetes cluster).

> **Note**
> The AppHost project containing the app model is used when starting and debugging the project during development. For deployment, the manifest of the app model is used. When running the solution in the production environment, the app host is no longer in action.

.NET Aspire is used in this book from the first to the last chapter. Let's look into what we are building.

Codebreaker – the solution

The Codebreaker solution is a traditional game to solve a set of colors. With one game type, the player needs to place four colors (which can be duplicates) from a list of six different colors. The correct colors are chosen randomly by the game service. With every move the player makes, an answer is returned: for every color that is correct and positioned at the correct place, a black peg is returned. For every color that is correct but wrongly positioned, a white peg is returned. The player now has up to 12 moves to find the correct solution. *Figure 1.4* shows a game run using a Blazor client application.

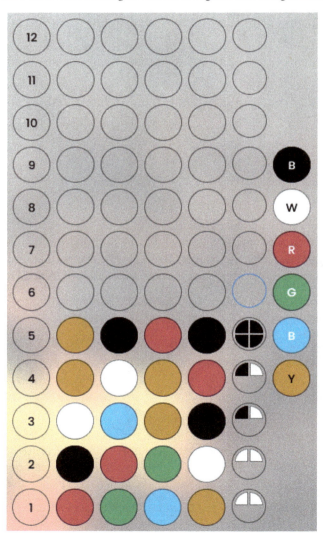

Figure 1.4 – Blazor client application

This gameplay shows that the solution was found after five moves. In this case, the correct result was yellow – black – red – black. The first selection was red – green – blue – yellow, with a result of two white pegs. With the fifth move, yellow – black – red – black was selected and four black pegs were returned, which means this is the correct move.

> **Note**
>
> Creating client applications is not part of this book (just a simple console application accessing the API is done in *Chapter 4*). However, the source code for several client applications is available at `https://github.com/codebreakerapp`.

Creating a service to run some game rules seems like a simple task that doesn't need a microservices architecture. However, there's more, as shown in the sequence diagrams in *Figure 1.5* and *Figure 1.6*.

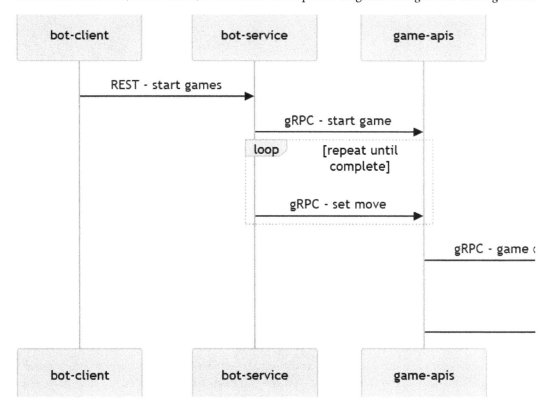

Figure 1.5 – The Codebreaker play games sequence

Multiple services are needed with the solution. The game API service is not only invoked by UIs used by human players; a bot service, which can be triggered on receiving a message, plays multiple games on its own, and the game API service writes information about games and every move set to a database.

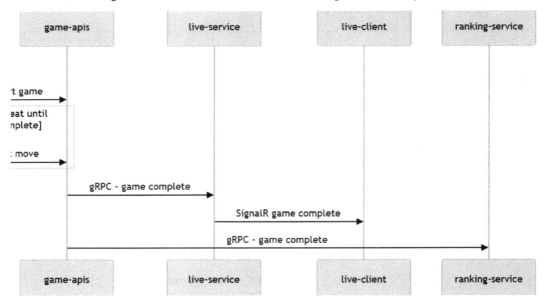

Figure 1.6 – Codebreaker game completion sequence

On completion of a game, the game API service not only writes this information to a database but also sends events. These events are received by a live service and a ranking service. The live service is used by live clients to monitor running games using ASP.NET Core SignalR. The ranking service writes completed games to its own database, which can be used by clients to get daily, weekly, and monthly game ranks. A service running Microsoft YARP is used as well to authenticate users and forward requests to the different services.

The Codebreaker solution makes use of several Azure services, as discussed next.

Using Microsoft Azure

To create and run the code from this book, you also need to have an Azure subscription. You can activate Microsoft Azure for free at https://azure.microsoft.com/free, which gives you an amount of about $200 Azure credits that are available for the first 30 days and several services that can be used for free for the time after.

What many developers miss is that if you have a Visual Studio Professional or Enterprise subscription, you also have a free amount of Azure resources every month. You just need to activate this with your Visual Studio subscription: https://visualstudio.microsoft.com/subscriptions/.

To create and manage resources, we use the Azure Portal, the Azure CLI, and the Azure Developer CLI. On Windows, you can install them with the following:

```
winget install Microsoft.AzureCLI
winget install Microsoft.Azd
```

To install these tools on Mac and Linux, check https://learn.microsoft.com/en-us/cli/azure/install-azure-cli and https://learn.microsoft.com/en-us/azure/developer/azure-developer-cli/install-azd.

Let's look at the resources used with Microsoft Azure.

Azure resources used by Codebreaker

To see what Azure resources are used, check *Figure 1.7.*

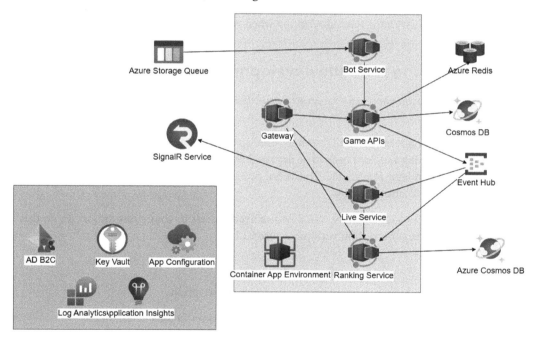

Figure 1.7 – Azure resources for Codebreaker

The compute services where the solution is running is an Azure Container App environment. This is a service that abstracts a Kubernetes cluster. The bot service, game APIs, live service, ranking service, and the gateway using YARP run within Azure Container Apps. The Azure Storage queue is used from the bot service: as a message arrives in the queue, the bot service is triggered to play a series of games. The bot service can be used from all client apps as well – indirectly via a gateway, implemented with

YARP. The game API service writes games to the Azure Cosmos DB and caches games with a Redis cluster. As games are completed, game events are pushed to the Azure Event Hub. The live service and ranking service are subscribers to the Event Hub. The live service uses ASP.NET Core SignalR, and to reduce the load of this service, the Azure SignalR service is used.

What's commonly used is the Azure App Configuration for application configuration values and feature management, Azure Key Vault to store secrets, Azure Active Directory B2C for user registrations, and Log Analytics and Application Insights to monitor the application.

> **Note**
> Starting with a small version of Codebreaker, not that many Azure services would be required to use. For a flexible and scalable solution, which might be accessed worldwide, and to learn about all the different aspects of microservices, all these services are in use. Don't be afraid of the cost when deploying the services with your Azure subscription. As long as you don't create a huge load (which we do in *Chapter 12*), the cost stays very small, and by far you will not use the full $200 available with the free subscription when you delete the resources again after use.

Azure provisioning from the development environment

Your .NET Aspire solution can easily integrate with Microsoft Azure and deploy resources while debugging the solution.

By debugging the solution locally, not all resources need to be deployed to Azure. The service projects can run locally while testing and don't need to be deployed. With Azure Cosmos DB, a Docker container or a locally installed emulator is available. This is not possible with all resources, for example, Azure Key Vault or Azure Application Insights.

To deploy these resources automatically, .NET Aspire needs access to your subscription. To do this, first, log in to your Azure subscription with the Azure CLI:

```
az login
```

This opens a browser, and you can log in with your Azure subscription.

In case you have multiple subscriptions, check the Azure CLI is set to the current one:

```
az account show
```

This shows the current active subscription. In case a different one should be used, use `az account list` to list all subscriptions, and `az account set -subscription <your subscription id>` to set the current subscription to a different one. Remember the value that's listed with `id` – this is the subscription ID that's needed with the next steps.

Now, we need to connect the project to the subscription and specify some settings. It's best to put this information within user secrets; this shouldn't be put into source code repositories.

In case user secrets are not configured yet with the `AppHost`, initialize it:

```
cd AspireSample.AppHost
dotnet user-secrets init
```

The configurations we need are the following:

```
dotnet user-secrets set Azure:SubscriptionId <your subscription id>
dotnet user-secrets set Azure:AllowResourceGroupCreation true
dotnet user-secrets set Azure:ResourceGroup rg-firstsample
dotnet user-secrets set Azure:Location westeurope
dotnet user-secrets set Azure:CredentialSource AzureCli
```

With `SubscriptionId`, you specify the subscription where resources are created. The resource group you specify with the value for `ResourceGroup` is used to create all the resources needed. The resource group will be created if you set `AllowResourceGroupCreation` to `true`. Otherwise, you need to create the resource group first. With the `Location` setting, specify your preferred location. To see the locations available with your subscription, use `az account list-locations -o table`.

Setting the `CredentialSource` setting to `AzureCli` specifies that you are using the same account you just used to log in with the Azure CLI to create the resources. Without this setting, `DefaultAzureCredential` will be used, which tries to use multiple account types with a predefined list until one succeeds. This includes Visual Studio, Azure CLI, PowerShell, Azure Developer CLI, and other credentials. Here, credentials might be used which don't have access to the subscription. In my experience, it's better to supply the credentials explicitly.

To see all the secrets, use the following:

```
dotnet user-secrets list
```

> **Note**
>
> Using Visual Studio, you can connect the project to Azure by using the Solution Explorer. Within the AppHost project, select **Connected Services**, open the context menu, and select **Azure Resource Provisioning Settings**. This opens a dialog to select the subscription, location, and resource group.

Next, let's add the `Aspire.Hosting.Azure.KeyVault` NuGet package to the AppHost project, and update the app model:

```
var builder = DistributedApplication.CreateBuilder(args);

var keyVault = builder.AddAzureKeyVault("secrets");

var apiService = builder.AddProject<Projects.AspireSample_
ApiService>("apiservice")
  .WithReplicas(3)
  .WithReference(keyVault);
```

The `AddAzureKeyVault` method creates a key vault named `secrets`. This is referenced from the `apiservice` project.

When you start the AppHost now, the key vault will be created within Azure. Opening the Azure portal at `https://portal.azure.com`, you will see the resource group, and within the resource group, the Azure Key Vault is created. If you check the user secrets again, an `Azure:Deployments` section is added, which contains links to the resources created. This information is used to find the resources again, and they don't need to be published again the next time you start the application.

When you are finished with this chapter, just delete the complete resource group from the portal, so no additional cost applies.

Note

To publish all the resources including the projects to Azure, you can use the Azure Developer CLI. This is covered in *Chapter 6*.

Summary

In this chapter, you learned about the core features of .NET Aspire, which includes tooling, orchestration, and Aspire components. You learned how resources are connected by the Aspire app model, and how service discovery is done. You've seen how to create a manifest describing the app model, which can be used by tools to deploy the solution.

With the Codebreaker solution, you learned about the rules of the game and the parts of the application that are created from the second to the last chapter.

Now, you know the different Microsoft Azure services that are used by the Codebreaker solution when running in Azure. An alternative to these services is offered as well to run the complete solution in an on-premises environment (which can also be hosted in the Azure cloud this way).

From the next chapter on, we'll start developing the Codebreaker solution. In *Chapter 2*, we will create REST services using ASP.NET Core minimal APIs to play games. We'll test this API using HTTP files.

Further reading

To learn more about the topics discussed in this chapter, you can refer to the following links:

- **.NET Aspire setup and tooling**: `https://learn.microsoft.com/en-us/dotnet/aspire/fundamentals/setup-tooling`

- **.NET Aspire components**: `https://learn.microsoft.com/en-us/dotnet/aspire/fundamentals/components-overview`

- **.NET Aspire manifest format**: `https://learn.microsoft.com/en-us/dotnet/aspire/deployment/manifest-format`

- **GitHub repository for Aspir8**: `https://github.com/prom3theu5/aspirational-manifests`

2
Minimal APIs – Creating REST Services

Since .NET 6, minimal APIs are the new way to create REST APIs. With later .NET versions, more and more enhancements have been made available, which makes them the preferred way to create REST services with .NET.

In this chapter, you'll learn how to create a data representation of the game with model types, use these types in a service to implement the game functionality, create a minimal API project to create games, update games by setting game moves, and return information about games.

You'll implement functionality to offer an OpenAPI description for developers accessing the service to get information about the service, and an easy way to create a client application.

In this chapter, you'll be exploring these topics:

- Creating models for the game
- Implementing an in-memory game repository
- Implementing the REST service of the game using minimal APIs
- Using OpenAPI to describe the service
- Testing the service using HTTP files
- Enabling .NET Aspire

By the end of this chapter, you'll have a running service implementing the Codebreaker Games API with an in-memory games store, accessible using HTTP requests.

Technical requirements

The code for this chapter can be found in the following GitHub repository: `https://github.com/PacktPublishing/Pragmatic-Microservices-with-CSharp-and-Azure`. The `ch02` source code folder contains the code samples for this chapter. You'll find the code for the following:

- `Codebreaker.GamesAPIs` – The Web API project
- `Codebreaker.GamesAPIs.Models` – A library for the data models
- `Codebreaker.GameAPIs.Analyzers` – A library containing game move analyzers for the game
- `Codebreaker.GamesAPIs.Analyzers.Tests` – Unit tests for the game move analyzers
- `Codebreaker.AppHost` – The host project for .NET Aspire
- `Codebreaker.ServiceDefaults` – A library used by the .NET Aspire configuration

> **Note**
>
> You don't implement the game move analyzers of the game in this chapter. The `Analyzers` project is just for reference purposes, but you can simply use a NuGet package for the analyzers (`CNinnovation.Codebreaker.Analyzers`) that has been made available for you to build the service.

For the installation of Visual Studio, Visual Studio Code, and .NET Aspire, check the README file for this chapter in the repository.

The game models

Before creating the REST API project, we start with a library that contains the models to represent the game along with a move of the game. This model will contain the main data part of the Codebreaker Game API service solution, which will be used to read and write to the database (in *Chapter 3*), while the model also serves as an implementation of the main functionality of the game.

The major types in a simplified version are shown in *Figure 2.1*. The Game class implements the `IGame` interface. The `IGame` interface is used by the `Analyzers` package. A game contains a list of moves. A single game move is represented by the `Move` class.

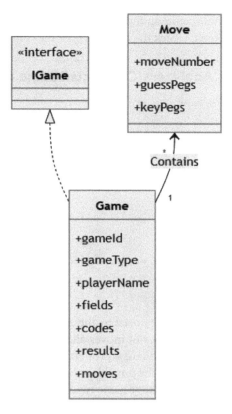

Figure 2.1 – The game model

Exploring the game analyzers library

As the focus of this book is not implementing the game rules with .NET, you can use the existing project, Codebreaker.GameAPIs.Analyzers, or just reference the NuGet package that has been made available on the NuGet server. This library contains game move analyzers for these game types:

- Game6x4 – Six colors with four codes to guess
- Game8x5 – Eight colors with five codes
- Game6x4Mini – Six colors with four codes with a *small children* mode
- Game5x5x4 – Five colors and five shapes with four codes

The analyzers work using the generic IGame interface, which you need to implement with the game models library.

The IGame interface specifies some common functionalities for the Codebreaker games, as you can see in the following code snippet. Check the repository for the complete interface definition:

Codebreaker.GameAPIs.Analyzers/Contracts/IGame.cs

```
namespace Codebreaker.GameAPIs.Contracts;

public interface IGame
{
  Guid Id { get; }
  string GameType { get; }
  int NumberCodes { get; }
  int MaxMoves { get; }
  DateTime StartTime { get; }
  // code removed for brevity
  IDictionary<string, IEnumerable<string>> FieldValues { get; }
  string[] Codes { get; }
}
```

The IGame interface defines game members used by the analyzers, such as the number of codes that must be set and the maximum number of moves allowed, which are used by the analyzer to verify correct input data. The FieldValues property of the IDictionary<string, IEnumerable<string>> type defines the possible values to select from. With the color game types, this will be a list of colors. With the shape game type, this will be a list of colors and a list of shapes. The Codes property contains an array of strings. How the string looks is different based on the game type. The string array contains the correct solution for a game run.

The analyzers library also contains record types that are supported by the analyzers implementation that you can use for the generic parameters of your game types.

One example of such a type used for the generic parameters is ColorField:

Codebreaker.GameAPIs.Analyzers/Fields/ColorField.cs

```
public partial record class ColorField(string Color)
{
  public override string ToString() => Color;

  public static implicit operator ColorField(string color) =>
    new(color);
}
```

The ColorField record class just contains a color string. This field type is used with all game types with the exception of the Game5x5x4 game type, which uses the ShapeAndColorField record.

To specify the results of a game move, three different types are available: `ColorResult`, `SimpleColorResult`, and `ShapeAndColorResult`. `SimpleColorResult` adds information for small children (which position has a correct color), whereas the `ColorResult` record struct just contains information on the number of colors placed in a correct hole, and the number of colors placed in an incorrect hole.

The following code snippet shows the `ColorResult` record struct:

Codebreaker.GameAPIs.Analyzers/Results/ColorResult.cs

```
public readonly partial record struct ColorResult(int Correct,
  int WrongPosition)
{
  private const char Separator = ':';
}
```

This record is implemented partially, separating parts of the implementation to simplify the source code. The other parts are defined in other source files and implement the `IParsable` and `IFormattable` interfaces. The `const` member called `Separator` is used with the other parts of the `ColorResult` type.

> **Note**
>
> `ColorField`, `ColorResult`, and the other classes representing a field and results are just used to analyze moves and return results. The `Game` and `Move` classes that you'll implement in this chapter are just data holders and don't contain any logic. The field guesses and results are all represented using strings, which makes them flexible for every game type available and easier to use with JSON serialization and database access. Converting the specific field and result types to and from strings is done using the `IParsable`, `ISpanParsable`, and `IFormattable` interfaces. The parsable interfaces are new since .NET 7 and are based on a C# 11 feature that allows static members with interfaces. These types are important in case you want to create your own game types and game analyzers.
>
> You can read the article at `https://csharp.christiannagel.com/2023/04/14/iparsable/` for more information about parsable interfaces.

Exploring game analyzers

The implementation of the game analyzers is done within the `GameGuessAnalyzer` base class and the `ColorGameGuessAnalyzer`, `SimpleGameGuessAnalyzer`, and `ShapeGameGuessAnalyzer` derived classes. The implementation of these analyzers is disconnected from the game model types. All these analyzers implement the `GetResult` method specified with the `IGameGuessAnalyzer<Tresult>` interface. After creating an instance of the analyzer passing the game, the guesses, and the move number, just the `GetResult` method needs to be invoked to calculate the result of the move.

If you are interested in checking the analyzers, dive into the `Codebreaker.GameAPIs.Analyzers` project within the book's source code repository. Working through the following steps, instead of referencing this project, you can add the `CNinnovation.Codebreaker.Analyzers` NuGet package. Just make sure to use the latest 3.x version of this package, as 4.x and newer versions might contain breaking changes.

If you choose to, you can create an analyzer on your own and also add more game types. Make sure to read the information on the game rules in *Chapter 1*.

Creating a .NET library

The model types are added to the `Codebreaker.GameAPIs.Models` .NET library. Having a library allows the creation of different data access libraries (in *Chapter 3*) to offer a flexible data store choice for hosting the service.

You can use the .NET CLI as shown to create the class library, or use Visual Studio to create a class library:

```
dotnet new classlib --framework net8.0 -o Codebreaker.GameAPIs.Models
```

Implementing classes for the model types

The Game class holding all the data needed by a game is shown here (check the GitHub repository for the complete implementation):

Codebreaker.GameAPIs.Models/Game.cs

```
public class Game(
  Guid id,
  string gameType,
  string playerName,
  DateTime startTime,
  int numberCodes,
  int maxMoves) : IGame
{
  public Guid Id { get; } = id;
  public string GameType { get; } = gameType;
  public string PlayerName { get; } = playerName;
  public DateTime StartTime { get; } = startTime;
  // code removed for brevity
  public ICollection<Move> Moves { get; } = [];
  public override string ToString() =>
    $"{Id}:{GameType} - {StartTime}";
}
```

This class implements the `IGame` interface to support the analyzer to check the correctness of a move and to set some game state. In addition to the members of the interface, the `Game` class also contains a collection of `Move` objects. Primary constructor syntax is used to reduce the number of needed code lines.

> **Primary constructors**
>
> Several classes created in this book make use of primary constructors. Primary constructors have been in use with records since C# 9. With C# 12, primary constructors can be used with normal classes and structs. However, while primary constructors with records create properties, with classes, these are just parameters. The parameters can be assigned to fields and properties, or just be used within members.

The `Move` class is simpler as it just represents the move of a player within a game along with its result:

Codebreaker.GameAPIs.Models/Move.cs

```
public class Move(Guid id, int moveNumber)
{
  public Guid Id { get; private set; } = id;
  public int MoveNumber { get; private set; } = moveNumber;
  public required string[] GuessPegs { get; init; }
  public required string[] KeyPegs { get; init; }

  public override string ToString() =>
    $"{MoveNumber}. {string. Join(':', GuessPegs)}";
}
```

The `Move` class contains string arrays for the guesses (`GuessPegs`) and the results (`KeyPegs`). String arrays can be used for every game type.

Defining the game repository contract

To be independent of the data store, the `IGamesRepository` interface defines all the members needed from a data store when playing a game. The `AddGameAsync` method is invoked when a new game is started. Setting a move, the game needs to be updated invoking the `UpdateGameAsync` method. The `GetGameAsync` and `GetGamesAsync` methods are specified by the interface to retrieve games:

Codebreaker.GameAPIs.Models/Data/IGamesRepository.cs

```
public interface IGamesRepository
{
  Task AddGameAsync(Game game,
        CancellationToken cancellationToken = default);
```

```
      Task AddMoveAsync(Game game, Move move,
        CancellationToken cancellationToken = default);
      Task<bool> DeleteGameAsync(Guid id,
        CancellationToken cancellationToken = default);
      Task<Game?> GetGameAsync(Guid id,
        CancellationToken cancellationToken = default);
      Task<IEnumerable<Game>> GetGamesAsync(GamesQuery gamesQuery,
        CancellationToken cancellationToken = default);
      Task<Game> UpdateGameAsync(Game game,
        CancellationToken cancellationToken = default);
}
```

This interface specifies asynchronous methods. This wouldn't be required with the implementation in this chapter. The next chapter will add asynchronous implementations, thus the contract should be ready for this.

Specifying asynchronous methods, it's a good practice to allow passing a `CancellationToken`. This allows canceling long-running operations even across network boundaries.

The `IGamesRepository` interface is specified in the `Codebreaker.GameAPIs.Models` library. This makes it possible to reference this interface from all the data store libraries that will be implemented later on. In this chapter, just an in-memory collection will be implemented as part of the next step.

The minimal APIs project

After having the models and the repository contract in place, we can move over to creating the project hosting the REST API.

Here, we'll use ASP.NET Core with minimal APIs to create a REST API, and store games and moves in an in-memory repository. To create the running games API, we need to do the following:

1. Create a Web API project.
2. Implement the games repository.
3. Create a games factory.
4. Create data transfer objects.
5. Create endpoints to run the game via HTTP requests.
6. Configure the JSON serializer.
7. Add endpoint filters.

To better understand how the different classes interact in creating the game and setting a move, the flow of the functionality we need to implement is shown in the next two figures.

Figure 2.2 shows the sequence when a new game is created. On invoking the API call, `GamesService` is invoked to start the game. This service class uses `GamesFactory` to create a new game based on the parameters received and returns random code values. For persistence, `GameMemoryRepository` is invoked from `GamesService` to add the game before the game is returned.

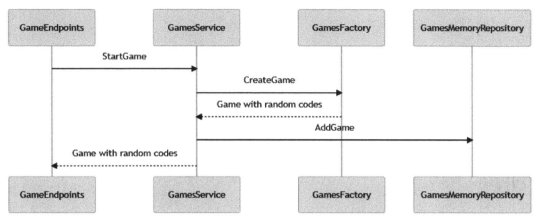

Figure 2.2 – Creating a game

Figure 2.3 shows the sequence when a game move is set. Because the client does not have the complete state of the game, `GamesService` retrieves the game from the repository. Then, `GamesFactory` is used again to select an analyzer based on the game type, and the analyzer is invoked to find the results of the game move. After the results are available with `GamesService`, the game is updated with the new move, and the results are returned to the game endpoint.

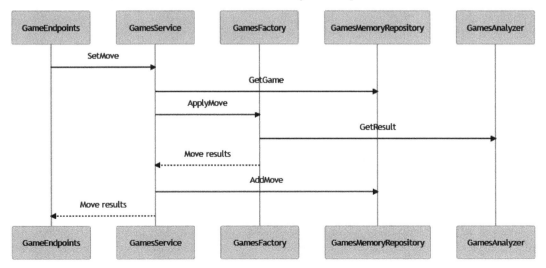

Figure 2.3 – Setting a game move

Let's implement this functionality with the API.

Creating the Web API project

To create the Web API project, you can use the .NET CLI as shown or use the Web API project template from Visual Studio:

```
dotnet new webapi --use-minimal-apis --framework net8.0 -o
Codebreaker.GameAPIs
```

The `--use-minimal-apis` option is used to create minimal APIs instead of the traditional controllers. Because the API will be hosted behind an Ingress server, the Ingress server will offer HTTPS, while HTTP is okay in the backend. To allow but not require HTTPS remove the codeline `app.UseHttpsRedirection();` from the code.

Exploring WebApplicationBuilder and WebApplication

With the created project, `Program.cs` is created and contains the configuration for the dependency injection container with the `WebApplicationBuilder` class and the configuration of middleware with the `WebApplication` class. The project template creates a random weather service. However, as weather information is not needed with the games service, some of the created code can be removed.

The part of the code that remains is shown in the next code snippet:

```
var builder = WebApplication.CreateBuilder(args);
builder.Services.AddEndpointsApiExplorer();
builder.Services.AddSwaggerGen();
var app = builder.Build();
if (app.Environment.IsDevelopment())
{
  app.UseSwagger();
  app.UseSwaggerUI();
}
app.Run();
```

`WebApplication.CreateBuilder` returns `WebApplicationBuilder` and configures the dependency injection container, the default configuration, and default logging. The `AddEndpointsApiExplorer` and `AddSwaggerGen` methods add services to the dependency injection container. The services registered by these two methods are needed for the `OpenAPI` description of the services.

Invoking the `Build` method returns a `WebApplication` instance. This instance is then used to configure the middleware. Middleware is invoked with every request to the service. The `UseSwagger` method registers the Swagger middleware to create an `OpenAPI` description of the service. The `UseSwaggerUI` method registers the `SwaggerUI` middleware to show a web page where the API is described and can be tested. As an API is not yet implemented, a description will not be generated yet.

> **What is the relationship between Swagger and OpenAPI?**
>
> The Swagger specification was created to describe HTTP APIs. In 2015, SmartBear Software acquired the Swagger API specification and started to form a new organization in 2016 together with Google, IBM, Microsoft, PayPal, and others under the sponsorship of the Linux Foundation – the **OpenAPI Initiative**. New versions of this specification are defined with OpenAPI.
>
> The SwaggerXX method names originate from the original specification and haven't been changed since. These methods are defined within the Swashbuckle.AspNetCore NuGet package that's referenced with the Web API project template.

Implementing the repository

In the earlier *Defining the game repository contract* section, we created the IGamesRepository interface to specify the methods that need to be implemented by every data store used with the Games API. We can implement this contract now. The GamesMemoryRepository class implements the IGamesRepository interface:

Codebreaker.GameAPIs/Data/GamesMemoryRepository.cs

```
namespace Codebreaker.GameAPIs.Data.InMemory;
public class GamesMemoryRepository(ILogger<GamesMemoryRepository>
    logger) : IGamesRepository
{
  private readonly ConcurrentDictionary<Guid, Game> _games = new();
  private readonly ILogger _logger = logger;

  public Task AddGameAsync(Game game,
    CancellationToken cancellationToken = default)
  {
    if (!_games.TryAdd(game.Id, game))
    {
      _logger.LogWarning("id {Id} already exists", game.Id);
    }
    return Task.CompletedTask;
  }
  // code removed for brevity
  public Task AddMoveAsync(Game game, Move move,
    CancellationToken cancellationToken = default)
  {
    _games[game.Id] = game;
    return Task.CompletedTask;
  }
}
```

With the implementation, the `ConcurrentDictionary` class is used for a thread-safe collection when multiple clients access the service concurrently. With the implemented `AddGameAsync`, `GetGameAsync`, and similar methods, games are added, updated, and removed from the dictionary. Here, all the games are just kept in memory.

> **Persistent state and multiple server instances**
>
> This chapter's implementation of the repository does not persist the state and also does not allow multiple server instances to run, as the state is only stored within the memory of the process. In the next chapter, other implementations of this interface will be used to store the games in a database.

Creating game objects initialized with random values

The `GamesFactory` class creates games with random values. The following code snippet shows creating a game with six colors and four holes; you can extend this to other game types as well:

Codebreaker.GameAPIs/Services/GamesFactory.cs

```
public static class GamesFactory
{
  private static readonly string[] s_colors6 =
    [ Colors.Red, Colors.Green, Colors.Blue, Colors.Yellow,
    Colors. Purple, Colors.Orange ];

  // code removed for brevity
  public static Game CreateGame(string gameType, string playerName)
  {
    Game Create6x4Game() =>
      new(Guid.NewGuid(), gameType, playerName, DateTime.Now, 4, 12)
      {
        FieldValues = new Dictionary<string, IEnumerable<string>>()
        {
          { FieldCategories.Colors, s_colors6 }
        },
        Codes = Random.Shared.GetItems(s_colors6, 4)
      };

    // code removed for brevity
    return gameType switch
    {
      GameTypes.Game6x4Mini => Create6x4SimpleGame(),
      GameTypes.Game6x4 => Create6x4Game(),
      GameTypes.Game8x5 => Create8x5Game(),
```

```
      GameTypes.Game5x5x4 => Create5x5x4Game(),
      _ => throw new CodebreakerException("
      Invalid game type") { Code = CodebreakerExceptionCodes.
      InvalidGameType }
    };
  }
}
```

With the code, pattern matching is used. For every game type, a local function such as `Create6x4Game` is defined, which specifies the available colors or shapes, the random code, and the maximum number of moves.

Creating data transfer objects

Data transfer objects (DTOs) are objects used for communication to define the data that should be transferred. When you create a microservice for creating, returning, updating, and deleting simple resources, there's no need to create DTOs that are just used for communication. For example, there's no need to have a `BookData` class to write information to the database and a `BookDTO` class for communication, implementing the same properties. With such a design in place, a change to the book would result in changing `BookData` and `BookDTO`, as well as the implementation to transform these objects. Using the same `Book` class with every of these scenarios reduces the programming effort, and also reduces the memory and CPU usage during runtime.

A reason to use different types for data access and communication is the requirements for the data mapping to the database and the requirements of the serializer used with communication. Nowadays, both EF Core and the `System.Text.Json` serializer support constructors with parameters, which might fill the requirements.

If there are other requirements, DTOs can become important. With the games API, different data should be used with the communication than used internally with the service. In creating a game, not all the data for the game is coming from the client. Much of this data, such as the list of available fields as well as the code, is generated on the server. When sending a move from the client to the server, a game gets updated on the server, but there's only a subset of a move needed that's sent from the client to the server. When returning information from the server to the client, again, only a subset of data is required. This makes creating DTOs important.

With the games API, to start a new game, we implement the `CreateGameRequest` and `CreateGameResponse` class record types:

Codebreaker.GameAPIs/Models/GameAPIModels.cs

```
public enum GameType
{
  Game6x4,
```

```
    Game6x4Mini,
    Game8x5,
    Game5x5x4
}

public record class CreateGameRequest(GameType GameType,
    string PlayerName);

public record class CreateGameResponse(Guid id, GameType GameType,
    string PlayerName, int NumberCodes, int MaxMoves)
{
    public required IDictionary<string, IEnumerable<string>>
        FieldValues { get; init; }
}
```

In creating a game, the client just needs to send the game type and the player's name. This is all that's needed from the client. In the backend, just a string is used for the game type. Using a string allows for easy enhancements of other game types. The API using an enum type allows showing the available game types with the OpenAPI description, as you'll see in the section about OpenAPI.

After the game is created, the client just needs to have the identifier of the game. For convenience, the game type and player name are also returned. The client should also know the possible fields that can be selected. This is specified by the FieldValues dictionary.

To set a game move, we implement the UpdateGameRequest and UpdateGameResponse class record types:

Codebreaker.GameAPIs/Models/GameAPIModels.cs

```
public record class UpdateGameRequest(Guid Id, GameType GameType,
string PlayerName, int MoveNumber, bool End = false)
{
    public string[]? GuessPegs { get; set; }
}

public record class UpdateGameResponse(
    Guid Id,
    GameType GameType,
    int MoveNumber,
    bool Ended,
    bool IsVictory,
    string[]? Results);
```

When sending a move, the client needs to send the list of guess pegs. With the API service, guess pegs and key pegs are represented with strings, such as the `GuessPegs` property. This makes it independent of any game type. With the analyzers, different types for every game type are implemented. The `IParsable` interface is used to convert string values.

Implementing the games service

To simplify the implementation of the endpoints, we create the `GamesService` class, which will be injected into the endpoints:

Codebreaker.GameAPIs/Services/GamesService.cs

```
public class GamesService(IGamesRepository dataRepository) :
IGamesService
{
  public async Task<Game> StartGameAsync(string gameType,
    string playerName, CancellationToken cancellationToken = default)
  {
    Game game = GamesFactory.CreateGame(gameType, playerName);

    await dataRepository.AddGameAsync(game, cancellationToken);
    return game;
  }
// code removed for brevity
```

`StartGameAsync` just invokes the `AddGameAsync` method with `IGamesRepository`. Such a simple implementation is the case with many other methods of `GamesService`. This will change in the next chapter when data will be cached in memory before accessing the database. When using the in-memory repository, this is not necessary.

Implementing `SetMoveAsync` is more complex, as here we have to decide to use one of the game analyzers to calculate the game. For the game type selection and the calculation, the `ApplyMove` method is defined as an extension method for the Game type within the `GamesFactory` class:

Codebreaker.GameAPIs/Services/GamesFactory.cs

```
public static Move ApplyMove(this Game game, string[] guesses, int
moveNumber)
{
  static TField[] GetGuesses<TField>(IEnumerable<string> guesses)
    where TField: IParsable<TField> => guesses
      .Select(g => TField.Parse(g, default))
      .ToArray();
```

```
string[] GetColorGameGuessAnalyzerResult()
{
  ColorGameGuessAnalyzer analyzer =
    new (game, GetGuesses<ColorField>(guesses), moveNumber);
  return analyzer.GetResult().ToStringResults();
}
// code removed for brevity
string[] results = game.GameType switch
{
  GameTypes.Game6x4 => GetColorGameGuessAnalyzerResult(),
  GameTypes.Game8x5 => GetColorGameAnalyzerResult(),
  // code removed for brevity
};
Move move = new(Guid.NewGuid(), moveNumber)
{
  GuessPegs = guesses,
  KeyPegs = results
}
game.Moves.Add(move);
return move;
}
```

The implementation of this method uses pattern matching with the `switch` expression to invoke the correct analyzer class to get the result of the game move.

Having this `Game` extension method in place let's us switch back to the implementation of the `SetMoveAsync` method of the `GamesService` class:

Codebreaker.GameAPIs/Services/GamesService.cs

```
public async Task<(Game game, string Result)> SetMoveAsync(
  Guid id, string[] guesses, int moveNumber,
    CancellationToken cancellationToken = default)
{
  Game? game = await dataRepository.GetGameAsync(id, cancellationToken);
  CodebreakerException.ThrowIfNull(game);
  CodebreakerException.ThrowIfEnded(game);

  Move move = game.ApplyMove(guesses, moveNumber);
  await dataRepository.AddMoveAsync(game, move, cancellationToken);
  return (game, move);
}
```

The `SetMoveAsync` method retrieves the game from the repository before invoking the `ApplyMove` method to do the calculation.

Converting transfer objects to and from the object model

In receiving a `CreateGameRequest` to create a new game, there's no need for a conversion. The members of `CreateGameRequest` can directly be used when using `GamesService`. We need to create a conversion from a `Game` type to `CreateGameResponse`. This is done as an extension method for the `Game` type:

Codebreaker.GameAPIs/Extensions/ApiModelExtensions.cs

```
public static partial class ApiModelExtensions
{
  public static CreateGameResponse ToCreateGameResponse(
    this Game game) =>
    new(game.Id, Enum.Parse<GameType>(game.GameType), game.PlayerName)
    {
      FieldValues = game.FieldValues;
    };
    // code removed for brevity
```

With the implementation, the needed data from the `Game` type is transferred to the `CreateGameResponse` type.

Creating endpoints for the Games API service

Before creating the endpoints, the services we created need to be registered with the dependency injection container to inject them within the endpoint:

Codebreaker.GameAPIs/Program.cs

```
builder.Services.AddSingleton<IGamesRepository,
GamesMemoryRepository>();
builder.Services.AddScoped<IGamesService, GamesService>();
```

`GamesMemoryRepository` was created to store game objects in memory. This is registered as a singleton for creating a single instance that's injected into the endpoints. Games should be kept as long as the server keeps running. `GamesMemoryRepository` implements `IGamesRepository`. In the next chapter, an EF Core context will be created to implement the same interface, which will allow changing the **Dependency Injection (DI)** configuration to use the database instead of the in-memory repository. The `GamesService` class implements the `IGamesService` interface. This service class is registered scoped; thus one instance is created with every HTTP request.

With the middleware, we invoke one extension method that defines all the games API endpoints:

Codebreaker.GameAPIs/Program.cs

```
app.MapGameEndpoints();
app.Run();
```

The `MapGameEndpoints` method is an extension method for the `IEndpointRouteBuilder` interface and is implemented next.

In creating a REST API, different HTTP verbs are used to read and write resources:

- GET – With an HTTP GET request, resources are returned from the service
- POST – The HTTP POST request creates a new resource
- PUT – PUT is usually used to update a complete resource
- PATCH – With PATCH, a partial resource can be sent for an update
- DELETE – The HTTP DELETE request deletes a resource

Creating games with HTTP POST

Let's start with the endpoint to create a new game by mapping an HTTP POST request:

Codebreaker.GameAPIs/Endpoints/GameEndpoints1.cs

```
namespace Codebreaker.GameAPIs.Endpoints;

public static class GameEndpoints
{
  public static void MapGameEndpoints(
    this IEndpointRouteBuilder routes)
  {
    var group = routes.MapGroup("/games");

    group.MapPost("/", async (
      CreateGameRequest request,
      IGamesService gameService,
      HttpContext context,
      CancellationToken cancellationToken) =>
    {
      Game game;
      try
      {
        game = await gameService.StartGameAsync(request.
```

```
        GameType.ToString(), request.PlayerName, cancellationToken);
    }
    catch (CodebreakerException) when (
      ex.Code == CodebreakerExceptionCodes.InvalidGameType)
    {
      GameError error = new(ErrorCodes.InvalidGameType,
        $"Game type {request.GameType} does not exist",
        context.Request.GetDisplayUrl(),
        Enum.GetNames<GameType>());
      return Results.BadRequest(error);
    }
  return Results.Created($"/games/{game.Id}",
    game.ToCreateGameResponse());
});
```

> **Note**
>
> In the source code repository, you will find the GameEndpoints.cs and GameEndPoints1.cs files. The current state of the endpoints is in the GameEndpoints1.cs file, but this will be changed later on when OpenAPI information is added. The file that gets compiled as defined in the project file is GameEndpoints.cs. If you want to compile the project from the repository with the current version, change the settings of the C# file in the project file.

The MapGameEndpoints method is – as previously mentioned – an extension method for the IEndpointRouteBuilder interface. The first method invoked is MapGroup to define common functionality for the endpoints, which, in turn, uses the returned group variable (a RouteGroupBuilder object). With this code, the common functionality is the /games URI, which will be prefixed. You can use this for common authorization needs or common logging, which will be shown in later chapters of this book. Common functionality for OpenAPI will be shown later in this chapter in the section on OpenAPI.

The created group is used with the MapPost method. The MapPost method will be invoked on an HTTP POST request. Similarly, MapGet, MapPut, and MapDelete are available as well. All these methods offer two overloads, where the overload with a pattern and the Delegate parameter are used. The Delegate parameter allows passing a lambda expression with any parameters and any return types – this is what minimal APIs take advantage of.

The parameter types specified with the MapPost method are listed here:

- CreateGameRequest – This comes from the HTTP body.

- IGamesService – The value is injected from the DI container.

- HttpContext and CancellationToken – These are special types bound with minimal APIs. Another special one is ClaimsPrincipal, which is used with authentication.

Other binding sources that can be used are route values, query strings, headers, and HTML form values. You can also add custom binding to map a route, query, or header binding to custom types.

You can add attributes such as `FromBody`, `FromRoute`, `FromServices`, and others, which can help with readability and also resolve issues if there's a conflict.

With the implementation of the lambda expression with the `MapPost` method, the injected `gameService` variable is used to start a game. Starting the game successfully, a Game-derived object is returned, which is converted to `CreateGameResponse` with the `ToCreateGameResponse` extension method. The method returns either `HTTP 201 Created` on success, or `HTTP 400 Bad Request` using the `Results` factory class. `Results` offers methods to return the different HTTP status codes.

Using `Results.Created`, a URI is assigned to the first parameter of the method, and the second parameter receives the created object. The `201 Created` status code is used to return the HTTP location header with a link that can be remembered by the client to retrieve the same resource at a later time. The link that's returned here can be used with an HTTP `GET` request to retrieve the game at a later time.

Defining an error object

In case of an error, the `MapPost` method returns `Results.BadRequest`. Here, we can define an object to return detailed error information to the client.

The following code snippet shows the `GameError` class:

Codebreaker.GameAPIs/Errors/GameError.cs

```
public record class GameError(string Code, string Message,
    string Target, string[]? Details = default);

public class ErrorCodes
{
    public const string InvalidGameType = nameof(InvalidGameType);
    // code removed for brevity
}
```

The `GameError` class defines the `Code`, `Message`, `Target`, and `Details` properties. With the `MapPost` method, in the case of requesting an invalid game type, the valid game types are returned with the details.

Returning games with HTTP GET

To fulfill a request to a single game, we use the `MapGet` method:

Codebreaker.GameAPIs/Endpoints/GameEndpoints1.cs

```
group.MapGet("/{id:guid}", async (
  Guid id,
  IGamesService gameService,
  CancellationToken cancellationToken
) =>
{
  Game? game = await gameService.GetGameAsync(id, cancellationToken);

  if (game is null)
  {
    return Results.NotFound();
  }

  return Results.Ok(game);
});
```

With the lambda parameters, here, the `Guid` is received from the route parameter. Within curly braces, the same variable name is used that matches the variable of `Guid`. The implementation simply returns the game with the status code of `OK`, or `not found` if the game ID is not found within the repository.

Updating games by setting a move with HTTP PATCH

To set a move, a game is updated with a move without sending the complete game. That's a partial update, thus we use the HTTP `PATCH` verb to invoke the `MapPatch` method:

Codebreaker.GameAPIs/Endpoints/GameEndpoints1.cs

```
group.MapPatch("/{id:guid} ", async (
  Guid id,
  UpdateGameRequest request,
  IGamesService gameService,
  HttpContext context,
  CancellationToken cancellationToken) =>
{
  try
  {
    (Game game, string result) = await gameService.SetMoveAsync(
        id, request.GuessPegs, request.MoveNumber, cancellationToken);
```

```
        return Results.Ok(game.AsUpdateGameResponse(result));
    }
    catch (CodebreakerException ex) when (
        ex.Code == CodebreakerExceptionCodes.GameNotFound)
    {
        return Results.NotFound();
    }
    // code removed for brevity
});
```

The complete route for this request is games/{id} – prefixed with the pattern specified by the group. The SetMoveAsync method of GamesService does the main work. On success, the method returns UpdateGameResponse, which is created from the Game object and the result string.

Configuring JSON serialization

To successfully run the application, the JSON serialization needs to be configured. The .NET JSON serializer has many configuration options – including polymorphic serialization (returning a hierarchy of objects). But with all these features supported from serialization and with ASP.NET Core, you need to pay attention to the client technologies you use, if the same support is available there. With the data types we transfer here, there's not a lot to do.

Just to serialize the enumeration value of game types, we prefer to use strings instead of just numbers, which would be returned by default. To return string values, the JSON options are configured:

Codebreaker.GameAPIs/Models/GameAPIModels.cs

```
[JsonConverter(typeof(JsonStringEnumConverter<GameType>))]
public enum GameType
{
    Game6x4,
    Game6x4Mini,
    Game8x5,
    Game5x5x4
}
```

The generic version of the JsonStringEnumConverter class is new since .NET 8 to support Native AOT. The non-generic version of this type uses reflection, which is incompatible with Native AOT.

> **Note**
>
> Instead of using the attribute with the type, you can also configure the JSON serialization behavior with the dependency injection container. The `ConfigureHttpJsonOptions` extension method with `JsonOptions` from the `Microsoft.AspNetCore.Http.Json` namespace can be used to configure the JSON serialization for minimal APIs. Be aware that the OpenAPI generation still uses the MVC serializer configuration, thus, here, you need to use the `Configure` method with `Microsoft.AspNetCore.Mvc.JsonOptions` as a generic parameter.

Creating endpoint filters

To simplify the code for endpoints, as you've seen, endpoints don't need to be specified within top-level statements. You can create multiple classes with extension methods to group endpoints together. Within one extension method, you can also group endpoints with common behaviors using the `MapGroup` method. We also used dependency injection to have the main functionality outside of the endpoint implementation in the `GameService` class.

There's another way to simplify the implementation of the endpoints – by using custom endpoint filters. Endpoint filters offer similar functionality as ASP.NET Core middleware, just with a different scope. By adding an endpoint filter to an endpoint, the filter code is invoked every time the endpoint is accessed. Adding multiple endpoint filters, the order of adding these is important: one filter is invoked after the other. You can also add an endpoint filter to a group; thus the filter is invoked with every endpoint specified with this group.

With `ValidatePlayernameEndpointFilter`, we create a filter to validate the minimum length of the player's name:

Codebreaker.GameAPIs/Endpoints/ValidatePlayernameEndpointFilter.cs

```
public class ValidatePlayernameEndpointFilter : IEndpointFilter
{
  public async ValueTask<object?>
    InvokeAsync(EndpointFilterInvocationContext context,
    EndpointFilterDelegate next)
  {
    CreateGameRequest request =
      context.GetArgument<CreateGameRequest>(0);
    if (request.PlayerName.Length < 4)
    {
      return Results.BadRequest("
        Player name must be at least 4 characters long");
    }
```

```
        return await next(context);
    }
}
```

An endpoint filter implements the IEndpointFilter interface. This interface defines the InvokeAsync method with EndpointFilterInvocationContext and EndpointFilterDelegate parameters. By using EndpointFilterInvocationContext, you can access HttpContext to access all the information about the request and also add responses, as well as the parameters passed to the endpoint. With the implementation of ValidatePlayernameEndpointFilter, the player's name is validated by accessing the first parameter of the endpoints lambda expression, accessing the CreateGameRequest object, and accessing the PlayerName property. An HTTP status code of 400 Bad Request is returned in the case that this does not succeed. To access the different parameters, the index of the parameter is needed. With a successful validation, the next endpoint is invoked using the next variable received with the second parameter.

Using an endpoint filter, you can also remove the exception handling code and create code with the filter that's invoked before and after the next filter or the endpoint code is invoked. We implement this functionality with CreateGameExceptionEndointFilter:

Codebreaker.GameAPIs/Endpoints/CreateGameExceptionEndpointFilter.cs

```
public class CreateGameExceptionEndpointFilter : IEndpointFilter
{
    private readonly ILogger _logger;
    public CreateGameExceptionEndpointFilter
        (ILogger<CreateGameExceptionEndpointFilter> logger)
    {
        _logger = logger;
    }
    public async ValueTask<object?>
        InvokeAsync(EndpointFilterInvocationContext context,
        EndpointFilterDelegate next)
    {
        CreateGameRequest request =
            context.GetArgument<CreateGameRequest>(1);
        try
        {
```

```
      return await next(context);
    }
    catch (CodebreakerException ex) when (
      ex.Code == CodebreakerExceptionCodes.InvalidGameType)
    {
      _logger.LogWarning("game type {gametype} not found",
        request.GameType);
      return Results.BadRequest("Gametype does not exist");
    }
  }
}
```

The `CreateGameExceptionEndpointFilter` endpoint filter defines a constructor to inject the `ILogger` interface. Everything registered with the dependency injection container can be injected with an endpoint filter. With this filter, the invocation of the next filter – or the endpoint itself, in this case – is surrounded by a `try/catch` block. Thus, `CodebreakerException` is caught with the endpoint invocation, logged, and a result returned. This way, the exception-handling code can be removed from the endpoint itself. The following code snippet shows the new code for the endpoint implementation with the configured filters:

Codebreaker.GameAPIs/Endpoints/GameEndpoints2.cs

```
group.MapPost("/", async (
  CreateGameRequest request,
  IGamesService gameService,
  CancellationToken cancellationToken) =>
{
  Game game = await gameService.StartGameAsync(request.
    GameType.ToString(), request.PlayerName, cancellationToken);
  return Results.Created($"/games/{game.Id}",
    game.ToCreateGameResponse());
      }).AddEndpointFilter<ValidatePlayernameEndpointFilter>()
        .AddEndpointFilter<CreateGameExceptionEndpointFilter>();
```

With this implementation, the important main functionality of this endpoint is easily visible. Validation and error handling is moved outside of the endpoint implementation. Calculating all the code together, the code is not becoming smaller, just the main functionality of the endpoint is becoming easier.

The real power of endpoint filters comes with shared functionality between different endpoints. In *Chapter 11*, *Logging and Monitoring*, we will create an endpoint filter that logs information for every endpoint that is configured with this filter.

Running the service

With the endpoints implemented, you can run and test the application and see the Open API user interface, as shown in *Figure 2.4*:

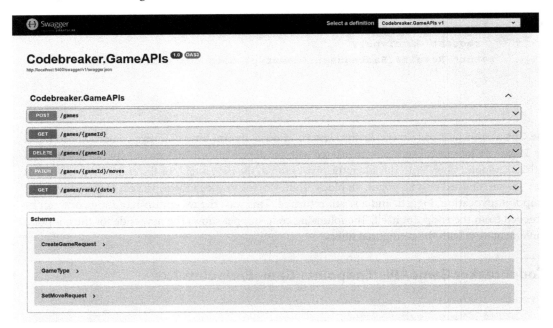

Figure 2.4 – OpenAPI UI without extra configuration

You can now run the application sending a POST request from the Swagger page. After getting the game information returned, copy the unique ID from the game, and send a PATCH request to set a move.

While the application is running, OpenAPI now lacks some information. The returned HTTP results are not listed, more description details should be shown, and the schema for return types is not available. This is solved in the next section.

OpenAPI information

Since there are many .NET versions, the Web API templates reference the Swashbuckle. AspNetCore NuGet package to create an OpenAPI description. Over the years, with later .NET versions, more and more functionality from OpenAPI has been added to ASP.NET Core itself, such as the OpenApiInfo class, which is now part of the Microsoft.OpenApis.Models namespace. Swashbuckle was changed to use the new classes and was also changed to be based on the System. Text.Json serializer instead of Newtonsoft.Json.

Adding OpenAPI documentation

Next, let's make use of classes from the `Microsoft.OpenApis.Models` namespace together with Swashbuckle to configure the OpenAPI documentation:

Codebreaker.GameAPIs/Program.cs

```
builder.Services.AddEndpointsApiExplorer();
builder.Services.AddSwaggerGen(options =>
{
  options.SwaggerDoc("v3", new OpenApiInfo
  {
    Version = "v3",
    Title = "Codebreaker Games API",
    Description =
      "An ASP.NET Core minimal APIs to play Codebreaker games",
    TermsOfService = new Uri("https://www.cninnovation.com/terms"),
    Contact = new OpenApiContact
    {
      Name = "Christian Nagel",
      Url = new Uri("https://csharp.christiannagel.com")
    },
    License = new OpenApiLicense
    {
      Name="API usage license",
      Url= new Uri("https://www.cninnovation.com/apiusage")
    }
  });
});
```

The Codebreaker Games API is already in its third major version. With the Swagger configuration of the dependency injection container, the `AddSwaggerGen` method supports receiving configuration options. The `SwaggerDoc` method of the `SwaggerGenOptions` options parameter allows specifying different document values, such as the version number, title, description, terms of service, contact, and license information, as shown in the code snippet. The `OpenApiInfo`, `OpenApiContact`, and `OpenApiLicense` classes are part of the `Microsoft.OpenApi.Models` namespace. All this configured information will show up in the generated OpenAPI documentation.

The middleware configuration needs to be updated as well:

Codebreaker.GameAPIs/Program.cs

```
app.UseSwagger();
app.UseSwaggerUI(options =>
```

```
{
  options.SwaggerEndpoint("/swagger/v3/swagger.json", "v3");
});
```

Here, the version number is included with the OpenAPI endpoint invoking the `SwaggerEndpoint` method using the `SwaggerUIOptions` parameter. In case you prefer a different look for the generated document, you can create your own style sheet, and pass the style sheet file by invoking the `options.InjectStylesheet` method.

Documentation for the endpoints

Several extension methods are available to add OpenAPI documentation to the endpoints, as we'll do within the `MapGameEndpoints` implementation:

Codebreaker.GameAPIs/Endpoints/GameEndpoints.cs

```
public static void MapGameEndpoints(this IEndpointRouteBuilder routes)
{
  var group = routes.MapGroup("/games")
    .WithTags("Games API");
```

The `WithTags` method could be added to every endpoint or, as shown here, to the endpoint group. `WithTags` adds a category name – or multiple names if the API should show up with multiple categories. With this, for the documentation for every tag name, a heading is used to show all the endpoints together that belong to the same category. If you don't supply a tag name, the project name is used, and all the endpoints are listed in this category.

Next, using `RouteHandlerBuilder` extension methods, documentation is added to every endpoint:

Codebreaker.GameAPIs/Endpoints/GameEndpoints.cs

```
group.MapPost("/", async Task<Results<Created<CreateGameResponse>,
BadRequest<GameError>>> (
// code removed for brevity
})
.WithName("CreateGame")
.WithSummary("Creates and starts a game")
.WithOpenApi(op =>
{
  op.RequestBody.Description = "
    The game type and the player name of the game to create";
  return op;
});
```

Using the WithName, WithSummary, and WithOpenApi methods, a name and description of the API – including a description of every parameter – can be added.

Adding return type information to OpenAPI

The Produces extension method could be used to define what types are returned from endpoints that need to be described. Since .NET 7, there's a better option: using the TypedResults class instead of the Results class that was used earlier in implementing the endpoints. TypedResults adds the class specified to the OpenAPI documentation. However, if there is more than one type returned, we need to specify this with the return types of the endpoint lambda expression.

The first method to be changed to TypedResults is MapDelete:

Codebreaker.GameAPIs/Endpoints/GameEndpoints.cs

```
group.MapDelete("/{id:guid}", async (
  Guid id,
  IGamesService gameService,
  CancellationToken cancellationToken
) =>
{
  await gameService.DeleteGameAsync(id, cancellationToken);

  return TypedResults.NoContent();
})
// code removed for brevity
```

The implementation of the lambda expression for the MapDelete method just returns the HTTP status code 204 and doesn't require a return type for the lambda expression.

This is different with the MapPatch implementation:

```
group.MapPatch("/{id:guid}/moves", async
Task<Results<Ok<UpdateGameResponse>, NotFound, BadRequest<GameError>>>
(
  Guid id,
  UpdateGameRequest request,
  IGamesService gameService,
  HttpContext context,
  CancellationToken cancellationToken) =>
{
  try
  {
    (Game game, string result) = await gameService.SetMoveAsync(id,
      request.GuessPegs, request.MoveNumber, cancellationToken);
```

```
    return TypedResults.Ok(game.AsUpdateGameResponse(result));
  }
  catch (ArgumentException ex) when (ex.HResult >= 4200 &&
    ex.HResult <= 4500)
  {
    string url = context.Request.GetDisplayUrl();
    return ex.HResult switch
    {
      4200 => TypedResults.BadRequest(new GameError(
        ErrorCodes.InvalidGuessNumber, "Invalid number of guesses
        received", url)),
      4300 => TypedResults.BadRequest(new GameError(
        ErrorCodes.InvalidMoveNumber, "Invalid move number received",
        url)),
    // code removed for brevity
    };
  }
  catch (GameNotFoundException)
  {
    return TypedResults.NotFound();
  }
})
// code removed for brevity
```

The implementation of the lambda expression contains several invocations of the TypedResults factory class. The BadRequest, NotFound, and Ok methods are invoked. With BadRequest, an object of the GameError type is returned. This is just a simple record class with a Message and other properties to return useful information for the clients. The Ok method returns an object of the UpdateGameResponse class. With two or more typed results returned, the lambda expression needs a return type. To specify the return type with the lambda expression, .NET 7 added generic Results types – for example, the type with two generic parameters: Results<TResult1, TResult2>. The generic types specify the constraint to require the IResult interface. The Microsoft.AspNetCore.Http.HttpResults namespace contains generic Results types from two to six generic parameters. By using Results<Ok<UpdateGameResponse>, NotFound, BadRequest<InvalidGameMoveError>>, it's defined that the method either returns an Ok result with a SetMoveError object, NotFound, or BadRequest with an InvalidGameMoveError object. Because the lambda expression makes use of the async keyword, the complete result is put into a task: Task<Results<Ok<UpdateGameResponse>, NotFound, BadRequest<InvalidGameMoveError>>>.

With all this OpenAPI configuration in place, we can start the service, generate the documentation to describe the API, and use it from a web interface, as shown in *Figure 2.5*.

Figure 2.5 – OpenAPI documentation

Opening an endpoint, you can see the different HTTP results returned and all the schemas created.

Testing the service

To run the service, the OpenAPI test page can be used. However, there's a better way without leaving Visual Studio or Visual Studio Code. Visual Studio offers the **Endpoints Explorer** window to show all the API endpoints in your solution, as shown in *Figure 2.6*. Open this window using **View | Other Windows | Endpoints Explorer**:

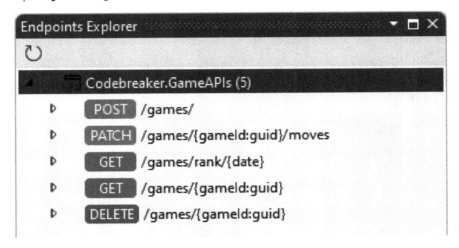

Figure 2.6 – Endpoints Explorer

By selecting an endpoint and opening the context menu, you can generate a request. This creates an HTTP file that you can use to send HTTP requests, including the body, and see the returned results beside it.

> **Using HTTP files with Visual Studio Code**
>
> If you use Visual Studio Code, install the REST Client extension from Huachao Mao.

First, we start a game by sending a POST request:

Codebreaker.GameAPIs/Codebreaker.GameAPIs.http

```
@HostAddress = http://localhost:9400
@ContentType = application/json

### Create a game
POST {{HostAddress}}/games/
Content-Type: {{ContentType}}

{
```

```
  "gameType": "Game6x4",
  "playerName": "test"
}
```

The first two lines in the HTTP file specify variables that are referenced later surrounded by two curly braces. Every request specified needs to be separated by three hash characters, ###. With this, using Visual Studio or the Visual Studio Code extension, you can see a green arrow that, when clicked, sends the request. To separate the HTTP header from the body, a blank line is required. The body to create a game contains the gameType and playerName JSON elements.

After sending the request to create a game, you can set a move:

Codebreaker.GameAPIs/Codebreaker.GameAPIs.http

```
### Set a move

@id = 1eae1e79-a7fb-41a6-9be8-39f83537b7f3

PATCH {{HostAddress}}/games/{{id}}/moves
Content-Type: {{ContentType}}

{
  "gameType": "Game6x4",
  "playerName": "test",
  "moveNumber": 1,
  "guessPegs": [
    "Red",
    "Green",
    "Blue",
    "Yellow"
  ]
}
```

Before sending the move, get the ID that was returned on creation of the game, and paste it into the id variable used with the HTTP file. There's no need to save this file; just clicking on the link will send the PATCH request. Remember to update the move number with every move.

To get information about the game, send a GET request:

Codebreaker.GameAPIs/Codebreaker.GameAPIs.http

```
### Get game information

GET {{HostAddress}}/games/{{id}}
```

When sending the GET request, a body is not supplied. This request gives complete information about the game, including its moves, and the result code.

By using HTTP files, you can easily debug the APIs without leaving Visual Studio. In case you're in a debug session, you might need to change the **Request timeout** setting in **Text Editor | Rest | Advanced settings**. The HTTP files also serve as good documentation as part of the project.

Enabling .NET Aspire

Let's add .NET Aspire to this solution. Using Visual Studio, you can select the Games API project in **Solution explorer** and use the context menu to select **Add | .NET Aspire Orchestrator Support…**.

> **Note**
> Instead of using Visual Studio, you can use the dotnet CLI to create a .NET Aspire project: dotnet new aspire. With the two projects created, you can connect the Games API project using the explanations from this section.

This creates two projects ({solution}.AppHost and {solution}.ServiceDefaults) and makes small changes to the Games API project. The AppHost project is a web application that runs a dashboard to monitor all the configured services. ServiceDefaults is a library to specify default configurations. This library is referenced by the Games API. Let's look into the details next.

Exploring the Aspire host

The source code of the startup code of the Aspire host is shown here:

Codebreaker.AppHost/Program.cs

```
var builder = DistributedApplication.CreateBuilder(args);

builder.AddProject<Projects.Codebreaker_GameAPIs>("gameapis");

builder.Build().Run();
```

This Aspire host runs a web application that's only used during development time. Here, the DistributedApplication class is used with the app builder pattern. The CreateBuilder method returns the IDistributedApplicationBuilder interface, which allows the configuration of all the services that should be orchestrated by the distributed application. Using this interface, similar to WebApplicationBuilder, the configuration and DI container can be configured. Contrary to WebApplicationBuilder, resources can be added that are orchestrated by the app host.

A project resource is added to the resources of the builder by invoking the AddProject extension method. Adding a project reference to the Codebreaker.GameAPIs project creates the Codebreaker_GameAPIs class in the Projects namespace. The name passed with the parameter – gameapis – can be used when one service references another one. Currently, we only have one service in the Codebreaker solution, but this will change going forward with the chapters of this book.

Exploring the ServiceDefaults library

The Codebreaker.ServiceDefaults project is here to be referenced by all service projects making use of .NET Aspire functionality. This project adds references to NuGet packages for HTTP resiliency, service discovery, and several OpenTelemetry packages used for logging, metrics, and distributed tracing.

This is the configuration defined by the Extensions class:

Codebreaker.ServiceDefaults/Extensions.cs

```
public static IHostApplicationBuilder AddServiceDefaults(this
IHostApplicationBuilder builder)
{
  builder.ConfigureOpenTelemetry();

  builder.AddDefaultHealthChecks();

  builder.Services.AddServiceDiscovery();

  builder.Services.ConfigureHttpClientDefaults(http =>
  {
    http.AddStandardResilienceHandler();
    http.AddServiceDiscovery();
  });

  return builder;
}
```

The AddServiceDefaults method configures the DI container with OpenTelemetry for logging, default health checks, and service discovery, and configures the HttpClient class with resiliency.

This complete configuration can be used from the Codebreaker Games API by invoking the AddServiceDefaults method with the application builder. This invocation is added from the Visual Studio .NET Aspire integration when adding Aspire orchestration. If you created the .NET Aspire projects from the command line, you need to add the invocation to the AddServiceDefaults method when configuring the DI container.

Another extension method defined by the service defaults library is `MapDefaultEndpoints`:

Codebreaker.ServiceDefaults/Extensions.cs

```
public static WebApplication MapDefaultEndpoints(this WebApplication
app)
{
  app.MapHealthChecks("/health");
  app.MapHealthChecks("/alive", new HealthCheckOptions
  {
    Predicate = r => r.Tags.Contains("live")
  });
  return app;
}
```

This is an extension method to configure the ASP.NET Core middleware with health checks. The `MapDefaultEndpoints` method needs to be invoked with the middleware configuration of the Games API.

All the functionality mentioned here will be discussed in detail in the following chapters. Now, you can already start the Aspire host project, which, in turn, starts up the Games API.

Running the .NET Aspire host

When running the .NET Aspire host, a dashboard is shown that shows the `codebreaker.gameapis` project with the resources where you can access endpoints, check environment variables, and open log information, as shown in *Figure 2.7*.

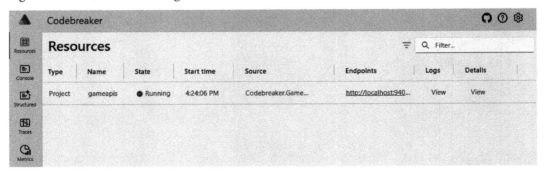

Figure 2.7 – Resources in the .NET Aspire dashboard

While the Aspire host is running, you can play games using the HTTP files created earlier, and now monitor the logs, metrics information, and traces (shown in *Figure 2.8*).

Figure 2.8 – Traces in the .NET Aspire dashboard

Details on logging, metrics, tracing, and adding custom information are covered in *Chapter 11*, *Logging and Monitoring*.

Summary

In this chapter, we created a Web API using ASP.NET Core minimal APIs. We covered creating services and an in-memory repository, configured them with the dependency injection container, created models, and used game analyzer classes to calculate moves.

We created endpoints to create, read, and update games, specified information to show up with the OpenAPI documentation, tested the service using HTTP files, and finally, added .NET Aspire for hosting and a dashboard.

After working through this chapter, you deserve a break to play a game. Use the HTTP files to create a game and set moves until the answer returned shows that you won. Don't cheat by making GET requests to the game before you find the answer!

In the next chapter, we'll replace the repository by using Entity Framework Core with SQL Server and Azure Cosmos DB to have a persistent games store.

Further reading

To learn more about the topics discussed in this chapter, you can refer to the following links:

- *Primary constructors*: `https://csharp.christiannagel.com/2023/03/28/primaryctors/`

- *Microsoft REST API Guidelines*: `https://github.com/microsoft/api-guidelines/blob/vNext/Guidelines.md`

- *The OpenAPI Initiative*: `https://www.openapis.org/`

- *Minimal APIs parameter binding*: `https://learn.microsoft.com/aspnet/core/fundamentals/minimal-apis/parameter-binding`

- *JSON serialization customization*: `https://learn.microsoft.com/dotnet/standard/serialization/system-text-json`

- The `Swashbuckle.AspNetCore` GitHub repo: `https://github.com/domaindrivendev/Swashbuckle.AspNetCore`

- .NET Aspire documentation: `https://learn.microsoft.com/en-us/dotnet/aspire/`

3

Writing Data to Relational and NoSQL Databases

After creating the first implementation of the service using minimal APIs, we build on that to read and write to databases. In this chapter, we will replace the in-memory repository built in *Chapter 2* using **Entity Framework Core (EF Core)** to access a relational database – Microsoft SQL Server – and an Azure Cosmos DB NoSQL database using EF Core.

You'll create two libraries to access these databases, create EF Core context classes, specify the mappings from model classes, and configure the minimal APIs service to use one or the other database. After adding these changes, the games will be persisted and you can continue game runs when the service is restarted.

In this chapter, you'll be exploring these topics:

- Exploring the models for the data to be stored in the databases
- Creating and configuring an EF Core context to access Microsoft SQL Server
- Creating migrations to update the database schema
- Creating and configuring an EF Core context to access Azure Cosmos DB

Technical requirements

The code for this chapter can be found in the following GitHub repository: `https://github.com/PacktPublishing/Pragmatic-Microservices-with-CSharp-and-Azure`.

The `ch03` source code folder contains the code samples for this chapter. The most important projects for this chapter are the following:

- `Codebreaker.Data.SqlServer` – This is the new library to access Microsoft SQL Server.
- `Codebreaker.Data.Cosmos` – This is the new library to access Azure Cosmos DB.
- `Codebreaker.GamesAPIs` – This is the web API project created in the previous chapter. In this chapter, the **dependency injection (DI)** container is updated to use .NET Aspire components to use SQL Server and Azure Cosmos DB.
- `Codebreaker.GameAPIs.Models` – This project just has a minimal change in this chapter, adding a property to the `Game` class.
- `Codebreaker.AppHost` – This project is updated with SQL Server and Azure Cosmos DB resources and forwarding configuration values.
- `Codebreaker.ServiceDefaults` – This project is unchanged from the previous chapter.

The `analyzer` library from the previous chapter is not included in this chapter. Here, we'll just use the `CNinnovation.Codebreaker.Analyzers` NuGet package.

If you worked through the previous chapter to create the models and implemented the minimal APIs project, you can continue from there. You can also copy the files from the `ch02` folder if you didn't complete the previous work and start from there. `ch03` contains all the updates from this chapter.

Other than a development environment, you need Microsoft SQL Server and Azure Cosmos DB. You don't need an Azure subscription at this point. SQL Server is installed together with Visual Studio. You can also download the SQL Server 2022 Developer Edition instead. This is easy via `winget` (but you can also download and install Windows installer packages instead):

```
winget install Microsoft.SQLServer.2022.Developer
```

If you use a Mac, you can use a Docker image for SQL Server. In *Chapter 5*, you can read more details on Docker and running SQL Server within a Docker container.

For easy use of SQL Server and Azure Cosmos DB, Docker images are used in this chapter. You can also use SQL Server, which is installed together with Visual Studio, and the Azure Cosmos DB emulator instead.

To run Azure Cosmos DB, an emulator to run it locally is available. You can install this NoSQL database emulator with the following command:

```
winget install Microsoft.Azure.CosmosEmulator
```

> **Note**
>
> The Azure Cosmos emulator is only available on Windows. With Linux environments (and also on Windows and the Mac), you can use a Docker image to run the emulator. See `https://learn.microsoft.com/en-us/azure/cosmos-db/how-to-develop-emulator` for information on running the emulator. Read *Chapter 5* for more information on Docker.

To read and write your SQL Server data, within Visual Studio you can use SQL Server Object Explorer. Outside of Visual Studio, and with more functionality, use **SQL Server Management Studio** (**SSMS**), which can be installed with the following command:

```
winget install Microsoft.SQLServerManagementStudio
```

The projects of this chapter and how they relate to each other are shown in *Figure 3.1* with a C4 component diagram. The `gamesAPI` and `models` components have been created in *Chapter 2*. In this chapter, two projects for accessing SQL Server and Azure Cosmos DB databases will be added (`sqlDatabase` and `cosmosDatabase`). Depending on the configuration, the games API will use either the in-memory repository (created in *Chapter 2*) or one of the other `IGamesRepository` implementations:

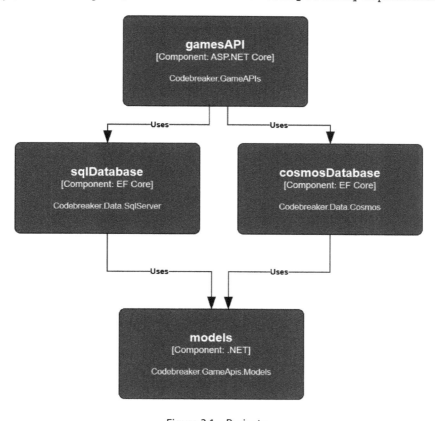

Figure 3.1 – Projects

Let's start exploring the project containing the models while making a small change to the models.

Exploring the models for database mapping

When creating services, different models can be used for the database, the functionality, and the APIs. The database layer might have different requirements than the other layers. When creating a monolithic application, this usually is the case, but it also means that while maintaining the application and adding a field, all the different layers need to be touched and updated. When creating microservices with smaller scopes, there's a good chance to use the same model with the database, the functionality of the application, and the APIs. This not only reduces maintenance costs but also improves performance in that not every layer creates new instances and copies the values around.

With the sample application, the Game and Move types with their generic counterparts created in the previous chapter are not that simple, but it's possible to use them directly with EF Core.

Let's look at the models and what must be mapped for the database, starting with the Game type:

Codebreaker.GameAPIs.Models/Game.cs

```
public class Game(
    Guid id,
    string gameType,
    string playerName,
    DateTime startTime,
    int numberCodes,
    int maxMoves)
{
    public Guid Id { get; private set; } = id;
    public string GameType { get; private set; } = gameType;
    public string PlayerName { get; private set; } = playerName;
    public DateTime StartTime { get; private set; } = startTime;
    // code removed for brevity
    public required IDictionary<string, IEnumerable<string>> FieldValues
        { get; init; }
    public required string[] Codes { get; init; }
    public ICollection<Move> Moves { get; } = new List<Move>();
    public override string ToString() => $"{Id}:{GameType} -
        {StartTime}";
}
```

The Game class contains properties of type Guid, string, DateTime, TimeSpan, int, and bool. All these properties can easily be mapped to database columns. Just the size of the string needs to be configured. With SQL Server, the convention to map a string is nvarchar(max). This can be reduced in size. What's more interesting is the constructor.

The class doesn't define a parameterless constructor. While some tools require a parameterless constructor, both JSON serialization as well as EF Core nowadays don't need one. EF Core supports constructors with parameters as long as the constructors map to simple properties – which is the case with the `Game` class. EF Core mapping supports properties with `get` and `set` accessors. If there's just a `get` accessor available, the mapping will fail. A way around this is to use `private` fields. EF Core supports explicit mapping to fields. Another option is to use private `set` accessors – which are used with the `Game` class.

Some other interesting members are available with the `Game` class: the `FieldValues` property is of type `IDictionary<string, IEnumerable<string>>`. The field values define the possible options the user can choose from. All games of one game type usually have the same field values, but this can change over time. We shouldn't expect these values to always stay the same. The application might change the colors or shapes to choose from over time. So, we can't just ignore the property to be stored – this should be stored with the game. As for this type, a default mapping is not available, so we need to add a conversion.

The `Codes` property is of type string array. EF Core 8.0 supports built-in mapping of collections of primitive types; that is, a list of arrays, integers, strings, and more. With the built-in functionality, the collection is stored in the JSON format in a string table. This fits the purpose. The `Codes` property contains a list of a solution. With the color-based game types, this is a list of up to five colors; with the shape game type, one string of the collection consists of a shape and a color with a delimiter. Using a version older than EF Core 8, a custom conversion would be required. With EF Core 8, this will map with the default functionality.

The `Moves` property is typically a relation with relational databases. Using SQL Server, we will use the `Moves` table to store every move. It would be possible to store moves with a JSON column, but we'll use a separate table and query for moves if needed. Using a NoSQL database, it's a natural way to store moves within the game.

Let's look into the `Move` type:

Codebreaker.GameAPIs.Models/Move.cs

```
public class Move(Guid id, int moveNumber)
{
  public Guid Id { get; private set; } = id;
  public int MoveNumber { get; private set; } = moveNumber;
  public required string[] GuessPegs { get; init; }
  public required string[] KeyPegs { get; init; }

  public override string ToString() => $"{MoveNumber}. " +
    $"{string.Join('#',GuessPegs)} : " +
    $"{string.Join('#', KeyPegs)}";
}
```

With the `Move` class, `GuessPegs` (the guess of the player for the move) and `KeyPegs` (the result from the analyzer) can be serialized similarly to the `Codes` property from the `Game` type. What's more interesting here is what's *not* here. That is, the absence of a foreign key property like `GameId` or a direct `Game` property to establish a relationship between the `Move` and `Game` types. While using the `Move` type so far, this relationship was not required. With EF Core, it's also not required to add this relationship to the model. EF Core supports a feature named **shadow properties**. These properties are not part of the model but are stored within the database and can be accessed while using the EF Core context.

Let's summarize what needs to be done to map the `Game` and `Move` types with EF Core:

1. For simple string properties, the size of the database string needs to be defined with SQL Server.

2. The `FieldValues` property of type `IDictionary<string, IEnumerable<string>>` needs a **value converter**.

3. The `Moves` property maps a collection to the `Move` type. With a relational database, `Move` objects should be stored in a separate `Moves` table. Because the `Move` type doesn't define a primary key, **shadow properties** are needed.

4. Using Azure Cosmos DB, the moves should be stored in the game JSON document.

> **Note**
>
> EF Core supports mapping via conventions, annotations, and the fluent API. Conventions are provider-specific. For example, a .NET string mapping to `nvarchar(max)` is a convention. Using nullability, a non-nullable property maps to a required database column, while a nullable property is not required. Conventions are overridden using annotations. Annotations are attributes such as `[StringLength(20)]`, which not only can be used to validate user input but also to specify the column to be `nvarchar(20)`. Using the fluent API, annotations can be overridden. The fluent API offers most options and overrides all the other settings. We'll use the fluent API in the next sections.

Let's define a mapping to work with these models, both with a relational database and NoSQL.

Using EF Core with SQL Server

Let's start with a relational database to store games and moves in multiple tables. We need to do the following:

1. Create a class library project

2. Create the EF Core context

3. Customize the mapping of simple properties

4. Create value conversions to map complex properties

5. Define relations between games and moves

6. Create shadow properties for the `Move` type

7. Implement the repository contract

8. Configure the application model with SQL Server

9. Configure the DI container with the minimal APIs project

Creating a data class library using SQL Server

To create the class library project, you can use the .NET CLI as shown or use the class library project template from Visual Studio:

```
dotnet new classlib --framework net8.0 -o Codebreaker.Data.SqlServer
```

To access SQL Server with EF Core, add the `Microsoft.EntityFrameworkCore.SqlServer` NuGet package. This project also has a dependency on the `Codebreaker.GameAPIs.Models` project.

Creating an EF Core context for SQL Server

The mapping to the database is specified with an EF Core context implemented with the `GamesSqlServerContext` class, as shown in the following code snippet:

Codebreaker.Data.SqlServer/GamesSqlServerContext.cs

```
public class
GamesSqlServerContext(DbContextOptions<GamesSqlServerContext> options)
: DbContext(options), IGamesRepository
{
  protected override void OnModelCreating(ModelBuilder modelBuilder)
  {
    modelBuilder.HasDefaultSchema("codebreaker");
    modelBuilder.ApplyConfiguration(new GameConfiguration());
    modelBuilder.ApplyConfiguration(new MoveConfiguration());
    // code removed for brevity
  }

  public DbSet<Game> Games => Set<Game>();
  public DbSet<Move> Moves => Set<Move>();
  // code removed for brevity
}
```

An EF Core context class needs to derive from the `DbContext` base class. Using the context from a DI container, the connection string can be configured outside of the context. This requires using the constructor with the `DbContextOption` parameter, which needs to be forwarded to the base class.

The `GamesSqlServerContext` class implements the `IGamesRepository` interface, which we defined in *Chapter 2*, to be used by the `GamesService` class. In *Chapter 2*, we implemented this interface with the `GamesMemoryRepository` in-memory repository class. The EF Core context class supports the repository pattern by implementing the same interface. This way, we can easily switch the in-memory repository by using `GamesSqlServerContext` instead.

The overridden `OnModelCreating` method allows customization to map the model types to the database. With SQL Server, the default schema name is `dbo`. This is changed by invoking `modelBuilder.HasDefaultSchema`.

To reduce the complexity of the `OnModelCreating` method, `GameConfiguration` and `MoveConfiguration` classes are created to customize the mapping with the `Game` and `Move` types.

One more thing that's required with the context class is for properties of type `DbSet<TEntity>` to allow using properties to access mapped database tables.

> **Note**
>
> Creating the `codebreaker` solution had several iterations. One time, an abstract base class and generically derived class were used to support all the different game types. Using EF Core, it's also possible to map inheritance, and this is also possible with the JSON serialization and the OpenAPI definition. EF Core can map an inheritance tree to a single table (**table-per-hierarchy** or **TpH**), to a table for every type (**table-per-type** or **TpT**), and a table for every concrete type (**table-per-concrete-type** or **TpC**).
>
> Instead of creating a complex class hierarchy for the model types to plan for future features that might be never needed, it often helps to have model types as simple as possible, depending on what's required with the current version. A complex model design adds complexity in multiple places.
>
> The Game class as it is defined now fulfills requirements for a group of different Game types as data holders. Functionality is abstracted away and done from the `analyzers` library, which just uses a contract to access the game.
>
> As microservices are used in a smaller scope, the **KISS** principle (**Keep It Simple, Stupid**) can help not only reduce work that needs to be done but also increase performance.

Customizing the mapping of simple properties

The mapping of the Game class is specified with the `GameConfiguration` class. A configuration class that is applied using `ApplyConfiguration` with the context configuration needs to implement the `IEntityTypeConfiguration` generic interface:

Codebreaker.Data.SqlServer/Configuration/GameConfiguration.cs

```
internal class GameConfiguration : IEntityTypeConfiguration<Game>
{
```

```
public void Configure(EntityTypeBuilder<Game> builder)
{
  builder.HasKey(g => g.Id);
  builder.Property(g => g.GameType).HasMaxLength(20);
  builder.Property(g => g.PlayerName).HasMaxLength(60);

  builder.Property(g => g.Codes).HasMaxLength(120);
  // code removed for brevity
```

With the implementation of this class, the key of the table is specified to map to the `Id` property. This would not be required, as convention defines that a property with the name `Id` or an ID prefixed with the class name maps to the primary key.

The fluent API `HasMaxLength` property is used to change the database types for the `GameType` and `PlayerName` properties. The `Codes` property is not such a simple property but can still be limited to a size of 120 characters.

This configuration for the `Games` table is applied by invoking `ApplyConfiguration` from the context configuration.

The `MoveConfiguration` class specifies a similar configuration for the `GuessPegs` and `KeyPegs` properties:

Codebreaker.Data.SqlServer/Configuration/MoveConfiguration.cs

```
internal class MoveConfiguration : IentityTypeConfiguration<Move>
{
  public void Configure(EntityTypeBuilder<Move> builder)
  {
    // code removed for brevity
    builder.Property(g => g.GuessPegs).HasMaxLength(120);
    builder.Property(g => g.KeyPegs).HasMaxLength(60);
  }
}
```

For the `Move` type, shadow properties will be specified later in the *Defining relations between games and moves* section.

Creating value conversion to map complex properties

To allow mapping of types not directly supported by EF Core, value conversion can be used. The `FieldValues` property of type `IDictionary<string, IEnumerable<string>>` is not directly supported with the default mapping. With the game, the content of this value is not really that huge and also doesn't need to be searched within. This allows us to map this to a `nvarchar`-type column.

Different options are available for the implementation. We will use a different one with SQL Server and with Azure Cosmos DB, but both options could be used with any of these providers.

Let's look at an example of what the data looks like. Let's imagine we have a dictionary such as the following with colors and shapes:

```
Dictionary<string, IEnumerable<string>> input = new ()
{
  { "colors", ["Red", "Green", "Blue"] },
  { "shapes", ["Rectangle", "Circle"] }
};
```

This should result in this string:

```
var expected =
"colors:Red#colors:Green#colors:Blue#shapes:Rectangle#shapes:Circle";
```

Every value is prefixed with the key of the value. With the source repository, you'll find a unit test to check for the implementation of this conversion.

To convert this dictionary to a string, the `ToFieldsString` extension method is implemented:

Codebreaker.Data.SqlServer/MappingExtensions.cs

```
public static class MappingExtensions
{
  public static string ToFieldsString(this IDictionary<string,
    IEnumerable<string>> fields)
  {
    return string.Join('#',
      fields.SelectMany(
        key => key.Value
          .Select(value => $"{key.Key}:{value}")));
  }
  // code removed for brevity
}
```

With the implementation, using the LINQ `SelectMany` method, for every key in the dictionary, a value is created that's prefixed by the key.

The reverse functionality converts the string to a dictionary with the `FromFieldsString` method:

Codebreaker.Data.SqlServer/MappingExtensions.cs

```
public static IDictionary<string, IEnumerable<string>>
  FromFieldsString(this string fieldsString)
```

```
{
  Dictionary<string, List<string>> fields = new();

  foreach (var pair in fieldsString.Split('#'))
  {
    var index = pair.IndexOf(':');
    if (index < 0)
    {
      throw new ArgumentException($"Field {pair} does not contain ':'
      delimiter.");
    }

    var key = pair[..index];
    var value = pair[(index + 1)..];

    if (!fields.TryGetValue(key, out List<string>? List))
    {
      list = [];
      fields[key] = list;
    }
    list.Add(value);
  }

  return fields.ToDictionary(
    pair => pair.Key,
    pair => (IEnumerable<string>)pair.Value);
}
```

With the implementation, the complete string is first split using the # separator. Each resulting string contains a key and value separated with :. These results are added to a pair, to finally return a dictionary.

These methods are now used with the configuration of the Game class:

Codebreaker.Data.SqlServer/Configuration/GameConfiguration.cs

```
public void Configure(EntityTypeBuilder<Game> builder)
{
  // code removed for brevity
  builder.Property(g => g.FieldValues)
    .HasColumnName("Fields")
    .HasColumnType("nvarchar")
    .HasMaxLength(200)
    .HasConversion(
      convertToProviderExpression: fields => fields.ToFieldsString(),
```

```
        convertFromProviderExpression: fields => fields.
          FromFieldsString(),
        valueComparer: new ValueComparer<IDictionary<string,
          Ienumerable<string>>>(
          equalsExpression: (a, b) => a!.SequenceEqual(b!),
          hashCodeExpression: a => a.Aggregate(0, (result, next) =>
            HashCode.Combine(result, next.GetHashCode())),
          snapshotExpression: a => a.ToDictionary(kv => kv.Key, kv =>
          kv.Value)));
  }
```

The column name and data type are specified using the fluent API's HasColumnName, HasColumnType, and HasMaxLength properties. The HasConversion method is used to convert a type to be mapped to the database representation. This method has several overloads for different use cases. Here, the first parameter references an expression to convert the .NET property type to the database type, while the second parameter does the reverse. Here, we invoke the previously created extension methods. With the third parameter, an instance of the ValueComparer class is invoked. This is used to compare the value for equality.

Defining relations between games and moves

With the relational database, the Games table has a relation to the Moves table. One game maps to a list of moves. To make this possible, with the Moves table a foreign key named GameId is defined to reference the primary key of the Games table:

Codebreaker.Data.SqlServer/Configuration/MoveConfiguration.cs

```
internal class MoveConfiguration : IEntityTypeConfiguration<Move>
{
  public void Configure(EntityTypeBuilder<Move> builder)
  {
    builder.Property<Guid>("GameId");

    builder.Property(g => g.GuessPegs).HasMaxLength(120);
    builder.Property(g => g.KeyPegs).HasMaxLength(60);
  }
}
```

Using EntityTypeBuilder for the Move type, invoking the Property method creates a **shadow property** in case the Move type doesn't have a property with this name. If there's not a property with the same name, specifying the type is required as done here using the generic parameter.

The following code snippet maps this relationship to database tables:

Codebreaker.Data.SqlServer/GameConfiguration.cs

```
public void Configure(EntityTypeBuilder<Game> builder)
{
  builder.HasKey(g => g.Id);
  builder.HasMany(g => g.Moves)
    .WithOne()
    .HasForeignKey("GameId");
  // code removed for brevity
}
```

EF Core supports one-to-one, one-to-many, and many-to-many relationships. With games and moves, a one-to-many relationship is defined with the `HasMany` and `WithOne` methods. The `HasForeignKey` method specifies the `GameId` value of the `Move` class to reference the ID of the game records.

After defining the mapping from the classes to the tables, let's implement the contract of the repository and add migrations to create the database.

Implementing the repository contract

In the previous chapter, we defined the `IGamesRepository` interface and implemented it with an in-memory representation. Now, let's implement this interface to read and write to the SQL Server database.

Adding and deleting games

Let's add the implementation of the `AddGameAsync` and `DeleteGameAsync` methods of the contract to the `GamesSqlServerContext` class:

Codebreaker.Data.SqlServer/GameSqlServerContext.cs

```
public async Task AddGameAsync(Game game, CancellationToken
cancellationToken = default)
{
  Games.Add(game);
  await SaveChangesAsync(cancellationToken);
}

public async Task<bool> DeleteGameAsync(Guid id, CancellationToken
  cancellationToken = default)
{
```

```
    var affected = await Games
      .Where(g => g.Id == id)
      .ExecuteDeleteAsync(cancellationToken);
      return affected == 1;
  }
```

With the AddGameAsync method, the passed Game object is added to the Games property of the
EF Core context, which marks the entity as *added* with the change tracker. The SaveChangesAsync
method creates INSERT statements in the database.

The DeleteGameAsync method receives the game ID with the parameter. Here, the
ExecuteDeleteAsync method is invoked on the record matching the ID. ExecuteDeleteAsync
and ExecuteUpdateAsync methods, available since EF Core 7, don't use tracking and directly
execute DELETE and UPDATE statements. This increases performance when change tracking is not
necessary. When the record was not found to be deleted, this method returns false.

Starting a 6x4 game creates this SQL statement to store the game:

```
INSERT INTO [codebreaker].[Games] ([Id], [Codes], [Duration],
[EndTime], [Fields], [GameType], [LastMoveNumber], [MaxMoves],
[NumberCodes], [PlayerName], [StartTime], [Won])
      VALUES (@p0, @p1, @p2, @p3, @p4, @p5, @p6, @p7, @p8, @p9, @p10,
@p11);
```

Let's set a move with the next implementation.

Updating a game

When a move is set, some game information such as the last move number is updated as well. The
implementation to add a move and update the game is shown here:

Codebreaker.Data.SqlServer/GameSqlServerContext.cs

```
public async Task AddMoveAsync(Game game, Move move, CancellationToken
cancellationToken = default)
{
  Moves.Add(move);
  Games.Update(game);

  await SaveChangesAsync(cancellationToken);
}
```

The Move object is added to the context with the Add method, and the Game object is added with the
Update method. This way, the change tracker is configured in that invoking the SaveChangesAsync
method creates SQL UPDATE and INSERT statements.

> **Note**
> By default, one invocation of `SaveChangesAsync` creates one transaction. If updating the game fails, there's a rollback for updating the move. In case you need multiple `SaveChangesAsync` instances within one transaction, the easiest option to use is ambient transactions (using the `TransactionScope` class from the `System.Transactions` namespace).

Querying games

To retrieve games, we need to implement `Getxx` methods. Let's start with `GetGameAsync` to get the game by using the game ID:

Codebreaker.Data.SqlServer/GamesSqlServerContext.cs

```
public async Task<Game?> GetGameAsync(Guid id, CancellationToken
cancellationToken = default)
{
  var game = await Games
    .Include("Moves")
    .TagWith(nameof(GetGameAsync))
    .SingleOrDefaultAsync(g => g.Id == id, cancellationToken);

  return game;
}
```

The `GetGameAsync` method uses the `SingleOrDefaultAsync` method to get either one or zero records. If the game ID is not found, `null` is returned. Behind the scenes, a query using SELECT TOP (2) is created to check if more than one record would be returned from this query. If this is the case, the `SingleOrDefaultAsync` method throws an exception.

The `Include` method is used to create a query that includes moves that relate to the returned query. Here, the SQL LEFT JOIN statement is used to join multiple tables. EF Core writes all the queries and updates to the log output. To better see which output maps to which LINQ methods, the `TagWith` method can be used. This tag is shown as a title with the log output.

> **Note**
> The `TagWith` method is of great help with debugging and troubleshooting. Checking the log outputs to see SQL queries sent, the tag gives a fast way to see where this query was generated.

The following snippet shows the log output from this query, including the title:

```
-- GetGameAsync

SELECT [t].[ Id], [t].[Codes], [t].[Duration], [t].[EndTime], [t].
```

```
[Fields], [t].[GameType], [t].[LastMoveNumber], [t].[MaxMoves], [t].
[NumberCodes], [t].[PlayerName], [t].[StartTime], [t].[Won], [m].[Id],
[m].[GameId], [m].[GuessPegs], [m].[KeyPegs], [m].[MoveNumber]
FROM (
  SELECT TOP(2) [g].[Id], [g].[Codes], [g].[Duration], [g].[EndTime],
[g].[Fields], [g].[GameType], [g].[LastMoveNumber], [g].[MaxMoves],
[g].[NumberCodes], [g].[PlayerName], [g].[StartTime], [g].[Won]
  FROM [codebreaker].[Games] AS [g]
  WHERE [g].[Id] = @__Id_0
) AS [t]
LEFT JOIN [codebreaker].[Moves] AS [m] ON [t].[Id] = [m].[GameId]
ORDER BY [t].[Id]
```

To query by date, player name, or some other query option, pass the GamesQuery object to the GetGamesAsync method:

Codebreaker.Data.SqlServer/GamesSqlServerContext.cs

```csharp
public async Task<IEnumerable<Game>> GetGamesAsync(GamesQuery?
gamesQuery, CancellationToken cancellationToken = default)
{
    IQueryable<Game> query = Games
        .TagWith(nameof(GetGamesAsync))
        .Include(g => g.Moves);

    if (gamesQuery.Date.HasValue)
    {
        DateTime begin = gamesQuery.Date.Value.ToDateTime(TimeOnly.
            MinValue);
        DateTime end = begin.AddDays(1);
        query = query.Where(g => g.StartTime < end && g.StartTime >
            begin);
    }
    if (gamesQuery.PlayerName != null)
        query = query.Where(g => g.PlayerName == gamesQuery.PlayerName);
    if (gamesQuery.GameType != null)
        query = query.Where(g => g.GameType == gamesQuery.GameType);
    if (gamesQuery.Ended)
    {
        query = query.Where(g => g.EndTime != null)
            .OrderBy(g => g.Duration);
    }
    else
    {
        query = query.OrderByDescending(g => g.StartTime);
```

```
    }

    query = query.Take(MaxGamesReturned);

    return await query.ToListAsync(cancellationToken);
}
```

The implementation of this method uses the `IQueryable` variable to add different LINQ query methods. Depending on the values passed with the `GamesQuery` parameter, multiple `Where` methods are added, in addition to `OrderBy` or `OrderByDescending`, to define the order of the result. To not return all the games played, only the first 500 games based on the filter are returned.

Calling this method passing the player's name and a date results in this SQL query:

```
SELECT [t].[Id], [t].[Codes], [t].[Duration], [t].[EndTime], [t].
[Fields], [t].[GameType], [t].[IsVictory], [t].[LastMoveNumber],
[t].[MaxMoves], [t].[NumberCodes], [t].[PlayerIsAuthenticated],
[t].[PlayerName], [t].[StartTime], [m].[Id], [m].[GameId], [m].
[GuessPegs], [m].[KeyPegs], [m].[MoveNumber]
FROM (
    SELECT TOP(@__p_3) [g].[Id], [g].[Codes], [g].[Duration], [g].
[EndTime], [g].[Fields], [g].[GameType], [g].[IsVictory], [g].
[LastMoveNumber], [g].[MaxMoves], [g].[NumberCodes], [g].
[PlayerIsAuthenticated], [g].[PlayerName], [g].[StartTime]
    FROM [codebreaker].[Games] AS [g]
    WHERE [g].[StartTime] < @__end_0 AND [g].[StartTime] > @__begin_1
AND [g].[GameType] = @__gamesQuery_GameType_2
    ORDER BY [g].[StartTime] DESC
) AS [t]
LEFT JOIN [codebreaker].[Moves] AS [m] ON [t].[Id] = [m].[GameId]
ORDER BY [t].[StartTime] DESC, [t].[Id]
```

The `Include` method results in a `LEFT JOIN` operation to access the `Moves` table. Because of the `Take` method, `SELECT TOP` is used. Multiple invocations of the LINQ `Where` method results in a `WHERE` clause.

Configuring user secrets

To access the database, we need to retrieve some configuration values. Some of these configuration values are secrets that shouldn't be part of the source code repository. During development time, you can use user secrets. User secrets are stored with the user profile.

To initialize user secrets, use this .NET CLI command:

```
cd Codebreaker.AppHost
dotnet user-secrets init
```

This creates a `UserSecretsId` property in the project file. Because all user secrets are stored with the user profile, this string is used to differentiate the configurations with multiple applications.

To set a configuration value with the secrets, use the `dotnet user-secrets set` command:

```
dotnet user-secrets set Parameters:sql-password [enter the password]
```

With the SQL Server Docker container we use, there are some requirements for the password. Be aware that you can't use simple passwords. Three of four sets need to match: uppercase letters, lowercase letters, base 10 digits, and symbols. You can check the log output to see if there's an issue with the password.

You can also use Visual Studio and a context menu with Visual Studio to configure user secrets.

Be aware that the provider to read configuration values from user secrets by default is only used if the secret ID is configured and the application is running in the `Development` environment.

Note

User secrets cannot be used in production. The idea of user secrets is to not store secrets with a configuration file that's pushed to a source code repository. Every developer working on this project needs to configure the configuration values for secrets. In production, you can use other services such as Azure Key Vault. This is covered in *Chapter 7*.

Configuring the application model with SQL Server

To run SQL Server, .NET Aspire makes it easy to run a Docker container. Just add this code to the application model in the `Codebreaker.AppHost` project:

Codebreaker.AppHost/Program.cs

```
var builder = DistributedApplication.CreateBuilder(args);

var sqlServer = builder.AddSqlServer("sql", sqlPassword)
  .AddDatabase("CodebreakerSql", "codebreaker");
```

The `AddSqlServer` method adds a SQL Server resource. Using this method, during development time, a Docker container is used. In *Chapter 5*, we'll get into the details of Docker and add more configuration with this SQL Server Docker container. The name of this resource is sql. Optionally, a password can be passed to the `AddSqlServer` method. If a configuration parameter value is set with the resource name postfixed with -password (as we did), this `password` is used. Otherwise a random `password` is generated.. The `AddDatabase` method adds a database to the resource with the first parameter, the name of the resource that is used as a name for the connection string name, and the database name.

To allow us to dynamically decide between different game repositories, we use the `DataStore` configuration to decide between in-memory, SQL Server, and Azure Cosmos DB on the startup of the application:

Codebreaker.AppHost/appsettings.json

```
{
  "Logging": {
    "LogLevel": {
      "Default": "Information",
      "Microsoft.AspNetCore": "Warning",
      "Aspire.Hosting.Dcp": "Warning"
    }
  },
  "DataStore": "SqlServer"
}
```

Depending on the database provider you want to use, change the value as needed.

> **Note**
>
> *Chapter 7, Flexible Configuration*, goes into the details of the `appsettings.json` file and environment-specific counterparts, as well as other options to store configuration values such as environmental variables, program arguments, and Azure App Configuration instances. In this chapter, all we need is to configure settings within `appsettings.json` as well as user secrets, which are covered with Azure Cosmos DB.

The configuration value is retrieved with the startup of the `AppHost` project:

Codebreaker.AppHost/Program.cs

```
string dataStore = builder.Configuration["DataStore"] ??
  "InMemory";
```

In case the value is not configured, it defaults to the in-memory provider we created in the previous chapter.

Now, we can change dependencies for the game APIs:

Codebreaker.AppHost/Program.cs

```
builder.AddProject<Projects.Codebreaker_GameAPIs>("gameapis")
  .WithEnvironment("DataStore", dataStore)
  .WithReference(sqlServer);
// code removed for brevity
```

The WithEnvironment method creates an environment variable for the game APIs project with the DataStore key and the value that's retrieved from the configuration. The WithReference method references the SQL Server resource and creates an environment variable for the connection string.

Next, let's configure the minimal APIs project to retrieve the configuration values from the AppHost project.

Configuring the DI container with the minimal APIs project

After the mapping of the model to the database is completed and the resource dependencies are defined with the Aspire AppHost project, the DI container can be configured to use the EF Core context.

The games API project needs a reference to the Codebreaker.Data.SqlServer project and the Aspire.Microsoft.EntityFrameworkCore.SqlServer NuGet package.

The configuration for the DataStore is retrieved with the following code snippet:

Codebreaker.GameAPIs/ApplicationServices.cs

```
public static void AddApplicationServices(this IHostApplicationBuilder
builder)
{
  // code removed for brevity
  string? dataStore = builder.Configuration.
    GetValue<string>("DataStore");
  switch (dataStore)
  {
    case "SqlServer":
      ConfigureSqlServer(builder);
      break;
    default:
      ConfigureInMemory(builder);
      break;
  }

  builder.Services.AddScoped<IGamesService, GamesService>();
}
```

Depending on the retrieved configuration value for DataStore, we configure Azure Cosmos DB, SQL Server, or the in-memory repository that we implemented in the previous chapter.

The configuration for SQL Server, which is called from the previous switch/case statement, is shown here:

Codebreaker.GameAPIs/ApplicationServices.cs

```
static void ConfigureSqlServer(IHostApplicationBuilder builder)
{
   builder.AddDbContextObjectPool<IGamesRepository,
GamesSqlServerContext>(options =>
   {
      var connectionString = builder.Configuration.
      GetConnectionString("CodebreakerSql") ?? throw new
      InvalidOperationException("Could not read SQL Server connection
      string");
      options.UseSqlServer(connectionString);
      options.UseQueryTrackingBehavior(
      QueryTrackingBehavior.NoTracking);
   }
   builder.EnrichSqlServerDbContext<GamesSqlServerContext>();
}
```

Using the .NET Aspire SqlServer EF Core component, we can invoke the AddSqlServerDbContext API to configure the EF Core context with .NET Aspire. However, this API doesn't provide the level of flexibility we need for working with different database providers. Thus, instead, we configure the EF Core context using EF Core APIs such as AddDbContext and AddDbContextPool and add Aspire functionality by using EnrichSqlServerDbContext. The AddDbContextObjectPool method configures to use the SQL Server EF Core provider, passing the connection string, which is passed via the AppHost project and thus needs to match the name configured with the top-level statements in the AppHost project.

Invoking the UseQueryTrackingBehavior method adds one interesting aspect when using EF Core. By default, all queries are tracked within the EF Core context to allow the context to know about changes. Within the API service, the context is newly created with every new HTTP request. Thus, keeping this tracking state for every context is not required. Adding and updating entities are explicitly marked with the Add and Update methods. Setting the query tracking behavior to QueryTrackingBehavior.NoTracking disables tracking with all queries (unless overwritten with a query using the AsTracking method) and thus reduces the overhead.

Instead of turning tracking off by default, you can also use the option to turn tracking off with a single query using the AsNoTracking method.

The EnrichSqlServerDbContext method adds health checks, logging, and telemetry configuration offered by the Aspire component. Logging and telemetry configuration are covered in *Chapter 11*, and health checks are covered in *Chapter 12*.

As the mappings and the repository contract are implemented, we can now continue to create the database using migrations.

Creating migrations with EF Core

Using EF Core, you can create the database with the `Database.EnsureCreatedAsync` context API method. However, this does not take schema changes into account. Over time, the database schema will change as new features are added – and it's best to do this automatically.

Chapter 8 describes how to automatically publish services to testing and production environments. With this, updating the database is important as well. When the database schema changes, updates should be published to the environments. EF Core offers **migrations** to record all schema changes and update the database schema programmatically.

Next, let's do the following:

1. Add the .NET EF Core tool
2. Add the EF Core tool and create initial migrations
3. Update the model and add migrations
4. Update the database programmatically

Adding the .NET EF Core tool

If you don't have the EF Core .NET command-line tool installed yet, you can install it with the `dotnet` CLI as a global or a local tool. Here, we install it as a local tool to have a specific version of this tool as part of the `Codebreaker.Data.SqlServer` project.

To install local tools, a `tool-manifest` file first needs to be created:

```
cd Codebreaker.Data.SqlServer
dotnet new tool-manifest
```

With the `tool-manifest` template, the `dotnet new` command creates a `.config` directory with a `dotnet-tools.json` file. This manifest file will contain all the tools that should be installed when working on the project.

As soon as this manifest file is available, we can install the `dotnet-ef` tool:

```
dotnet tool install dotnet-ef
```

This command configures this tool with the tool manifest file and installs it locally. In case you've installed another version of this tool globally, while the current directory of your command prompt is within the project folder, you use the tool version that's specified with the tool manifest file.

To get all the tools installed and configured with a tool manifest file, you can use the `tool restore` command:

```
dotnet tool restore
```

The `restore` command can be practically used when you clone a repository containing a tool manifest file. Using `dotnet tool restore`, all tools specified with the project are restored.

Let's use this tool to create an initial migration for the actual context:

```
dotnet ef migrations add InitGames -s ..\Codebreaker.GameAPIs
```

`migrations` is a command of the `dotnet ef` tool. Using `add`, a new migration is added with the name that follows the `add` command (here, `InitGames`). The `-s` (or `--startup-project`) option specifies the project where the EF Core context is configured with the DI container, and the connection string to the database is specified. This is a different project than the project where the EF Core context is implemented (`Codebreaker.Data.Cosmos` and `Codebreaker.Data.SqlServer`); that's why this option is needed.

> **Note**
>
> In case creating the migration fails, check the error message. An error could be that you failed to specify a mapping, and here the errors are very detailed. While working around issues, you can temporarily ignore the properties of the model to see if the error is really based on a property mapping.

After a successful run of this tool, you'll see a `Migrations` folder with the project. This folder contains a snapshot of the current state of the database, including all the table mappings, the property mappings, and the relations. This class is named based on the EF Core context suffixed by `ModelSnapshot`; for example, `GameSqlServerContextModelSnapshot`.

Every time you add a new migration, the snapshot will be updated and a new `Migration`-derived class created that includes all the schema changes based on the previous migration. The migration is named with the migration name prefixed with the time. The generated class contains an `Up` method that will be invoked when the migration is applied to the SQL Server database and a `Down` method that will be invoked when the migration is dropped from the database.

Next, we'll use the `dotnet ef` tool to apply the migration to the database and create the database if it doesn't exist yet. This can be done using the `dotnet ef database update` command:

```
dotnet ef database update -s ..\Codebreaker.GameAPIs.
```

This command now uses the connection string from the startup project to apply migrations to the database. Using migrations to create the database, you'll see all the games and moves tables created – along with the `_EFMigrationsHistory` table. Reading the content of this table, you'll see all the migration names applied to the database. This information is checked when doing another update to the database schema using migrations.

> **Note**
>
> There are some cases where creating the database fails while creating the migration succeeds. Mapping errors can be the reason here as well. Checking the error message again gives good details on the reason for the failure.

Creating or updating the database programmatically

Instead of using the command line to apply a migration, migrations can be started programmatically invoking the EF Core context with `context.Database.MigrateAsync`. Let's implement this functionality with the `CreateOrUpdateDatabaseAsync` method, which is called from the application startup code for easy use of the solution:

Codebreaker.GameAPIs/ApplicationServices.cs

```
public static async Task CreateOrUpdateDatabaseAsync(this
WebApplication app)
{
    var dataStore = app.Configuration["DataStore"] ?? "InMemory";
    if (dataStore == "SqlServer")
    {
        try
        {
            using var scope = app.Services.CreateScope();

            var repo = scope.ServiceProvider.
GetRequiredService<IGamesRepository>();
            if (repo is GamesSqlServerContext context)
            {
                await context.Database.MigrateAsync();
                app.Logger.LogInformation("SQL Server database updated");
            }
        }
        catch (Exception ex)
        {
            app.Logger.LogError(ex, "Error updating database");
            throw;
        }
    }
}
```

With the implementation, we check if the solution is configured to use SQL Server. In that case, the `MigrateAsync` method is invoked to update the database to the newest version.

With the `codebreaker` solution, this is really convenient – running the solution, everything is ready, including the database. From a security standpoint, in the production environment, the service running should not have a database connection string that is allowed to change the schema. Instead, a separate program can be used to update the database. This could be invoked from a GitHub action with automatic deployment. Using the `dotnet ef` tool, you can even create a standalone application to update the database schema: `dotnet ef migrations bundle` creates an application with the .NET runtime included, thus you can start this application from clients who don't have the .NET runtime installed. You can also create a SQL script to start the migration if this is the preferred option by the database administrator: `dotnet ef migrations script`.

Next, let's make a change to the model that influences the database schema.

Updating the database schema

In *Chapter 9*, we'll differentiate anonymous from authenticated users. With this, game information will be stored when the game is played by an authenticated user. For this, we'll add a `PlayerIsAuthenticated` flag to the Game class:

Codebreaker.GameAPIs.Models/Game.cs

```
public class Game(
  Guid id,
  string gameType,
  string playerName,
  DateTime startTime,
  int numberCodes,
  int maxMoves) : IGame
{
  public Guid Id { get; private set; } = id;
  public string GameType { get; private set; } = gameType;
  public string PlayerName { get; private set; } = playerName;
  public bool PlayerIsAuthenticated { get; set; } = false;
  // code removed for brevity
```

This new property is not defined to be ignored from the database. To update the database schema, we add a new migration:

```
cd Codebreaker.Data.SqlServer
dotnet ef migrations add AddPlayerIsAuthenticated -s ..\Codebreaker.
GameAPIs
```

The new migration is named `AddPlayerIsAuthenticated`. This change updates the snapshot in the `Migrations` folder and adds a new migration, as shown in this code snippet:

Codebreaker.Data.SqlServer/Migrations/ 20231225095931_AddPlayerIs-Authenticated.cs

```
public partial class AddPlayerIsAuthenticated : Migration
{
    protected override void Up(MigrationBuilder migrationBuilder)
    {
        migrationBuilder.AddColumn<bool>(
            name: "PlayerIsAuthenticated",
            schema: "codebreaker",
            table: "Games",
            type: "bit",
            nullable: false,
            defaultValue: false);
    }

    protected override void Down(MigrationBuilder migrationBuilder)
    {
        migrationBuilder.DropColumn(
            name: "PlayerIsAuthenticated",
            schema: "codebreaker",
            table: "Games");
    }
}
```

With the `Up` method, updating a database from the previous version, the column is added to the database schema (`AddColumn`), and the `Down` method removes the column (`DropColumn`).

During development, you might often update the schema and create many migrations. Before publishing a new version of the application, it's a good idea to combine migrations into one. Just pay attention to the version installed in production or staging environments. You should keep the migrations already deployed. Migrations that have been used only in your development environment can be removed with `dotnet ef migrations remove` (possibly called multiple times to always remove the last migration) – and finally, one invocation of `dotnet ef migrations add <Name of the migration>`, which then creates one migration with all schema changes since the last migration.

Now, let's run the solution using SQL Server.

Running the application with SQL Server

Starting the host application now, not only the game APIs service is running, but also SQL Server in a Docker container, as shown in *Figure 3.2*:

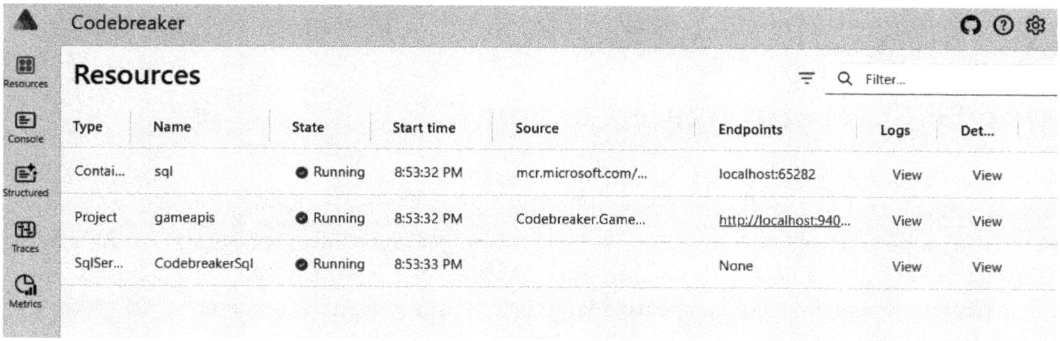

Figure 3.2 – .NET Aspire resources with SQL Server

You can access the OpenAPI endpoint description to start games. Make sure to check into the details of the games API service. Details give information about the resource, the endpoints, and the environment variables, as shown in *Figure 3.3*:

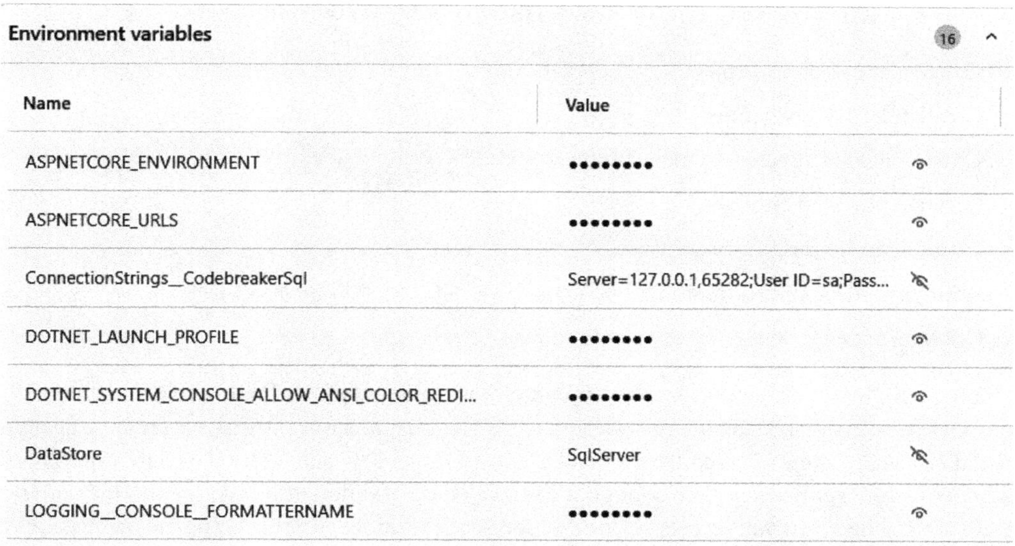

Figure 3.3 – Environment variables

With the environment variables set for this service, check `DataStore` and `ConnectionStrings__ CodebreakerSql`, which are set by the `AppHost` project.

Try to play a game using the HTTP files. Verify how records are added to the SQL Server database. However, when you stop the project and run the application again, the database is created from a fresh state. With Docker, we need volumes to map storage outside of the Docker container. This is covered in *Chapter 5*.

After this, let's move over to Azure Cosmos DB.

Using EF Core with Azure Cosmos DB

With Azure Cosmos DB, Microsoft offers different databases with several APIs that make use of the same infrastructure. Most of these database offerings are NoSQL databases for different purposes. Azure Cosmos DB offers a JSON document store that can be accessed with the Mongo DB API. The Apache Cassandra API offers a wide column store where each row can have different columns. The Apache Gremlin query language can be used to access a graph version of the database. This is great to query for relations using vertices and edges. Azure Cosmos DB for PostgreSQL is a distributed high-performance relational database using the same infrastructure to read and write from a database network worldwide.

For the `codebreaker` solution, we'll use Azure Cosmos DB for NoSQL. Here, an EF Core provider is available. This allows us to use the same API as with SQL Server, but the mapping will be different.

Writing the games and moves to Azure Cosmos DB, we need to do the following:

1. Create a class library project
2. Create the EF Core context
3. Create a value converter to map complex types
4. Create embedded entities
5. Implement the repository contract
6. Configure the application model
7. Configure the DI container

We created a SQL Server database when applying the migrations first. With Azure Cosmos DB, migrations are not available and are not needed. As JSON documents are stored, we are very flexible in the data to write. There is no concept of tables and relations between tables – we just store JSON documents within a container. One container can keep data of different kinds. A container can be used as a scaling unit, but you can also decide to specify the scaling with the complete database and share **request units** (**RU/s**) with different containers in a database.

With containers, you also need to know about partitions. Partitions are used to scale containers for performance. Before specifying partitions, you need to know about some attributes of Azure Cosmos DB:

*   A partition is limited to 20 GB storage. The size limit of a container is 1 TB.

- Writing to the database, a transaction can only span writing to a single partition. If different data should be written within a transaction, this data should use the same partition key.

- 10,000 RU/s is the maximum limit for a partition. With a container, the limit is 1,000,000 RU/s (with serverless, the container RU/s limit is 20,000). For the best-performance parallel reading of data, different partition keys should be used.

- The maximum length of a partition key is 2048 bytes.

- The maximum size of one item to store is 2 MB.

- There's no limit on the distinct values of partition keys.

We will use the game ID for the partition key. Games are created and updated independently of other games. It's not necessary to write multiple games within one transaction. Running Azure Cosmos DB with a multi-region write configuration allows us to create games from different Azure regions with high performance. This makes the `Id` value of the game a good candidate for the partition key.

With this information, we'll create a class library next.

Creating a class library project for EF Core with NoSQL

Similar to creating a library for SQL Server, we use a library to access Azure Cosmos DB:

```
dotnet new classlib --framework net8.0 -o Codebreaker.Data.Cosmos
```

This library makes use of the `Microsoft.EntityFrameworkCore.Cosmos` NuGet package – and of course, a reference to the `Codebreaker.GameAPIs.Models` project is needed.

Creating an EF Core context for Azure Cosmos DB

Let's create a context class to access Azure Cosmos DB, as shown with the following code snippet:

Codebreaker.Data.Cosmos/GamesCosmosContext.cs

```
public class GamesCosmosContext(DbContextOptions<GamesCosmosContext>
optoins) : DbContext(options), IGamesRepository
{
  private const string PartitionKey = nameof(PartitionKey);
  private const string ContainerName = "GamesV3";
  private const string DiscriminatorValue = "GameV3";

  protected override void OnModelCreating(ModelBuilder modelBuilder)
  {
    modelBuilder.HasDefaultContainer(ContainerName);
    var gameModel = modelBuilder.Entity<Game>();
```

```
    gameModel.Property<string>(PartitionKey);
    gameModel.HasPartitionKey(PartitionKey);
    gameModel.HasKey(nameof(Game.Id), PartitionKey);

    gameModel.HasDiscriminator<string>("Discriminator")
      .HasValue<Game>(DiscriminatorValue);
    // code removed for brevity
}
public DbSet<Game> Games => Set<Game>();

public static string ComputePartitionKey(Game game) =>
    game.GameId.ToString();

public void SetPartitionKey(Game game) =>
    Entry(game).Property(PartitionKey).CurrentValue =
      ComputePartitionKey(game);
// code removed for brevity
```

Similar to before, the custom context class derives from the DbContext base class and defines a constructor with context options, which allows us to configure the DI container with the connection string. The differences start now. With SQL Server, we defined the default schema name. This is not available with Azure Cosmos DB, but we can define the default container name using the HasDefaultContainer method. In case you have entities that should not be stored with the default container, these can be configured to use a different container with the help of the ToContainer method. The previously discussed partition key is configured by invoking the HasPartitionKey method. Using the SetPartitionKey and ComputePartitionKey methods, the partition key is configured as a **shadow property** with the same value as the game ID.

While Id is a good option for the partition key, other types that could be stored in the same container might not have an Id value. Thus, for the partition key, PartitionKey is used. With games, the Id value will be mapped to PartitionKey.

Writing different object types to a single container requires the use of a discriminator value. By default, the discriminator value is the name of the class. By invoking the HasDiscriminator method, the default discriminator configuration is overridden by specifying the Discriminator shadow property. For Game types, the GameV3 value is written. This allows us to differentiate game objects stored with incompatible new versions.

Azure Cosmos DB stores JSON documents, thus only the Game type needs to be specified with a DbSet property, and not the Move type, as we did with SQL Server. Defining maximum sizes for string properties is not needed as well – there's no schema describing this.

Creating a value converter to convert complex types

In the SQL Server section, we converted the `Idictionary`-typed property, passing expressions to the `HasConversion` method to convert the dictionary to a string. The same could be done with Azure Cosmos DB, but now we'll create a class deriving from `ValueConverter` and convert the dictionary to and from JSON, as shown in the following code snippet:

Codebreaker.Data.Cosmos/Utilities/FieldValueValueConverter.cs

```
internal class FieldValueValueConverter :
ValueConverter<IDictionary<string, IEnumerable<string>>, string>
{
  static string GetJson(IDictionary<string, IEnumerable<string>>
    values) => return JsonSerializer.Serialize(values);
  static IDictionary<string, IEnumerable<string>> GetDictionary(string
    json) => JsonSerializer.Deserialize<IDictionary<string,
    IEnumerable<string>>>(json) ??
      new Dictionary<string, IEnumerable<string>>();

  public FieldValueValueConverter() : base(
    convertToProviderExpression: v => GetJson(v),
    convertFromProviderExpression: v => GetDictionary(v))
  { }
}
```

An EF Core value converter derives from the `ValueConverter` base class and specifies with the generic parameter what type to convert. With the `FieldValues` property, this is `IDictionary<string, IEnumerable<string>>`. The constructor of the base class requires parameters to convert to the database data type and to convert from the database data type. With the implementation, the `JsonSerializer` class from the `System.Text.Json` namespace is used to do the serialization and deserialization.

An instance of this value converter is now passed to an overload of the `HasConversion` method with the `FieldValues` property configuration:

Codebreaker.Data.Cosmos/GamesCosmosContext.cs

```
public class GamesCosmosContext(DbContextOptions<GamesCosmosContext>
options) : DbContext(options), IGamesRepository
{
  private static FieldValueValueConverter s_fieldValueConverter =
new();
  private static FieldValueComparer s_fieldValueComparer = new();
```

```
protected override void OnModelCreating(ModelBuilder modelBuilder)
{
   // code removed for brevity
   gameModel.Property(g => g.FieldValues)
      .HasConversion(s_fieldValueConverter, s_fieldValueComparer);
}
```

Similar to `FieldValueValueConverter`, a `FieldValueComparer` instance is created. Instances of these two types are created to pass to the `HasConversion` method.

When creating the mapping for a relational database, much more was needed to configure. We reduced the code from the data context by creating configuration classes with every mapped table. This is not worthwhile doing here. The complete EF Core configuration, as well as the implementation of the repository interface, is done with the context class.

Creating embedded entities

What about the relationship between games and moves? EF Core defines the `OwnsOne` and `OwnsMany` methods to define an owned relationship. With a relational database, `OwnsOne` adds columns of the owned type to the owner type. With the Azure Cosmos DB provider, invoking `OwnsMany` from `gameModel` and referencing the `Moves` property, moves will be stored as JSON within the game.

Since EF Core 7, this is the default behavior with related entity types with the Azure Cosmos DB provider. So, nothing needs to be configured to make this happen.

Implementing the repository contract

With the implementation of the repository, there are many similarities to the implementation with SQL Server, but because of the different storage, some changes are necessary. Here, we'll concentrate on the differences:

Codebreaker.Data.Cosmos/GamesCosmosContext.cs

```
public async Task AddGameAsync(Game game, CancellationToken
cancellationToken = default)
{
   SetPartitionKey(game);
   Games.Add(game);
   await SaveChangesAsync(cancellationToken);
}
```

When adding or updating the game, the partition key needs to be set. Other than that, the code is the same as with SQL Server.

There's a difference in what happens at runtime. Instead of using SQL `INSERT` and `UPDATE` statements, the Azure Cosmos DB provider executes `CreateItem` and `ReplaceItem` functions. When you check the log output, you can see the number of RUs required for every statement done.

The `GetGamesAsync` method defined previously also works with the Cosmos DB provider. This is the query created:

```
SELECT c
FROM root c
WHERE ((((c["Discriminator"] = "Game") AND ((c["StartTime"] < @__
end_0) AND (c["StartTime"] > @__begin_1))) AND (c["GameType"] = @__
gamesQuery_GameType_2))
ORDER BY c["StartTime"] DESC
OFFSET 0 LIMIT @__p_3
```

Comparing this query to the query to the SQL Server database, with Cosmos DB, it's a lot simpler: joining of tables is not required. An interesting part of this query is the filtering on `Discriminator`. By default, every object stored in a container has a `Discriminator` filter that includes the type name. This allows the storage of different documents in the container. Queries for a specific type include the `Discriminator` filter.

In case you only store objects of the same type within one container, you can turn off storing with the `Discriminator` filter with the `HasNoDiscriminator` model definition method.

Be aware that not all LINQ queries translate successfully from the Cosmos DB provider. For example, `Include` and `Join` methods are not translated. While the `Include` method was used with SQL Server to include the moves with a query for a game, with a JSON document where the moves are stored within a game, this is not required. As there are no tables with NoSQL, `Join` is usually not required as well. In case you want to combine a list of different object types, create two queries and combine the results with the caller.

Configuring the application model with Azure Cosmos DB

With SQL Server, we've been using a Docker container for SQL Server. Using Azure Cosmos DB, a Docker container is available as well. However, with Cosmos DB, this is just an emulator and should not be used for production. In *Chapter 5*, we'll use the database running with Microsoft Azure.

To add Azure resources to the `AppHost` project, we need to add the `Aspire.Hosting.Azure` NuGet package. Let's add Azure Cosmos DB to the Aspire `AppHost` application model:

Codebreaker.AppHost/Program.cs

```
var cosmos = builder.AddAzureCosmosDB("codebreakercosmos")
  .AddDatabase("codebreaker");
  .RunAsEmulator();
```

```
builder.AddProject<Projects.Codebreaker_GameAPIs>("gameapis")
.WithEnvironment("DataStore", dataStore)
.WithReference(cosmos)
.WithReference(sqlServer);
```

Invoking the `AddAzureCosmosDB` method registers the Azure Cosmos DB resource. `codebreakercosmos` is the resource name, which needs to be lowercase and is used as the connection string to the Azure Cosmos DB account. Here, the name of the database is not part of the connection string. The database is specified by invoking the `AddDatabase` method and defines the name of the database. The `RunAsEmulator` method specifies a Docker image to run the database within a Docker container, but only within the development environment. Similar to before, the Cosmos DB resource is referenced from the games API project, which forwards the connection string with the `codebreakercosmos` key to this project. Be aware it's not the name passed to `AddDatabase` (which was the case with SQL Server) because the database name is not part of the connection string.

Configuring the DI container

To configure the DI container with the games API project, we have to add the `Aspire.Microsoft.EntityFrameworkCore.Cosmos` NuGet package to use this Aspire component. The configuration of the DI container was already prepared with the configuration of the relational database. All that's needed now is to add the Cosmos DB EF Core context, as shown in the following code snippet:

Codebreaker.GameApis/ApplicationServices.cs

```
static void ConfigureCosmos(IHostApplicationBuilder builder)
{
  builder.AddDbContext<IGamesRepository, GamesCosmosContext>(options
=>
  {
    var connectionString = builder.Configuration.
      GetConnectionString("codebreakercosmos") ??
      throw new InvalidOperationException("Could not read Cosmos
      connection string");
    options.UseCosmos(connectionString, "codebreaker");
    options.UseQueryTrackingBehavior(
    QueryTrackingBehavior.NoTracking);
  });
  builder.EnrichCosmosDbContext<GamesCosmosContext>();
}
```

The .NET Aspire Azure Cosmos DB EF Core component offers the `AddCosmosDbContext` method, but similar to before, because we need the registration of the `IGamesRepository` interface, we use the EF Core `AddDbContext` method and add the Aspire component features by invoking the

EnrichCosmosDbContext method. The UseCosmos method registers to use the EF Core provider for Azure Cosmos DB and assigns the connection string that is passed from the application model definition.

To create the database and the Cosmos DB container, we add the else part to the CreateOrUpdateDatabaseAsync method:

Codebreaker.GameApis/ApplicationServices.cs

```
public static async Task CreateOrUpdateDatabaseAsync(this
WebApplication app)
{
  // code removed for brevity
  else if (dataStore == "Cosmos")
  {
    try
    {
      using var scope = app.Services.CreateScope();
      var repo = scope.ServiceProvider.
        GetRequiredService<IGamesRepository>();
      if (repo is GamesCosmosContext context)
      {
        bool created = await context.Database.EnsureCreatedAsync();
        app.Logger.LogInformation("Cosmos database created:
          {created}", created);
      }
    }
    catch (Exception ex)
    {
        app.Logger.LogError(ex, "Error updating database");
      throw;
    }
  }
}
```

The Database.EnsureCreatedAsync method creates the database and the Azure Cosmos DB container with the partition key specified.

Having the configuration in place, let's start the application as before with SQL Server and check the stored games with your Azure Cosmos DB database while you set moves. Just make sure that the DataStore configuration is set to the correct database type. Using the HTTP files, don't forget to use the returned game ID that is returned after creating the game.

Summary

In this chapter, we changed to using persistent storage with the API service using a relational and a NoSQL database. We created the database context to map the `Game` and `Move` types to tables in a relational database and to a JSON document with a NoSQL database – both using EF Core.

To select which database to use in your environment, if you have relational data with a fixed schema, select SQL Server. If the schema is not required in your scenario, and changes to the data happen often, a NoSQL database can be the best option.

You learned about how to map objects and how to deal with special mapping requirements based on the object model. Using a relational database, you also learned how to create migrations to update the database schema and to initially create the database.

You learned how to use database resources with the .NET Aspire application model specified with the `AppHost` project.

Before starting the next chapter, it's well deserved to play another game round. Just use the HTTP files to make your game run. With the state of the current implementation, the game can continue to run after you restart the service – the games and moves are persisted.

In the next chapter, we create a library that can be used by client applications to invoke the web API so that it becomes more convenient to play the game.

Further reading

To learn more about the topics discussed in this chapter, you can refer to the following links:

- Value conversions: `https://learn.microsoft.com/ef/core/modeling/value-conversions`

- Inheritance with EF Core: `https://learn.microsoft.com/ef/core/modeling/inheritance`

- Owned entity types: `https://learn.microsoft.com/ef/core/modeling/owned-entities`

- Transactions: `https://learn.microsoft.com/ef/core/saving/transactions`

- Migrations: `https://learn.microsoft.com/ef/core/managing-schemas/migrations`

- Azure Cosmos DB intro: `https://learn.microsoft.com/azure/cosmos-db/introduction`

- Limitations with the EF Core Cosmos provider: `https://learn.microsoft.com/ef/core/providers/cosmos/limitations`

- .NET Aspire SqlServer EF Core component: `https://learn.microsoft.com/en-us/dotnet/aspire/database/sql-server-entity-framework-component`

- .NET Aspire Microsoft EF Core Cosmos DB component: `https://learn.microsoft.com/en-us/dotnet/aspire/database/azure-cosmos-db-entity-framework-component`

4

Creating Libraries for Client Applications

With the updates from the last chapter, the Game API is available to use – including access to the database. In this chapter, we'll create a .NET library that can be used by all .NET client applications to access the service. Instead of the need to create HTTP requests with every client application, we create a library that can be shared.

In this chapter, you'll do the following:

- Create a library to send HTTP requests
- Create a client console application to play a game using the library
- Use the Microsoft Kiota tool to generate code based on the OpenAPI document

Technical requirements

The code for this chapter can be found in the GitHub repository at `https://github.com/PacktPublishing/Pragmatic-Microservices-with-CSharp-and-Azure`. The source code folder `ch04` contains the code samples for this chapter.

The service implementation from the previous chapter is stored in the `server` folder. There's just a small change to the previous chapter with the models. The models contain annotations (`Required`, `MinLength`, and `MaxLength` attributes). This information shows up in the OpenAPI document and can be used on creating the client. You can use the file `Chapter04.server.sln` to open and run the solution. You need to start the service when running the client application. Based on your preference, you need to configure SQL Server or Azure Cosmos DB as discussed in the previous chapter. You can also use the in-memory repository instead so that you don't need to have a database running. Change the configuration with the `appsettings.json` file based on your needs.

The new code is in the `client` folder. Here, you will find these projects:

- `Codebreaker.GameApis.Client`: This is the new library that includes custom models and the `GamesClient` class, which sends HTTP requests to the service
- `Codebreaker.Client.Console`: This is a new console application that references the client library and can be used to play the game
- `Codebreaker.GamesApis.Kiota`: This is a client library that can be used as an alternative to `Codebreaker.GameApis.Client` with generated code
- `Codebreaker.Kiota.Console`: This is a console application that uses the Kiota client library

Creating a library to create HTTP requests

With a microservices team, a good practice is when the team is not only responsible to develop the complete service including the database access code, but also at least one of the client applications. With traditional development teams, client and server development is often spread across different teams. The issue with that is that the client and service are best created in collaboration. Creating the client, you'll find answers missing from the services API. Here, a fast communication between the client and service developers helps.

Creating a library for the client allows us to reuse this functionality from all .NET clients; you can create clients with any .NET client technology, such as Blazor, WinUI, .NET MAUI, and others. In this chapter, we will just create a console application, but you can find clients using Blazor, WinUI, .NET MAUI, WPF, and Platform Uno in the GitHub organization at `https://github.com/codebreakerapp`.

To create the library to be used by client applications, we do the following:

1. Create a library with multi-targeting support to support different .NET versions with the clients
2. Inject the `HttpClient` with the main class the client interacts with.
3. Send HTTP requests to the games service.
4. Create a NuGet package for easier use.

Creating a library with multi-targeting support

We create the library for the client using the `dotnet` CLI:

```
dotnet new classlib --framework net8.0 -o Codebreaker.GameApis.Client
```

With this library, we need the model types for the data to transfer between the client and the service and a client class that does the HTTP requests to invoke the services API.

To support clients using different .NET versions, the library is configured with multi-targeting support:

Codebreaker.GameAPIs.Client/Codebreaker.GameAPIs.Client.csproj

```
<PropertyGroup>
  <TargetFrameworks>net7.0;net8.0</TargetFrameworks>
  <!-- code removed for brevity -->
</PropertyGroup>
```

Instead of the default entry, `TargetFramework`, an `s` is appended to contain a list of frameworks. With multiple frameworks added, multiple binaries are added when creating a NuGet package. It might be okay for you to create a library with .NET 6, which can also be used from .NET 7 and .NET 8 clients. Using multiple frameworks you can create optimized code based on the client version.

An optimization is shown with the following code snippet:

Codebreaker.GameAPIs.Client/Models/CreateGameRequest.cs

```
#if NET8_0_OR_GREATER
[JsonConverter(typeof(JsonStringEnumConverter<GameType>))]
#else
[JsonConverter(typeof(JsonStringEnumConverter))]
#endif
public enum GameType
{
    Game6x4,
    Game6x4Mini,
    Game8x5,
    Game5x5x4,
}
```

Generic attributes are new since C# 11. The generic type of the `JsonStringEnumConverter` is new with .NET 8. This generic version supports Native AOT compilation. The older version and the non-generic `JsonStringEnumConverter` uses reflection. Using the C# preprocessor directive `#if` and the predefined symbol `NET8_0_OR_GREATER`, different code gets compiled based on the framework version.

The models are mainly the same between the client and the service. Here, you might choose to move the models from the service-only library to a common library that is referenced both from the client and service applications. However, with client technologies, you might have other requirements based on validating and change notification. With the models of the client library, you can implement the interface `INotifyPropertyChanged`, which is used by different client technologies to update the user interface automatically if a change is notified. Later in this chapter, we'll also create a library from the OpenAPI document created in *Chapter 2*, which can be another reason not to create a shared library.

`CreateGameRequest` is the class we need to send the request when starting the game:

Codebreaker.GameAPIs.Client/Models/CreateGameRequest.cs

```
[JsonConverter(typeof(JsonStringEnumConverter<GameType>))]
public enum GameType
{
  Game6x4,
  Game6x4Mini,
  Game8x5,
  Game5x5x4,
}

public record class CreateGameRequest(
  GameType,
  string PlayerName);
```

The `CreateGameRequest` contains the properties `GameType` and `PlayerName`, which are required to start the game. The `GamesQuery` class is used to send different query parameters to retrieve a filtered list of games based on the query:

Codebreaker.GameAPIs.Client/Models/GamesQuery.cs

```
public record class GamesQuery(
  GameType? GameType = default,
  string? PlayerName = default,
  DateOnly? Date = default,
  bool? Ended = false)
{
  public string AsUrlQuery()
  {
    var queryString = "?";

    if (GameType != null)
    {
      queryString += $"gameType={GameType}&";
    }

    if (PlayerName != null)
    {
      queryString += $"playerName={Uri.EscapeDataString(PlayerName)}&";
    }
    // code removed for brevity
```

```
    queryString = queryString.TrimEnd('&');

    return queryString;
  }
}
```

The `AsUrlQuery` method converts the properties of the record to create HTTP query parameters as specified with the games API service and returns the combined query string. You might think about adding this method to the `Game` class. The `Game` class just defines the structure of the data representing a game. The `GamesQuery` class controls how its data can be converted into a URL query string.

Additionally, `CreateGameResponse`, `UpdateGameRequest`, `UpdateGameResponse`, `Game`, and `Move` are needed with this library. Check these types with the GitHub repo.

Injecting the HttpClient class

The `GamesClient` class we create next is used to send requests to the games service. To use the `HttpClient` class, an object of this class can be injected. With the application using the library, this `HttpClient` needs to be configured with the base address (in *Chapter 9*, this will be extended with authentication).

The implementation of the constructor of the `GamesClient` class is shown in the following code snippet:

Codebreaker.GameAPIs.Client/GamesClient.cs

```
public class GamesClient(HttpClient httpClient)
{
  private readonly HttpClient _httpClient = httpClient;
  private readonly JsonSerializerOptions _jsonOptions = new()
  {
    PropertyNameCaseInsensitive = true
  };
  // code removed for brevity
```

With the constructor of the `GamesClient`, the injected `HttpClient` instance is assigned to a variable, and `JsonOptions` is configured. ASP.NET Core maps properties on JSON serialization to lowercase. With the options as defined here, casing is ignored, so the lowercase map is transferred to uppercase properties.

> **Note**
>
> Don't create a new instance of the `HttpClient` class with every request. Instead, injecting the client will shift the responsibility to creating instances to the calling application. With the dependency injection container, we'll configure the `HttpClient` to be created from a factory.

Sending HTTP requests

Let's send some requests to the service to retrieve information about games, start games, and set moves.

The first methods are used to retrieve game information:

Codebreaker.GameAPIs.Client/GamesClient.cs

```
public async Task<Game?> GetGameAsync(bool id, CancellationToken
cancellationToken = default)
{
  Game game = default;
  try
  {
    game = await _httpClient.GetFromJsonAsync<Game>(
      $"/games/{id}", _jsonOptions, cancellationToken);
  }
  catch (HttpRequestException ex) when (ex.StatusCode =
  HttpStatusCode.NotFound)
  {
    return default;
  }
  return game;
}


public async Task<IEnumerable<Game>> GetGamesAsync(GamesQuery query,
CancellationToken cancellationToken = default)
{
  IEnumerable<Game> games = (
    await _httpClient.GetFromJsonAsync<IEnumerable<Game>>(
      $"/games/{query.AsUrlQuery()}", _jsonOptions,
      cancellationToken)) ?? Enumerable.Empty<Game>();
  return games;
}
```

The GetGameAsync method retrieves one game passing the identifier of the game. GetGamesAsync uses the previously created GamesQuery to create the URI for the service to send the HTTP GET request. GetFromJsonAsync is an extension method for the HttpClient class to send an HTTP GET request, checks for a successful status code using EnsureSuccessStatusCode with HttpResponseMessage (which throws an HttpRequestException if not successful), and uses the System.Text.Json deserializer to deserialize the stream from the response. When the game-id passed was not found, we want to return null instead of throwing an exception so this exception is caught.

Sending a request to start a game is implemented with the `StartGameAsync` method:

Codebreaker.GameAPIs.Client/GamesClient.cs

```
public async Task<(Guid id, int numberCodes, int maxMoves,
IDictionary<string, string[]> FieldValues)>
  StartGameAsync(GameType gameType, string playerName,
  CancellationToken cancellationToken = default)
{
  CreateGameRequest createGameRequest = new(_gameType, _playerName);
  var response = await _httpClient.PostAsJsonAsync("/games",
    createGameRequest, cancellationToken);
  response.EnsureSuccessStatusCode();
  var gameResponse = await response.Content.
    ReadFromJsonAsync<CreateGameResponse>(
    _jsonOptions, cancellationToken);
  if (gameResponse is null)
    throw new InvalidOperationException();
  return (gameResponse.Id, gameResponse.NumberCodes, gameResponse.
    MaxMoves, gameResponse.FieldValues);
}
```

The `StartGamesAsync` method sends an HTTP POST request after creating the data that should be sent with the HTTP body: `CreateGameRequest`. After receiving a success response, the `ReadFromJsonAsync` extension method deserializes the returned HTTP body and returns the methods result using a tuple.

To send a games move, the game is updated using an HTTP PATCH request:

Codebreaker.GameAPIs.Client/GamesClient.cs

```
public async Task<(string[] Results, bool Ended, bool IsVictory)>
SetMoveAsync(Guid id, string playerName, GameType gameType, int
moveNumber, string[] guessPegs, CancellationToken cancellationToken =
default)
{
  UpdateGameRequest updateGameRequest = new(id, gameType, playerName,
    moveNumber)
  {
    GuessPegs = guessPegs
  };
  var response = await _httpClient.PatchAsJsonAsync($"/games/{id}",
    updateGameRequest, _jsonOptions, cancellationToken);
  response.EnsureSuccessStatusCode();
  var moveResponse = await response.Content.
```

```
ReadFromJsonAsync<UpdateGameResponse>(_jsonOptions, cancellationToken)
    ?? throw new InvalidOperationException();
  (_, _, _, bool ended, bool isVictory, string[] results) =
    moveResponse;
  return (results, ended, isVictory);
}
```

Sending an HTTP PATCH request is very similar to sending a POST request: the UpdateGameRequest object is created to send this JSON-serialized information to the server. Receiving the result, the body is deserialized to an UpdateGameResponse object.

> **Note**
>
> With REST APIs, a HTTP PUT request is usually used to update a resource, while HTTP PATCH is used for a partial update. Here, the game resource is updated, but not by sending the complete game and just some partial data.

Creating a NuGet Package

To create a NuGet package from the library, you can use the dotnet CLI:

```
cd Codebreaker.GameAPIs.Client
dotnet pack --configuration Release
```

To see the content of the NuGet package, you can rename it to zip. For easy use of this package, you can add it to a shared folder and configure the Visual Studio NuGet Package manager to reference this folder. You can also publish the package to Azure DevOps Artifacts. Referencing this, you can create a nuget.config file:

```
dotnet new nugetconfig
```

With the generated nuget.config file, you need to specify the shared folder or the link to your Azure DevOps Artifacts feed.

This is a nuget.config file created using dotnet new with one additional entry for a custom feed:

```
<?xml version="1.0" encoding="utf-8"?>
<configuration>
  <packageSources>
    <clear />
    <add key="nuget" value="https://api.nuget.org/v3/index.json" />
    <add key="custom" value="https://pkgs.dev.azure.com/
      MyOrganization/_packaging/MyFeed/nuget/v3/index.json" />
  </packageSources>
</configuration>
```

With this NuGet configuration file, the entry `<clear />` removes all the default feeds. The `nuget` key with the first `add` element references the default feed of the NuGet server. Similarly, you can add custom feeds with other keys and links to the package feeds on the server.

For the Codebreaker solution, you can look for the NuGet package `Cninnovation.Codebreaker.Client`, which is available on the NuGet server. After making NuGet packages available on NuGet, a readme file, a license, and some more metadata should be added to the package. See further readings for more information.

Creating a client application

Having the library in place, let's create a client application. A simple console application fulfils the purpose to play the game. With the sample application of this chapter, the NuGet packages `Microsoft.Extensions.Hosting`, `Microsoft.Extensions.Http.Resiliency`, and `Spectre.Console.Cli` are added. Navigate to the folder of the solution file before invoking these commands:

```
dotnet new console -framework net8.0 -o Codebreaker.Console
cd Codebreaker.Console
dotnet add package Microsoft.Extensions.Hosting
dotnet add package Microsoft.Extensions.Http.Resilience
dotnet add package Spectre.Console.Cli
dotnet add reference ../Codebreaker.GameAPIs.Client
```

`Microsoft.Extensions.Hosting` will be used for a dependency injection container and configuration support and `Microsoft.Extensions.Http.Resilience` is the package offering an `HttpClientFactory`. Of course, the library created previously needs to be referenced as well.

To interact with the user, you can use simple `Console.ReadLine` and `Console.WriteLine` statements. With the sample application available in the books GitHub repo, the NuGet package `Spectre.Console.Cli` is used. Just check the source code for more information.

Configuring the dependency injection container

The top-level statements of the application are shown with the following code snippet:

Codebreaker.Console/Program.cs

```
var builder = Host.CreateApplicationBuilder(args);
builder.Services.AddHttpClient<GamesClient>(client =>
{
  string gamesUrl = builder.Configuration["GamesApiUrl"] ?? throw new
    InvalidOperationException("GamesApiUrl not found");
  client.BaseAddress = new Uri(gamesUrl);
});
```

```
builder.Services.AddTransient<Runner>();
var app = builder.Build();

var runner = app.Services.GetRequiredService<Runner>();
await runner.RunAsync();
```

The CreateApplicationBuilder of the Host class configures the dependency injection container and has default configuration for the application configuration providers and the logging providers. The AddHttpClient extension method is implemented with the HttpClient factory. Here, the generic method overload is used to specify the GamesClient class that will receive the HttpClient injected as specified with the configureClient lambda expression. The BaseAddress of the HttpClient is configured to have the GamesApiUrl configuration value.

For the configuration, we create the appsettings.json configuration file:

Codebreaker.Console/appsettings.json

```
{
  "Logging": {
    "LogLevel": {
      "Default": "Warning"
    }
  },
  "GamesApiUrl": "http://localhost:9400"
}
```

The GamesApiUrl key is configured to contain the address of the Games API service. To not mess up logging with the console output for the game play, the log level is configured to only log warning, error, and critical error messages.

Interacting with the user

The interaction with the user and the invocation of the service happens via the Runner class. Here, the previously created GamesClient is injected into the primary constructor:

Codebreaker.Console/Runner.cs

```
internal class Runner(GamesClient client)
{
  private readonly CancellationTokenSource _cancellationTokenSource =
new();
```

```
public async Task RunAsync()
{
  bool ended = false;
  while (!ended)
  {
    var selection = Inputs.GetMainSelection();
    switch (selection)
    {
      case MainOptions.Play:
        await PlayGameAsync();
        break;
      case MainOptions.Exit:
        ended = true;
        break;
      case MainOptions.QueryGame:
        await ShowGameAsync();
        break;
      case MainOptions.QueryList:
        await ShowGamesAsync();
        break;
      case MainOptions.Delete:
        await DeleteGameAsync();
        break;
      default:
        throw new ArgumentOutOfRangeException();
    }
  }
  // code removed for brevity
}
```

The RunAsync method first asks the user what to do next. The main options are to play a game, to show the status of a single game, to show a list of games, or to delete a game. This code snippet makes use of the Inputs class, which in turn uses the AnsiConsole class from the mentioned NuGet package Spectre.Console.Cli. With this, you get a nice console user interface, as shown in *Figure 4.1*, with easy selections. Depending on what you used for interacting with the user, your user interface might look different.

1. To run a game, first start the server before starting the client. With the client, select **Play** (*Figure 4.1*).

```
Select

> Play
  QueryGame
  QueryList
  Delete
  Exit
```

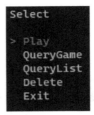

Figure 4.1 – Console output to select a main task

2. Next, select a game type (*Figure 4.2*), for example, **Game6x4**.

```
Select the game type

> Game6x4
  Game6x4Mini
  Game8x5
  Game5x5x4
```

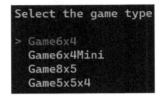

Figure 4.2 – Selecting the game type

3. Enter a player name (*Figure 4.3*) and enter all the colors needed for a single move.

```
Enter player name Christian
colors 1

  Red
> Green
  Blue
  Yellow
  Purple
  Orange
```

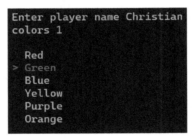

Figure 4.3 – Entering the name and selecting the colors depending on the game type

4. *Figure 4.4* shows the result of the move (here with three colors correct but in the wrong positions) and the start of the next move. Repeat this until you solve the codes.

```
Enter player name Christian
1. Green    Red      Blue     Yellow   ** White White White
colors 1

> Red
  Green
  Blue
  Yellow
  Purple
  Orange
```

Figure 4.4 – Moving the result and the next move

The result of a successful move is shown in *Figure 4.5*:

```
Enter player name Christian
1. Green      Red       Blue      Yellow    ** White White White
2. Red        Green     Yellow    Purple    ** White White
3. Blue       Orange    Green     Red       ** Black White White
4. Blue       Blue      Red       Green     ** Black Black Black Black
Victory: True
You won after 4 moves
Select

> Play
  QueryGame
  QueryList
  Delete
  Exit
```

Figure 4.5 – Game results

From there, you can repeat this to play another game or query for the list of games. From the list of games, you can get a game identifier and query for a single game passing the identifier.

Using Microsoft Kiota to create a client

Running the API service generating the OpenAPI document (this was done in *Chapter 2*), we can leverage this information, and create the client code automatically. With the sample code of this chapter, the OpenAPI document is stored with the file `gamesapi-swagger.json`, which you can reference without starting the service.

One option with Visual Studio is to use **Add | Connected Client** and add a service reference to an OpenAPI document. But this option (at the time of this writing) has some limitations:

- It still uses the Newtonsoft Json serializer, whereas the new `System.Text.Json` one is faster and uses less memory

- The client implementation makes use of strings instead of streams, which can result in objects in the large object heap

As you've seen in this chapter, creating a custom library to create HTTP requests is not that hard and can be optimized for your own domain.

But now there's another option that should be considered: Microsoft Kiota (`https://learn.microsoft.com/openapi/kiota/`). Microsoft Kiota is a command-line tool that offers code generation from the OpenAPI for several languages including Java, PHP, Python, Typescript, C#, and many others. Let's give this a chance.

Installing Kiota

Kiota is available as a `dotnet` tool. We install this tool as part of the new library for another class library project.

The library is created with the following commands. Run these commands from the `solution` folder:

```
dotnet new classlib --framework net8.0 -o Codebreaker.GameApis.
KiotaClient
cd Codebreaker.GameApis.KiotaClient
dotnet add package Microsoft.Kiota.Http.HttpClientLibrary
dotnet add package Microsoft.Kiota.Serialization.Json
dotnet add package Microsoft.Kiota.Serialization.Form
dotnet add package Microsoft.Kiota.Serialization.Multipart
dotnet add package Microsoft.Kiota.Serialization.Text
```

Using Kiota, we also need to add some Kiota NuGet packages for different serializers and a Kiota HTTP client library. The Kiota tool is installed with the project:

```
dotnet new tool-manifest
dotnet tool install microsoft.openapi.kiota
```

Generating Code with Kiota

After the Kiota tool is installed, we can generate the code using the OpenAPI document `gamesapi-swagger.json`. This file is available in the `ch04` folder:

```
dotnet kiota generate --openapi ..\..\gamesapi-swagger.json --output
codebreaker --language CSharp --class-name GamesAPIClient --namespace-
name Codebreaker.Client
```

Using these options, source code is generated using the referenced OpenAPI document `gamesapi-swagger.json`, the generated files are stored in the subdirectory `codebreaker`, C# is used for the code generation, the main class to do the HTTP requests is named `GamesAPIClient`, and the namespace for all the generated code is `Codebreaker.Client`.

> **Note**
>
> Looking at the generated code, you'll see that Kiota-generated code is not using the same coding convention as used in this book or as the .NET team is using. For example, the opening of curly braces is done in the same line as the method name, which is a convention used with many JavaScript programs. If you use Visual Studio, you can easily change this with the complete program using the context menu in Solution Explorer, navigating to **Analyze and Code Cleanup | Run Code Cleanup**. You might need to configure your preferences with code cleanup first.

The types created using Kiota are the models (using the `schemas` within OpenAPI) and request builders (using the `paths` where the requests are defined). Check the book repository for the generated code files.

Exploring the Kiota-generated models

For all the requests and responses and all the types specified within the schemas, Kiota generates classes in the `Models` directory. Let's have a look at the `CreateGameRequest` class:

Codebreaker.GamesAPIs.KiotaClient/codebreaker/Models/Create-GameRequest.cs

```
public class CreateGameRequest : IParsable
{
  public Codebreaker.Client.Models.GameType? GameType { get; set; }
  public string? PlayerName { get; set; }
  public static CreateGameRequest
    CreateFromDiscriminatorValue(IParseNode parseNode)
  {
    _ = parseNode ?? throw new ArgumentNullException(nameof(parseNode));
    return new CreateGameRequest();
  }
  public virtual IDictionary<string, Action<IParseNode>>
    GetFieldDeserializers()
  {
    return new Dictionary<string, Action<IParseNode>> {
      {"gameType", n => { GameType = n.GetEnumValue<GameType>(); } },
      {"playerName", n => { PlayerName = n.GetStringValue(); } },
    };
  }

  public virtual void Serialize(ISerializationWriter writer)
  {
    _ = writer ?? throw new ArgumentNullException(nameof(writer));
    writer.WriteEnumValue<GameType>("gameType", GameType);
    writer.WriteStringValue("playerName", PlayerName);
  }
}
```

The model types implement the `IParsable` interface. This is not the `System.IParsable` interface, but a version from the Kiota library in the namespace `Microsoft.Kiota.Abstractions.Serialization`. This interface defines instance members `GetFieldDeserializers` and `Serialize`. With this, Kiota offers an abstraction layer, which allows for the use of different serializers.

Another important aspect that needs mentioning is that all the properties of the model types are declared to be nullable. While EF Core supports nullability to map non-nullable members to be required in the database, this annotation is not used when generating the OpenAPI document using the minimal APIs. Adding the `Required` attribute to the models on the server adds `required`. Other annotations such as `MaxLength` and `MinLength` are mapped as well with `maxLength` and `minLength`, as you can see with `gamesapi-swagger.json`.

However, many APIs don't pay attention to nullability. With the OpenAPI definition too, it's also not exlpicitly specified how strict nullability should be enforced. Depending on the context where the model is used, information can still be left out from the server, and the data is not sent.

Here is a discussion about the Kiota implementation: `https://github.com/microsoft/kiota/issues/2594`

With the next major version of the OpenAPI specification, this might change. With the decisions made for the current Kiota implementation, Kiota is on the safe side to declare all the model properties as nullable, but this also means that we need to check for `null`.

Exploring the Kiota-generated request builders

The goal of request builders is to easily create requests. Let's look at some of the generated code:

Codebreaker.GamesAPIs.KiotaClient/codebreaker/GamesAPIClient.cs

```
public class GamesAPIClient : BaseRequestBuilder
{
  public GamesRequestBuilder Games
  {
    get => new GamesRequestBuilder(PathParameters, RequestAdapter);
  }

  public GamesAPIClient(IRequestAdapter requestAdapter) :
base(requestAdapter, "{+baseurl}", new Dictionary<string, object>())
  {
    ApiClientBuilder.
      RegisterDefaultSerializer<JsonSerializationWriterFactory>();
    ApiClientBuilder.
      RegisterDefaultSerializer<TextSerializationWriterFactory>();
    ApiClientBuilder.
      RegisterDefaultDeserializer<JsonParseNodeFactory>();
  // code removed for brevity
  }
}
```

All the request builders derive from the base class, BaseRequestBuilder. The GamesApiClient class, where the name was specified with the code generation, is the request builder that needs to be initiated to communicate with the Games API. In the constructor, you can see default serializers and deserializers configured. Here, Kiota gives another flexibility.

The Games property of GamesApiClient returns another request builder: GamesRequestBuilder. This builder is in the GamesRequestBuilder.cs source file:

Codebreaker.GamesAPIs.KiotaClient/codebreaker/Games/GamesRequestBuilder.cs

```
public class GamesRequestBuilder : BaseRequestBuilder
{
  public GamesItemRequestBuilder this[Guid position]
  {
    get
    {
      // code removed for brevity
      return new GamesItemRequestBuilder(urlTplParams,
        RequestAdapter);
    }
  }

  public async Task<List<Game>?> GetAsync(Action<RequestConfiguration
    <GamesRequestBuilderGetQueryParameters>>? requestConfiguration =
    default, CancellationToken cancellationToken = default)
  {
    var requestInfo = ToGetRequestInformation(requestConfiguration);
    var collectionResult = await RequestAdapter.
      SendCollectionAsync<Game>(requestInfo, Game.
      CreateFromDiscriminatorValue, default, cancellationToken).
      ConfigureAwait(false);
    return collectionResult?.ToList();
  }

  public async Task<CreateGameResponse?> PostAsync(CreateGameRequest
    body, Action<RequestConfiguration<DefaultQueryParameters>>?
    requestConfiguration = default, CancellationToken
    cancellationToken = default)
  {
    // code removed for brevity
  }
```

This request builder is then used to invoke requests of the games API. Methods implemented by this request builder are GetAsync and PostAsync. The GetAsync method is used to retrieve a list of games with query parameters. PostAsync sends a POST request with the generated CreateGameRequest model.

To get a single game, update a game by sending a game move, and delete a game, the games API needs a game identifier. With Kiota, this is solved by offering an indexer with GamesRequestBuilder, which in turn returns another request builder, GameItemsRequestBuilder. Here, a fluent API can be used to pass a game identifier and invoke the GetAsync and PutAsync methods.

Let's implement another console application to use this generated code in the next section.

Using the Kiota-generated code

The Kiota-generated code is used with the console application Codebreaker.KiotaConsole. With big parts, the code of this application is similar to the console application from before. Mainly the invocations to the service, which are done with the Runner class, are now replaced and the dependency injection container configuration is changed.

The HttpClient factory is no longer registered with the DI container, as shown in the following code snippet:

Codebreaker.KiotaConsole/Program.cs

```
var builder = Host.CreateApplicationBuilder(args);

builder.Services.Configure<RunnerOptions>(options =>
{
  options.GamesApiUrl = builder.Configuration["GamesApiUrl"] ??
    throw new InvalidOperationException("GamesApiUrl not found");
});

builder.Services.AddTransient<Runner>();
var app = builder.Build();

var runner = app.Services.GetRequiredService<Runner>();
await runner.RunAsync();
```

Other than removing the code for the HttpClient configuration, the base address is now configured to supply values for the RunnerOptions class. This class just defines the GamesApiUrl property to specify the base address for the games service.

The constructor of the Runner class where the options are passed is shown in the next code snippet:

Codebreaker.KiotaConsole/Runner.cs

```
internal class Runner
{
  private readonly GamesAPIClient _client;
  private readonly CancellationTokenSource _cancellationTokenSource =
new();

  public Runner(IOptions<RunnerOptions> options)
  {
    AnonymousAuthenticationProvider authenticationProvider = new();
    HttpClientRequestAdapter adapter = new(authenticationProvider)
    {
      BaseUrl = options.Value.GamesApiUrl ?? throw new
        InvalidOperationException("Could not read GamesApiUrl")
    };
    _client = new GamesAPIClient(adapter);
  }
```

With the implementation of the Runner constructor, the GamesAPIClient class is instantiated. This class receives HttpClientRequestAdapter, which has the base address of the service configured. The constructor of HttpClientRequestAdapter receives an object implementing the IAuthenticationProvider interface. Here, AnonymousAuthenticationProvider is used as no authentication is needed. Kiota offers various authentication providers.

Sending a GET request with query parameters to get a list of games, you invoke the GetAsync method of GamesRequestBuilder:

Codebreaker.KiotaConsole/Runner.cs

```
private async Task ShowGamesAsync()
{
  var games = await _client.Games.GetAsync(config =>
  {
    config.QueryParameters.Date = new Date(DateTime.Today);
  }, _cancellationTokenSource.Token);
  // code removed for brevity
```

The Games property returns the generated GamesRequestBuilder, which allows us to invoke the GetAsync method passing query parameters. Kiota offers its own Date type within the Microsoft.Kiota.Abstractions namespace, which represents the date-only part of DateTime. Today, .NET offers DateOnly, but this type is not available with .NET Standard 2.0, which is also supported by Kiota.

Starting a game is done by sending a POST request, as shown in the next snippet:

Codebreaker.KiotaConsole/Runner.cs

```
private async Task PlayGameAsync()
{
  // code removed for brevity
  CreateGameRequest request = new()
  {
    PlayerName = playerName,
    GameType = gameType
  };
  var response = await _client.Games.PostAsync(request,
    cancellationToken: _cancellationTokenSource.Token);
```

Starting the game, the user input for the player name and the game type are assigned to the `CreateGameRequest` object. This model type is then passed with the invocation of the `PostAsync` method to start a game and to receive `CreateGameResponse`.

Retrieving a single game passing the game identifier is shown in the next code snippet:

Codebreaker.KiotaConsole/Runner.cs

```
private async Task ShowGameAsync()
{
  // code removed for brevity
  var game = await _client.Games[id.ToString()].GetAsync(
    cancellationToken: _cancellationTokenSource.Token);
  // code removed for brevity
```

Retrieving a single game, updating the game with a HTTP PATCH request, and deleting a game with the HTTP DELETE request all need the game identifier as the query parameter. To use this, Kiota offers an indexer passing `game-id` and continuing with a fluent API. To retrieve a single game, the `GetAsync` method is used. Patching and deleting games are very similar.

With this information, you can use the Kiota-generated code and write the implementation to update the game by sending a game move with the `PostAsync` method.

Using the new client, you can run the game in the same way as shown before!

Summary

Working through this chapter, you'll have a running client console application to run the game. We used the `HttpClient` class to send requests to the games service. To reuse this with different client technologies, we created a library. For efficient use of the `HttpClient` class, you learned to use the HttpClient factory.

Instead of implementing the models on your own, you learned using Microsoft Kiota to create code from the OpenAPI definition. With your own scenarios, you can now decide what's the best option for you.

Before reading the next chapter, you can reuse this newly created library and create clients of your choice such as Blazor, WinUI, or .NET MAUI. While these frameworks are outside of the scope of this book, you can check `https://github.com/codebreakerapp` for more clients available.

No matter what client you implement, before diving into the next chapter, it's well deserved to play one more game—this time with your own created client application.

In the next chapter, the focus will be on the services again; we'll host the service application (and another service) with a Docker container. This new service will also use the HTTP client created in this chapter.

Further reading

To learn more about the topics discussed in this chapter, you can refer to the following links:

- `HttpClient` guidelines: `https://learn.microsoft.com/dotnet/fundamentals/networking/http/httpclient-guidelines`
- NuGet packages best practices: `https://learn.microsoft.com/nuget/create-packages/package-authoring-best-practices`
- The large object heap on Windows systems: `https://learn.microsoft.com/dotnet/standard/garbage-collection/large-object-heap`
- Microsoft Kiota documentation: `https://learn.microsoft.com/openapi/kiota/`
- Kiota GitHub repository: `https://github.com/microsoft/kiota`

Part 2:
Hosting and Deploying

This part focuses on essential aspects of hosting and deploying microservices. You will begin by gaining a comprehensive understanding of Docker fundamentals, such as creating Dockerfiles, building Docker images using the .NET CLI, and running Docker containers with .NET Aspire on your development environment. You will then proceed to publish Docker images to the Azure Container Registry, deploy them to the Azure Container Apps environment (based on Kubernetes), and incorporate Azure services like Azure App Configuration and Azure Key Vault.

Throughout this part, you will utilize Azure resources for local application execution, deploy applications to Azure using the Azure Developer CLI, and establish GitHub Actions for automated deployment to Azure upon code updates in the repository. To ensure seamless operation in both on-premises and Azure environments, authentication will be implemented with Azure Active Directory B2C and Microsoft Entra, alongside ASP.NET Core Identities.

This part has the following chapters:

- *Chapter 5, Containerization of Microservices*

- *Chapter 6, Microsoft Azure for Hosting Applications*

- *Chapter 7, Flexible Configuration*

- *Chapter 8, CI/CD – Publishing with GitHub Actions*

- *Chapter 9, Authentication and Authorization with Services and Clients*

5

Containerization of Microservices

After building clients and services with the previous chapters, now is the time to make the services ready for publishing. With Docker, we can prepare images that have everything included to run the complete solution.

In this chapter, you'll start learning the most important parts of Docker, building Docker images, running containers, and using .NET Aspire to run a solution consisting of multiple services locally on your developer system, including SQL Server running in a Docker container, as well as making use of native **ahead-of-time** (**AOT**) to create platform-specific native applications.

In this chapter, you'll learn about the following topics:

- Working with Docker
- Building a Docker image
- Running the solution with .NET Aspire
- Using native AOT with ASP.NET Core

Technical requirements

What you need to go through this chapter is **Docker Desktop**. Docker Desktop is free for individual developers and education and open source communities. You can download Docker Desktop from `https://www.docker.com/products/docker-desktop/`, best used with the **Windows Subsystem for Linux** (**WSL**). Check the README file of this chapter to install WSL 2 and Docker Desktop.

> **Note**
>
> The `dotnet publish` command supports building and publishing Docker images. While some features of `dotnet publish` can be used without Docker Desktop being installed, we start using Docker directly, as this also helps in understanding what can be done using the .NET CLI, and often you need a lot more in regard to Docker than offered by the .NET CLI.

The code for this chapter can be found in the following GitHub repository: `https://github.com/PacktPublishing/Pragmatic-Microservices-with-CSharp-and-Azure`.

The `ch05` source code folder contains the code samples for this chapter. For different sections of this chapter, different subfolders are available. For a start, working through the instructions, you can use the `StartXX` folders. `StartDocker` contains the projects before creating Docker containers have been added and the `FinalDocker` folder contains the project in the final state after building the Docker container.

The `StartAspire` folder contains multiple projects that the .NET Aspire-specific projects we created in the previous chapters are already part of. Use this as a starting point to work through the .NET Aspire part of this chapter. `FinalAspire` contains the complete result, which you can use as a reference. The `NativeAOT` folder contains the code for the games API that compiles with .NET native AOT.

In the subfolders of the `ch05` folder, you'll see these projects:

- `Codebreaker.GameAPIs` – The games API project we used in the previous chapter from our client application. In this chapter, we make minor updates to specify the connection string to the SQL Server database. This project has a reference to NuGet packages with implementations of the `IGamesRepository` interface for SQL Server and Azure Cosmos DB.

- `Codebreaker.Bot` – This is a new project that implements a REST API and calls the games API to automatically play games with random game moves. This project makes use of the client library we created in *Chapter 4* – it has a reference to the `CNinnovation.Codebreaker.Client` NuGet package to call the games API.

- `Codebreaker.AppHost` – This project is enhanced to orchestrate the different services.

- `Codebreaker.ServiceDefaults` – This project is unchanged in this chapter.

- `Codebreaker.GameAPIs.NativeAOT` – A new project that offers the same games API with some changes to support native AOT with .NET 8.

Working with Docker

Although nowadays, it's possible just to work with .NET tools to create microservices and run Docker containers, it helps to know about Docker. Thus, here, we look at the most important concepts about Docker, starting up a SQL Server instance running within a Docker container, creating a Dockerfile to build a Docker image for the games API service, and running these containers on the local system.

In case you already know all about Docker, you can skip and move over to the *.NET Aspire* section, which does not need the Docker containers created here.

Before diving into building Docker images, why do we need containers at all? When deploying an application, it often occurs that the application fails to run. Often, a reason for this is a missing runtime on the target system or wrong or missing configuration settings. One way to resolve this is to prepare **virtual machines** (**VMs**) where everything is preinstalled. The disadvantage of this is the resources that the VM needs. A VM comes with an operating system and allocates CPU and memory resources. Docker is a lot more lightweight. A Docker image can be small as an operating system is not part of the image – and multiple Docker containers can share the same CPU and memory.

Here's a brief list of important terms when using Docker before going into the details:

- A Docker **image** is an executable package that contains everything to run an application
- One image might have different versions, which are identified by Docker **tags**
- A **Dockerfile** is a text file with instructions to build a Docker image
- A Docker **container** is the running instance of a Docker image
- The Docker **registry** is where Docker images are stored
- A Docker **repository** is a collection of different versions of a Docker image in the registry

Using Docker Desktop

Docker Desktop for Windows offers an environment to build Docker images and run Docker containers on Windows. You can configure it to use Windows or Linux containers. With a previous edition of Docker Desktop for Windows, it was required to install Hyper-V. Docker Desktop then used a Linux VM to run all Linux containers on this VM. Because Windows now supports Linux more natively with WSL, Docker Desktop can use WSL and doesn't need a VM. With the Docker Desktop configuration, you can select WSL distros (see *Figure 5.1*) that should use the same Docker environment as on the Windows system itself. Using these Linux distributions, you can use the same Docker commands to manage your Docker environment:

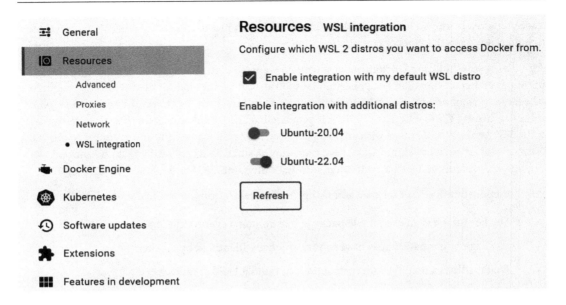

Figure 5.1 – WSL integration in Docker Desktop

Instead of configuring a VM with the amount of CPU and memory you allocate, WSL shares CPU and memory with Windows – but there are some limits for WSL. With the Windows build version 20176 and later, the memory is limited to 50% and 8 GB (whichever is less); with builds before that, WSL can use 80% of the total memory.

For the number of logical processors, by default, all available can be used. You can change memory and CPU limits globally for the complete subsystem but also define different limits for each Linux distribution you install. Check the settings configuration in WSL at `https://learn.microsoft.com/ windows/wsl/wsl-config`.

Running a Docker container

After you've installed Docker Desktop and you are running Windows, you can select to run Windows or Linux containers. While Windows containers are great for legacy applications that only run on Windows (for example, using .NET Framework), Linux containers offer more features, and Linux Docker images are smaller. The solution we build will run with Linux containers.

The Docker Desktop environment needs to be started, then to run the first container, use this command:

```
docker run hello-world
```

On the first run, the `hello-world` Docker image is downloaded from the Docker registry and started. This container just writes a message to the screen to verify that everything is running. Starting it a second time, you'll see that the image is no longer downloaded but started immediately.

To see all images downloaded, you can use this command:

```
docker images
```

To see running containers, use the following command:

```
docker container ls
```

There's also a shorthand notation to show all running containers: `docker ps`.

You will not see the `hello-world` container because this already stopped immediately after writing the output.

Running an image again and again, you start fresh again. But there's also a state kept with a running image. This allows you to continue a previously stopped container with the same state as before. The `docker container ls -a` command not only shows running but also stopped containers. Using `docker container prune`, you can delete the state from all stopped containers.

Running SQL Server in a Docker container

In *Chapter 3*, we used SQL Server and an Azure Cosmos DB emulator on the local system to access it from the games service. Instead of installing these products on your local system, you can also use Docker images.

Let's start downloading the Docker image for SQL Server:

```
docker pull mcr.microsoft.com/mssql/server:2022-latest
```

Previously, we used `docker run` to start a container and implicitly download it from a registry. `docker pull` just downloads the image from the registry. `mcr.microsoft.com` is the Microsoft repository where Microsoft stores images. `mssql/server` is the name of the image. You can read information about this image at `https://hub.docker.com/_/microsoft-mssql-server`. This is an Ubuntu-based image. `2022-latest` is a tag name. This is the actual version of SQL Server 2022. With SQL Server, other tags are `2019-latest`, `2017-latest`, and `latest`. These correspond to SQL Server 2019 and 2017. The `latest` tag is the latest version of SQL Server. At the time of this writing, the image is the same with both the `2022-latest` and the `latest` tags. If you download both images, a second download is not needed, and you will see images with the same image ID but with different tags.

> **Note**
>
> The default configuration to use SQL Server with the Docker image is the SQL Server Developer Edition. You can also configure to use the Express, Standard, Enterprise, and Enterprise Core editions by setting an environmental variable. Pay attention to required licenses with non-developer editions. Read the image documentation for setting environment variables for the different editions.
>
> Another option to use a SQL Server edition within a Docker environment is Azure SQL Edge. Check `https://learn.microsoft.com/en-us/azure/azure-sql-edge/disconnected-deployment` for running Azure SQL Edge.

To run the SQL Server image, you can use the following command:

```
docker run -e "ACCEPT_EULA=Y" -e "MSSQL_SA_PASSWORD=Pa$$w0rd" -p
14333:1433 --name sql1 --hostname sql1 -d mcr.microsoft.com/mssql/
server:2022-latest
```

With this command, these options are used:

- `-e` specifies an environmental variable. With the two variables, the license is accepted, and a password for the `sa` account is defined. `sa` is a privileged account configured, and a short name for *system administrator*.
- The `-p` option maps the port from the host to the container. On the target host, the same port cannot be used for multiple applications; for example, by having a local SQL Server running, the first value cannot use 1433. Make sure to use an available port.
- The `--name` option specifies a name for the container. By default, a random name combined from two lists is used.
- The `--hostname` option specifies the hostname for the container.
- The `-d` option runs the container in the background.

For some useful information to find out what the container does, use `docker container logs`:

```
docker container logs sql1
```

This command needs the name of the container. To connect and wait for all the logs to come, add the `-f` option (for *follow*).

To open a command prompt within the container and see what's there, use `docker exec -it sql1 bash`, which allocates a terminal, keeps `stdin` open (interactive mode), and executes the Bash shell within the container.

After the container with SQL Server is running, we can publish the database we created in *Chapter 3*.

Using volumes with a Docker container

The Docker container for SQL Server contains state (the database files). We can start a previous running container again (using `docker start`). When using `docker run`, we start fresh again, and the previous state is not used. Using `docker commit`, you can create a new image from a container. This keeps the database and the state together, and the Docker images grow in size. A better practice is to keep the state outside of the Docker container. You can mount external directories, files, and Docker volumes within a container. Docker volumes are completely managed by Docker. Let's use this for SQL Server.

First, create a volume:

```
docker volume create gamessqlstorage
```

This creates a volume with the name `gamessqlstorage`. To check the volumes available, use `docker volume ls`. To get more information about a volume, execute `docker volume inspect`.

Let's run the container with the database using this volume:

```
docker run -e "ACCEPT_EULA=Y" -e "MSSQL_SA_PASSWORD=Pa$$w0rd" -p
14333:1433 --name codebreakersql1 --hostname codebreakersql1 -v
gamessqlstorage:/var/opt/mssql -d mcr.microsoft.com/mssql/server:2022-
latest
```

The `-v` option mounts the `/var/opt/mssql` folder within the container to the `gamessqlstorage` volume. All data written by SQL Server to this folder now goes into this volume. The state is now kept externally from the container. So, let's create a database next.

> **Note**
>
> When creating backups of the database, you should also use volumes.

Creating a database in the Docker container

As the container is running, you can access it with a tool such as **SQL Server Object Explorer** from Visual Studio or **SQL Server Management Studio**. The connection string you can use is shown in the following .NET configuration file:

```
{
   "ConnectionStrings": {
     "GamesSqlServerConnection": "server=host.
 docker.internal,14333;database=CodebreakerGames;user
 id=sa;password=Pa$$w0rd;TrustServerCertificate=true"
   }
}
```

With the `appsettings.json` file, you also need to change the `DataStore` key to the `SqlServer` value.

Using Docker with a Linux host system, you can use the IP address of the Docker container with the port number to access services within the Docker container. This might not work with Windows. That's why the `host.docker.internal` hostname was introduced: to map to the service via a gateway using the local port number. With the connection string to the database, you need to add the port number after the hostname, separated by a comma. To pass the user and password, use the `user id` and `password` keys. Because the certificate from the server Docker container might not be from a trusted authority on the Windows system, add the `TrustServerCertificate` setting to the connection string.

In *Chapter 3*, we published a database using the `dotnet ef` command line. Now is the time to create this database within the Docker container. With the following command, your current directory needs to be the directory of the games API service (`Codebreaker.GameAPIs`), the `DataStorage` configuration value in `appsettings` set to `SqlServer`, and the connection string specified as shown before:

```
cd Codebreaker.GameAPIs
dotnet ef database update -p ..\Codebreaker.Data.SqlServer -c
GamesSqlServerContext
```

With this, the database is created – or migrated to the latest version. The `-p` option is needed because the EF Core context is in a different project than the **dependency injection** (**DI**) container configuration. With the solution, we have multiple EF Core contexts; that's why you need to specify the name of the context class with `-c`.

> **Note**
>
> With the games service project, you have another option to create the database. For an easier way to create a SQL Server database, the `/createsql` API is offered now in addition to the other APIs. Sending a `POST` request creates or upgrades the database (if SQL Server is configured) using the EF Core `MigrateAsync` method.

Next, let's create a custom Docker image for the games API service. *Figure 5.2* shows a C4 container diagram to give you an overview picture of the containers we use. The first container we create is the one on the right, hosting SQL Server. Next, we create a Docker image for the game APIs, which accesses the SQL Server container. The container on the left is the new bot project, which invokes the services running in the game APIs' container to automatically play games:

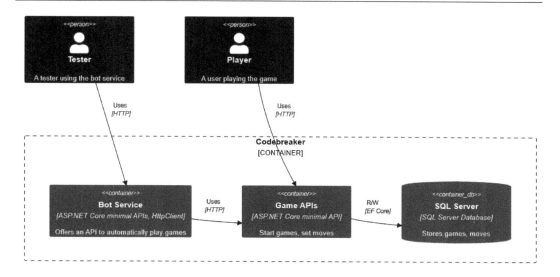

Figure 5.2 – C4 container diagram

Building a Docker image

The .NET CLI `dotnet publish` command supports creating Docker images without using a Dockerfile. However, to understand Docker, we need to know about Dockerfiles. That's why we start building a Docker image defining a Dockerfile first.

In this section, we will do the following:

- Create a Dockerfile for the games API
- Build a Docker image using the Dockerfile
- Run the games API with a Docker container
- Create a Docker image using `dotnet publish`

Creating a Dockerfile

Docker images are created using instructions in Dockerfiles. Using Visual Studio, you can easily create a Dockerfile from Solution Explorer, using **Add | Docker Support**. Make sure to select **Dockerfile** for the **Container build type** option. Adding a Dockerfile to the `Codebreaker.GamesAPI` project creates a multi-stage Dockerfile. A multi-stage Dockerfile creates interim images for different stages.

Base stage

With the following code snippets, the different stages are explained. The first stage prepares a Docker image for the production environment:

Codebreaker.GameAPIs/Dockerfile

```
FROM mcr.microsoft.com/dotnet/aspnet:8.0 AS base
USER app
WORKDIR /app
EXPOSE 8080
```

Every Dockerfile starts with a FROM instruction. The FROM instruction defines the base image that is used. mcr.microsoft.com/dotnet/aspnet is an image optimized for production. With .NET 8, this image is based on Debian 12 (Bookworm). The Debian:12-slim image is defined by a Dockerfile with FROM scratch, so this is the root of the hierarchy of instructions. The dotnet/aspnet image contains the .NET runtime and the ASP.NET Core runtime. The .NET SDK is not available with this image. The AS instruction defines a name that allows using the result of this stage with another stage. The USER instruction defines the user that should be used with the instructions of the stage. The WORKDIR instruction sets the working directory for following instructions. If the directory does not exist in the image, it's created. The last step in the first stage is the EXPOSE instruction. With EXPOSE, you define the ports that the application is listening to. By default, TCP is used, but you can also specify to have an UDP receiver.

> **Note**
> .NET 8 has some changes with Docker image generation: the default container images run with non-root users (the app user), and the default port is no longer port 80. Port 80 is a privileged port that requires the root user. The new default port is now 8080.

Build stage

The second stage builds the ASP.NET Core application:

Codebreaker.GameAPIs/Dockerfile

```
FROM mcr.microsoft.com/dotnet/sdk:8.0 AS build
ARG BUILD_CONFIGURATION=Release
WORKDIR /src
COPY ["Codebreaker.GameAPIs/Codebreaker.GameAPIs.csproj",
"Codebreaker.GameAPIs/"]
RUN dotnet restore "./Codebreaker.GameAPIs/Codebreaker.GameAPIs.
csproj"
COPY . .
```

```
WORKDIR "/src/Codebreaker.GameAPIs"
RUN dotnet build "./Codebreaker.GameAPIs.csproj" -c $BUILD_
CONFIGURATION -o /app/build
```

With the second stage, we ignore the first stage for a moment and use a different base image: `dotnet/sdk`. This image contains the .NET SDK and is used to build the application. First, a `src` directory is created, and the current directory is set to `src`. The `ARG` instruction specifies an argument that can be passed when invoking building the Docker image. If this argument is not passed, the default value is `Release`. Next, you'll see multiple `COPY` instructions to copy the project files to subfolders within the current directory. The project files contain the package references. In case the `dotnet restore` command that is started using the `RUN` instruction fails, there's no need to continue with the next steps. `dotnet restore` downloads the NuGet packages. In case you use a different NuGet feed, the Dockerfile needs some changes to copy `nuget.config` as well. When `dotnet restore` succeeds, the complete source code from `.` is copied to the current directory (which is `src` at this time) with the `COPY` instruction. Next, the working directory is changed to the directory of the games API project, and the `dotnet build` command is invoked to create release code for the application. With `dotnet build`, the `BUILD_CONFIGURATION` argument is used, which was specified with `ARG`. Having the release build in the `src/app/build` folder is the result for the interim image after the last command.

> **Note**
>
> To make sure that unnecessary files are not copied with instructions such as `COPY . .`, the `.dockerignore` file is used. Similar to the `.gitignore` file where files are specified to be ignored, with a `.dockerignore` file, you specify what files should not be copied to the image.

Publish stage

Next, the second stage is used as a base for the third stage:

Codebreaker.GameAPIs/Dockerfile

```
FROM build AS publish
ARG BUILD_CONFIGURATION=Release
RUN dotnet publish "./Codebreaker.GameAPIs.csproj" -c $BUILD_
CONFIGURATION -o /app/publish /p:UseAppHost=false
```

The `FROM build` instruction uses the result from the previous stage and continues here. The `dotnet publish` command results in code that's needed to publish the application. The files needed for publication are copied to the `/src/app/publish` folder. While the working directory was configured with the previous stage, continuing on the build image, the working directory is still set.

Final stage

With the final stage, we continue with the first stage, which was named `base`:

Codebreaker.GameAPIs/Dockerfile

```
FROM base AS final
WORKDIR /app
COPY --from=publish /app/publish .
ENTRYPOINT ["dotnet", "Codebreaker.GameAPIs.dll"]
```

The first instruction with this stage is to set the working directory to `app`. Then, referencing the third state with `--from=publish`, the `/app/publish` directory from the `publish` stage is copied to the current directory. The `ENTRYPOINT` instruction defines what should be done on running the image: the `dotnet bootstrapper` command starts and receives `Codebreaker.GameAPIs.dll` as an argument. You can do this from the command line as well: `dotnet Codebreaker.GameAPIs.dll` starts the entry point of the application to kick off the Kestrel server, and the application is available to receive requests.

Before building the Dockerfile, make sure the `DataStore` configuration in `appsettings.json` is set to `InMemory` to use the in-memory provider by default when starting the container.

Building a Docker image with a Dockerfile

To build the image, set the current directory to the directory of the solution, and use this command:

```
docker build . -f Codebreaker.GameAPIs\Dockerfile -t codebreaker/
gamesapi:3.5.3 -t codebreaker/gamesapi:latest
```

With simple Dockerfiles, using just `docker build` can be enough to build the image. However, the Dockerfile we use contains references to other projects. With this, we need to pay attention to the context. With the games API service, multiple projects need to be compiled, and the paths used as specified with the Dockerfile use the parent directory. Setting the current directory to the directory of the solution, the context is set to this directory with the first argument after the build command (`.`). The `-f` option next references the location of the Dockerfile. With the `-t` option, the image is tagged. The repository name (`codebreaker/gamesapi`) needs to be lowercase, followed by the `3.5.3` tag name and the `latest` tag name. Tag names can be strings; there's no requirement on the version. It's just a good practice to always tag the latest version with the `latest` tag. Specifying the `-t` option two times, we get two image names that reference the same image with the same image identifier.

To list the images built, use `docker images`. To restrict the output to `codebreaker` images, you can define a filter:

```
docker images codebreaker/*
```

To check a Docker image for how it was built, we can use the `docker history` command:

```
docker history codebreaker/gamesapi:latest
```

Figure 5.3 shows the result of the `docker history` command. This shows every instruction from the Dockerfile that was used to build the image. Of course, what you don't see with the codebreaker/gamesapi image is the `dotnet build` and `dotnet restore` commands. The `build` and `publish` stages have only been used for interim images to build the application and to create the files needed. Comparing the output to the Dockerfile we created, you see `ENTRYPOINT` on top, followed by `COPY`, `WORKDIR`, and so on. With each of these instructions, you can also see the size result of the instruction. The `COPY` command copied 3,441 MB into the image. The `USER` instruction was the first instruction coming from our Dockerfile; the instructions before that (in the lines below the `USER` instruction) show the instructions when the base image was created:

```
IMAGE          CREATED        CREATED BY                                        SIZE      COMMENT
2dd48eb88443   6 minutes ago  ENTRYPOINT ["dotnet" "Codebreaker.GameAPIs.d…     0B        buildkit.dockerfile.v0
<missing>      6 minutes ago  COPY /app/publish . # buildkit                    34.4MB    buildkit.dockerfile.v0
<missing>      8 days ago     WORKDIR /app                                      0B        buildkit.dockerfile.v0
<missing>      13 days ago    EXPOSE map[8080/tcp:{}]                           0B        buildkit.dockerfile.v0
<missing>      13 days ago    WORKDIR /app                                      0B        buildkit.dockerfile.v0
<missing>      13 days ago    USER app                                          0B        buildkit.dockerfile.v0
<missing>      2 weeks ago    COPY /shared/Microsoft.AspNetCore.App /usr/s…     23.7MB    buildkit.dockerfile.v0
<missing>      2 weeks ago    ENV ASPNET_VERSION=8.0.0-rc.1.23421.29            0B        buildkit.dockerfile.v0
<missing>      2 weeks ago    RUN /bin/sh -c ln -s /usr/share/dotnet/dotne…     24B       buildkit.dockerfile.v0
<missing>      2 weeks ago    COPY /dotnet /usr/share/dotnet # buildkit         72.3MB    buildkit.dockerfile.v0
<missing>      2 weeks ago    ENV DOTNET_VERSION=8.0.0-rc.1.23419.4             0B        buildkit.dockerfile.v0
<missing>      2 weeks ago    RUN /bin/sh -c groupadd        --gid=$APP_U…      8.46kB    buildkit.dockerfile.v0
<missing>      2 weeks ago    RUN /bin/sh -c apt-get update       && apt-get…   46.2MB    buildkit.dockerfile.v0
<missing>      2 weeks ago    ENV APP_UID=1654 ASPNETCORE_HTTP_PORTS=8080 …     0B        buildkit.dockerfile.v0
<missing>      2 weeks ago    /bin/sh -c #(nop)  CMD ["bash"]                   0B
<missing>      2 weeks ago    /bin/sh -c #(nop) ADD file:a1398394375faab8d…     74.8MB
```

Figure 5.3 – Result of docker history command

To see the exposed ports, environmental variables, the entry point, and more of an image, use the following command:

```
docker image inspect codebreaker/gamesapi:latest
```

The result is presented in JSON information and shows information such as exposed ports, environment variables, the entry point, the operating system, and the architecture this image is based on. When running the image, it helps to know what port number needs to be mapped and which environment variables could be useful to override.

Running the games API using Docker

You can start the Docker image of the games API service with the following command:

```
docker run -p 8080:8080 -d codebreaker/gamesapi:latest
```

Then, you can check the log output using `docker logs <container-id>` (get the ID of the running container with `docker ps`). You can interact with the games service using any client using the following HTTP address: `http://localhost:8080`.

With the default configuration of the data store, games are just stored in memory. To change this, we need to do the following:

1. Configure a network to let multiple Docker containers directly communicate

2. Start the Docker container for the SQL Server instance

3. Pass configuration values to the Docker container for the games API to use the SQL Server instance

Configuring a network for Docker containers

To let containers communicate with each other, we create a network:

```
docker network create codebreakernet
```

Docker supports multiple network types, which can be set using the `--driver` option. The default is a **bridge network** where containers within the same bridge network can directly communicate with each other. Use `docker network ls` to show all networks.

Starting the Docker container with SQL Server

To start the Docker container for SQL Server where the database already exists, you can either use the Docker image we committed or access the state of the previous running container. Here, we do the latter, where we define the name `sql1` as the name of the container:

```
docker start sql1
```

Use `docker ps` to check for the running container and see the port mapping as it was defined earlier.

Using the command adds the running container to the `codebreakernet` network:

```
docker network connect codebreakernet sql1
```

Starting the Docker container with the games API

Now, we need to start the games API but override the configuration values. This can be done by setting environmental variables. Passing environmental variables at the start of a Docker container can not only be done by setting the `-e` option but also by using `--env-file` and passing a file with environmental variables. This is the content of the `gameapis.env` file:

gameapis.env

```
DataStore=SqlServer
ConnectionStrings__
```

```
GamesSqlServerConnection=sql1,1433;database=CodebreakerGames;user
id=sa;password=<enter your password>;TrustServerCertificate=true
```

Creating the DI container of the application with the default container registers multiple configuration providers. A provider that uses environmental variables wins against the JSON file providers because of the order the providers are configured with the `WebApplicationBuilder` class (or the `Host` class). The `DataStore` key is used to select the storage provider. `GamesSqlServerConnection` is a key within the hierarchy of `ConnectionStrings`. Using command-line arguments to pass configuration values, you can specify a hierarchy of configuration values using : as a separator; for example, pass `ConnectionStrings:GamesSqlServerConnection`. Using : does not work everywhere; for example, with environmental variables on Linux. Using __ translates to this hierarchy.

With the environment variables file, the connection to the SQL Server Docker container is using the hostname of the Docker container and the port that is used by SQL Server. Being in the same network, the containers can communicate directly.

Starting the container, the environmental variables file is passed using `--env-file`:

```
docker run -p 8080:8080 -d --env-file gameapis.env -d --name gamesapi
codebreaker/gamesapi:latest
docker network connect codebreakernet gamesapi
```

> **Note**
> In case you get an error that the container name is already in use, you can stop running containers with `docker container stop <containername>`. To remove the state of the container, you can use `docker container rm <containername>` or the `docker rm <containername>` shorthand notation. To delete the state of all stopped containers, use `docker container prune`. When starting a new container with `dotnet run`, you can also add the `--rm` option to remove a container after exit.

Specifying the name of the `gamesapi` container, the container is added to the `codebreakernet` network. Now, we have two containers running communicating with each other, and can play games using the games service.

Let's create another Docker image, but this time without using a Dockerfile.

Building a Docker image using dotnet publish

In this chapter, we use multiple containers running at the same time. A project we didn't use in previous chapters is `Codebreaker.Bot`. This project offers an API and is also a client to the games service – to automatically play games in the background after requested by some API calls. In this section, we'll build a Docker image for this project – but without creating a Dockerfile first.

Since .NET 7, the `dotnet publish` command directly supports creating Docker images without a Dockerfile. Using the `docker build` command, we had to pay attention to some specific .NET behaviors, such as the need to compile multiple projects. With this, it was necessary to specify the context when building the image. The .NET CLI knows about the solutions and project structure. The .NET CLI also knows about the default Docker base images that are used to build an application with ASP.NET Core.

Options to configure the generation can be specified in the project file and using the parameters of `dotnet publish`. Here are some of the configurations specified in the project file:

```
<PropertyGroup>
  <ContainerRegistry>codebreaker/bot</ContainerRegistry>
  <ContainerImageTags>3.5.3;latest</ContainerImageTags>
</PropertyGroup>

<ItemGroup>
  <ContainerPort Include="8080" Type="tcp" />
</ItemGroup>
```

`ContainerImageTags` and `ContainerPort` are just two of the elements that are used by `dotnet publish`. You can change the base image with `ContainerBaseImage`, specify container runtime identifiers (`ContainerRuntimeIdentifier`), name the registry (`ContainerRegistry`), define environmental variables, and more. See `https://learn.microsoft.com//dotnet/core/docker/publish-as-container` for details.

With `dotnet publish`, the name of the repository is specified:

```
cd Codebreaker.Bot
dotnet publish Codebreaker.Bot.csproj --os linux --arch x64
/t:PublishContainer -c Release
```

The bot API service communicates with the games service, and the games service uses a container running SQL Server. Now, we already have three Docker containers collaborating. The games API needs a connection to the database, and the bot needs a link to the games API.

Running the solution with .NET Aspire

In *Chapter 2*, we added .NET Aspire to the solution, which contained only the games API service. In *Chapter 3*, we added a SQL Server database running in a Docker container without keeping state. Here, we'll extend the .NET Aspire configuration by using SQL Server running in a Docker container using a volume and configure the bot service to access the games API.

Configuring a Docker container for SQL Server

Using a Docker container with .NET Aspire can be orchestrated using .NET code – with extension methods of the IDistributedApplication interface:

Codebreaker.AppHost/Program.cs

```
var builder = DistributedApplication.CreateBuilder(args);
var sqlPassword = builder.AddParameter("SqlPassword", secret: true);
var sqlServer = builder.AddSqlServer("sql", sqlPassword)
  WithDataVolume("codebreaker-sql-data", isReadOnly: false)
  .AddDatabase("CodebreakerSql");

  var gameAPIs = builder.AddProject<Projects.
    Codebreaker_GameAPIs>("gameapis")
    .WithReference(sqlServer);
  // code removed for brevity
```

The AddSqlServer method adds SQL Server as a .NET Aspire resource to the app model. On the developer system, a Docker container is used to run SQL Server. When we created the container before in *Chapter 3*, we assigned a password using the default password configuration. Here we explicitly create an app model parameter using the AddParameter method. With this, parameters are retrieved from the Parameters configuration section. Thus, the parameter named SqlConfiguration is retrieved from Paramters:SqlConfiguration. This parameter resource is passed with the second parameter of AddSqlServer. If the password is not assigned, and a parameter named with the resource (sql) postfixed with -password does not exist, a random password is created. This is great if no volume mounts are used, and the database is created newly every time the container starts up. Using a persistent volume, the same password needs to be used with every run of the container. This is done by supplying the password configuration. The password is needed with the database connection string to access the database.

The WithDataVolume method defines the use of a Docker volume for the SQL Server container. Within the container, the database is stored within the /var/opt/mssql folder. Remember – without using a volume, the database is stored within the container itself, and the state is not kept when running a new container instance. While invoking WithDataVolume, there's no need to know what directories of the container need to mapped. This is known by the WithDataVolume method. We just pass the optional name of the volume, and if the volume should be read-only. With this volume, the database files are written to, thus it's not read-only. In *Chapter 11*, when adding Grafana and Prometheus Docker containers, we'll use read-only mounts.

The AddDatabase method adds the SQL Server database as a child resource to SQL Server. The name passed here defines the name of the resource and the name of the database.

So that .NET Aspire adds implicit service discovery, the database is referenced from the game APIs project using the `WithReference` method. With this, the `CodebreakerSql` database name can be used to reference the connection string.

Using the `AddSqlServer` method, there's no need to know the Docker image name for SQL Server or the environment variables that need to be specified, as we've done this earlier using the SQL Server Docker image. All this is done within the implementation of this method. To see how any Docker image can be added to the .NET Aspire app model, let's look at the implementation of this method:

```
public static IResourceBuilder<SqlServerServerResource>
  AddSqlServer(this IDistributedApplicationBuilder builder, string
  name, string? password = null, int? port = null)
{
var passwordParameter = password?.Resource
  ??  ParameterResourceBuilderExtensions.
  CreateDefaultPasswordPArameter(builder, $"{name}-password",
  minLower: 1, minUpper: 1, minNumeric: 1);
  var sqlServer = new SqlServerServerResource(name,
    passwordParameter);
  return builder.AddResource(sqlServer)
  .WithEndpoint(1433, port, null, "tcp")
  .WithImage("mssql/server", "2022-latest")
  .WithImageRegistry("mcr.microsoft.com")
  .WithEnvironment("ACCEPT_EULA", "Y")
  .WithEnvironment(context =>)
  {
    context.EnvironmentVariables["MSSQL_SA_PASSWORD"] =
      sqlServer.PasswordParameter;
  });

}
```

With this code, an endpoint is defined to map a passed port number to the port number `1433` that is used by SQL Server running in the container; the container registry, image name, and tag values are specified, and environment variables are created to accept the **end-user license agreement (EULA)** and to specify the password. All this is specified when invoking the `AddSqlServer` method. Using .NET Aspire, we don't need to know this with available components.

Before starting the application, make sure to set configuration via a user secret to store the password for SQL Server within the `Codebreaker.AppHost` project; for example, using the following command:

```
dotnet user-secrets set Parameters:SqlPassword "Password123!"
```

Now, we have to configure the .NET Aspire SQL Server component.

Configuring the .NET Aspire SQL Server component

With the games API, the EF Core context is configured with the DI container using the .NET Aspire SQL Server EF Core component:

Codebreaker.GameAPIs/ApplicationServices.cs

```
public static class ApplicationServices
{
  public static void AddApplicationServices(this
    IHostApplicationBuilder builder)
  {
    static void ConfigureSqlServer(IHostApplicationBuilder builder)
    {
      builder.AddDbContextPool<IGamesRepository,
        GamesSqlServerContext>(options =>
        {
          var connectionString = builder.Configuration.
            GetConnectionString("CodebreakerSql")
            ?? throw new InvalidOperationException("Could not read SQL
            Server connection string");
          options.UseSqlServer(connectionString);
          options.UseQueryTrackingBehavior(
            QueryTrackingBehavior.NoTracking);
        });
        builder.EnrichSqlServerDbContext<GamesSqlServerContext>();
    }
    // code removed for brevity
    string? dataStore = builder.Configuration.
      GetValue<string>("DataStore");
    switch (dataStore)
    {
      case "SqlServer":
        ConfigureSqlServer(builder);
        break;
      default:
        ConfigureInMemory(builder);
        break;
    }

    builder.Services.AddScoped<IGamesService, GamesService>();
  }
}
```

The first part of the configuration is just the usual .NET EF Core configuration to specify the EF Core database provider and the connection string. Just make sure to use the connection string key that was used with the app model definition. The connection string is forwarded from the `AppHost` project to the games API service. The parameter with the value `CodebreakerSql` matches the name of the database that has been configured with the app model. Using .NET Aspire orchestration (which uses `Microsoft.Extensions.ServiceDiscovery`) gets the connection string with this name from the orchestrator configuration.

The `EnrichSqlServerDbContext` method adds .NET Aspire configuration for retries, logging, and metrics.

With this, the games API service with the database configuration is in place. Next, add interaction between the bot service and the games API service.

Configuring interaction with multiple services

Both the games API service and the bot service are configured by adding a project to the distributed application:

Codebreaker.AppHost/Program.cs

```
var gameAPIs = builder.AddProject<Projects.Codebreaker_
GameAPIs>("gameapis")
  .WithReference(sqlServer);

builder.AddProject<Projects.CodeBreaker_Bot>("bot")
  .WithReference(gameAPIs);
```

The SQL Server Docker container was added using the `AddSqlServer` method. All .NET projects that should be orchestrated with .NET Aspire need to be added using the `AddProject` method. The class definition for the project is created from a source generator when adding a project reference to the `AppHost` project. Adding a reference to the `Codebreaker.Bot` project created a class named `Codebreaker_Bot`. All project classes are defined within the `Projects` namespace.

The bot service needs a reference to the games API, thus the `WithReference` method is used to generate implicit service discovery. The bot can use the name `http://gameapis` to reference the games API service.

The configuration to retrieve the link to the games API is configured with the `HttpClient` configuration:

Codebreaker.Bot/ApplicationServices.cs

```
public static void AddApplicationServices(this IHostApplicationBuilder
  builder)
{
```

```
builder.Services.AddHttpClient<GamesClient>(client =>
{
  client.BaseAddress = new Uri("http://gameapis");
});

  builder.Services.AddScoped<CodeBreakerTimer>();
  builder.Services.AddScoped<CodeBreakerGameRunner>();
}
```

The `BaseAddress` of `HttpClient` is configured with the name of the games API, as defined with the `gameapis` orchestration configuration – prefixed with `http://`. There's no need to specify .NET configuration to configure the link.

With just these few updates in place, we can start the `Codebreaker.AppHost` project in the next step.

Running the solution with .NET Aspire

Using `dotnet run` to build and run the `Codebreaker.AppHost` project starts up the Docker container for SQL Server, the games API, the bot service, and the .NET Aspire dashboard. *Figure 5.4* shows all started resources in the .NET Aspire dashboard. From here, you can see if services started successfully and access configured environment variables for services, logs, as well as accessible endpoints:

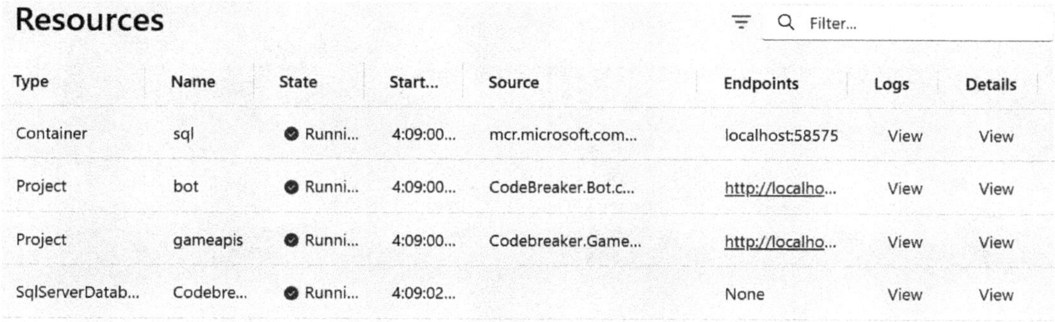

Resources

Type	Name	State	Start...	Source	Endpoints	Logs	Details
Container	sql	● Runni...	4:09:00...	mcr.microsoft.com...	localhost:58575	View	View
Project	bot	● Runni...	4:09:00...	CodeBreaker.Bot.c...	http://localho...	View	View
Project	gameapis	● Runni...	4:09:00...	Codebreaker.Game...	http://localho...	View	View
SqlServerDatab...	Codebre...	● Runni...	4:09:02...		None	View	View

Figure 5.4 – .NET Aspire resources in the dashboard

With resources, make sure to open the **Details** column to see the environment variables configured. When checking logs with the SQL Server Docker container, you can see issues when using an invalid password – for example, a password that doesn't fulfill the requirements or a password that doesn't match the volume that was initially set.

Clicking on the endpoint of the bot to show the OpenAPI test page, you can start multiple game runs, as shown in *Figure 5.5*. Specify the number of games the bot should play in a sequence, the delay time between the games, and the delay time with every move, and click the **Execute** button:

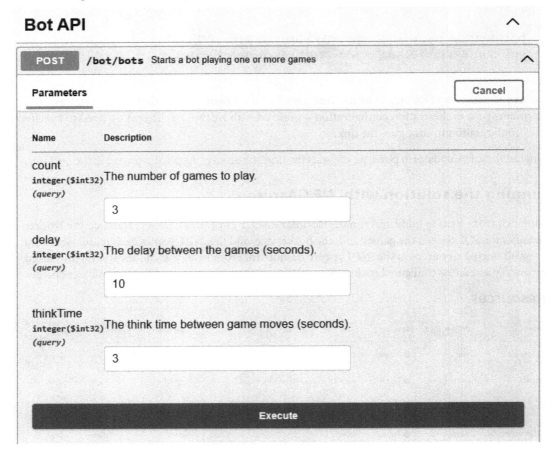

Figure 5.5 – OpenAPI test page for the bot service

With the logs of the bot service, you can monitor live information about games played by the bot. With every move set, the bot logs information about how good the result from the move was and how many possible options are remaining to find the solution:

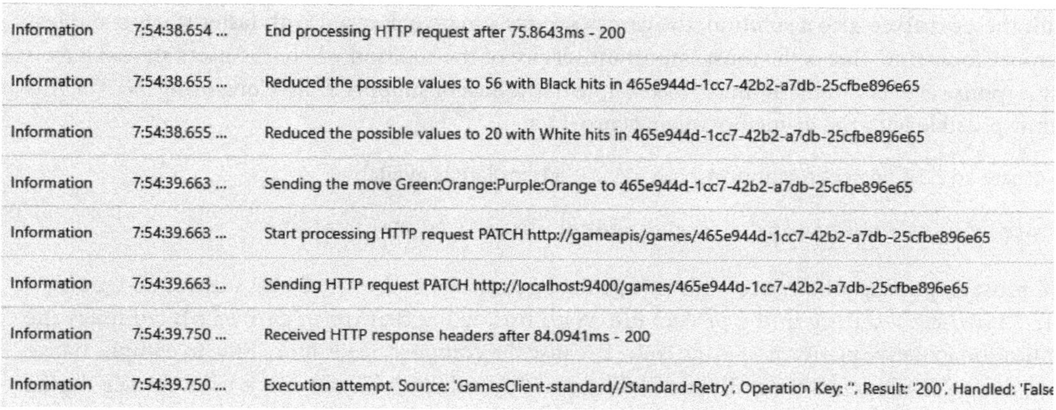

Information	7:54:38.654 ...	End processing HTTP request after 75.8643ms - 200
Information	7:54:38.655 ...	Reduced the possible values to 56 with Black hits in 465e944d-1cc7-42b2-a7db-25cfbe896e65
Information	7:54:38.655 ...	Reduced the possible values to 20 with White hits in 465e944d-1cc7-42b2-a7db-25cfbe896e65
Information	7:54:39.663 ...	Sending the move Green:Orange:Purple:Orange to 465e944d-1cc7-42b2-a7db-25cfbe896e65
Information	7:54:39.663 ...	Start processing HTTP request PATCH http://gameapis/games/465e944d-1cc7-42b2-a7db-25cfbe896e65
Information	7:54:39.663 ...	Sending HTTP request PATCH http://localhost:9400/games/465e944d-1cc7-42b2-a7db-25cfbe896e65
Information	7:54:39.750 ...	Received HTTP response headers after 84.0941ms - 200
Information	7:54:39.750 ...	Execution attempt. Source: 'GamesClient-standard//Standard-Retry'. Operation Key: ''. Result: '200'. Handled: 'False

Figure 5.6 – Structured logging of the bot service

After the services are running, you can also access the games API service, see the games played today, and use the client created in the previous chapter to play some games. Monitor the running Docker container using the Docker CLI, and also check the volume created.

Let's get into an exciting feature with .NET 8 – native AOT – next.

Using native AOT with ASP.NET Core

Comparing Docker images with VM images, Docker images are a lot smaller as they don't need to contain the operating system. With ASP.NET Core applications, the Docker image contains the application – and the .NET runtime. Over the last years, images have become smaller because more and more optimization has been done. Having smaller images means faster startup of the application.

Since .NET 7, it's possible to create native applications with C# using **native AOT**. With this, many changes are required with .NET. With .NET 7, the native AOT functionality was very limited. With .NET 8, we can already create ASP.NET Core services, which results in faster startup and less memory footprint.

Using native AOT, an AOT compiler is used to compile **Intermediate Language** (IL) code to native code. At build time, we still use the normal build process and the .NET runtime because native compilation takes some time. The native AOT compilation happens with `dotnet publish`.

Not all applications can be changed to use native AOT: libraries cannot be dynamically loaded, runtime code generation is not possible… With native AOT, the code is trimmed, and all libraries need to be native AOT compatible. With .NET 8, EF Core is not part of the libraries supporting native AOT. It's on the roadmap, and partial support is planned with EF Core 9.

By creating a solution based on microservices, it's possible to differentiate technologies with different services. What are the most used services where native AOT can give improvements?

With the `codebreaker` solution, the games service can be enhanced with faster startup and less memory footprint. This is the most important service of the solution where the users should have a fast response at every point in time. However, with the lack of support in EF Core, using .NET 8, this is only possible with the in-memory games provider.

To create an API project to support native AOT, a template is available:

```
dotnet new webapiaot -o Codebreaker.GameAPIs.NativeAOT
```

The most important difference with this project generated is the `<PublishAot>true</PublishAot>` setting in the project file. With this, using `dotnet publish` compiles the application to native platform-specific code. Because the compiler needs more time to compile native code, during development time, IL code is still generated, and the .NET runtime is used. As help during development to build native code, analyzers run and give compiler errors and warnings if code might not be compatible with native AOT.

With the `Codebreaker.GameAPIs.NativeAOT` project, you can start copying the code from the `Codebreaker.GameAPIs` project, but some changes are required. We'll focus on the changes needed for native AOT here.

OpenAPI document generation is removed – including methods to enhance OpenAPI documentation. This feature makes use of reflection and dynamic code generation, which is not supported. The EF Core SQL Server and Cosmos providers are removed from this project as well. Instead, the project only uses the in-memory games repository.

> **Note**
>
> .NET 7 included an extremely limited functionality for native AOT. .NET 8 brings many more features, but many libraries are not supported yet. With .NET 8, you can't use ASP.NET Core controllers, OpenAPI documentation is not available, the authentication library cannot be used, and most EF Core providers don't support native AOT. Over time, more features will be added to support native AOT.
>
> Native AOT doesn't allow the creation of code dynamically at runtime. Here, source generators are of big use. Instead of using reflection emit to create code at runtime, with source generators, code is created at compile time. This is not only an advantage with native AOT; even without using native AOT, source generators can improve the runtime performance.

Using the slim builder

Native AOT services make use of a slim application builder, as shown in the following code snippet:

Codebreaker.GameAPIs.NativeAOT/Program.cs

```
var builder = WebApplication.CreateSlimBuilder(args);
```

Contrary to the default builder, the number of services registered with the DI container is reduced. Logging is reduced as well. The only logging provider that is configured with the slim builder is the simple console logging provider. In case more functionality is needed, additional services can be added. To reduce the number of registered services even further, the `CreateEmptyBuilder` method can be used.

Using the JSON serializer source generator

The use of the `System.Text.Json` serializer needs to be changed to use a source generator. Without a source generator, the serializer uses reflection and creates code at runtime. This is not supported with native AOT. To generate code at compile time, source generators are used. To use the `System.Text.Json` source generator, the `AppJsonSerializerContext` class is added:

Codebreaker.GameAPIs.NativeAOT/Program.cs

```
[JsonSerializable(typeof(IEnumerable<Game>))]
[JsonSerializable(typeof(UpdateGameRequest))]
[JsonSerializable(typeof(UpdateGameResponse))]
[JsonSerializable(typeof(CreateGameResponse))]
[JsonSerializable(typeof(CreateGameRequest))]
[JsonSerializable(typeof(Game[]))]
internal partial class AppJsonSerializerContext :
JsonSerializerContext
{
}
```

The class is declared partial to allow the source generator to create additional sources to extend the class with additional members. For every type that's serialized with JSON, the `JsonSerializable` attribute is added.

This class is used with the DI configuration of the JSON serialization:

Codebreaker.GameAPIs.NativeAOT/Program.cs

```
builder.Services.ConfigureHttpJsonOptions(options =>
{
    options.SerializerOptions.TypeInfoResolverChain.Insert(0,
        AppJsonSerializerContext.Default);
});
```

With this code, the default instance of the context class is added to the type resolvers of the `System.Text.Json` serializer.

Building for Windows

After removing the code for the OpenAPI, removing SQL Server and Cosmos library references, and adding the `PublishAot` element to the project file, after a successful build, `dotnet publish` can be used to create a native application. This is the command to create a native image for Windows:

```
cd Codebreaker.GameAPIs.NativeAOT
dotnet publish -r win-x64 -c Release -o pubwin
```

Using `dotnet publish` with the `win-x64` runtime identifier starts the native compiler and writes the binary to the `pubwin` directory. The code is trimmed to remove not-used types and members from the binary. As a result, you receive a trimmed native executable that doesn't need to have the .NET runtime installed on the target system. Starting the application, you can use any client to play games with the in-memory provider.

Creating a Linux Docker image

This is the new Dockerfile we need for the native AOT games service:

Codebreaker.GameAPIs.NativeAOT/Dockerfile

```
FROM mcr.microsoft.com/dotnet/sdk:8.0 AS build
RUN apt-get update \
    && apt-get install -y --no-install-recommends \
    clang zlib1g-dev
ARG BUILD_CONFIGURATION=Release
WORKDIR /src
COPY ["Codebreaker.GameAPIs.NativeAOT.csproj", "."]
RUN dotnet restore "./Codebreaker.GameAPIs.NativeAOT.csproj"
COPY . .
WORKDIR "/src/."
RUN dotnet build "./Codebreaker.GameAPIs.NativeAOT.csproj" -c $BUILD_
CONFIGURATION -o /app/build

FROM build AS publish
ARG BUILD_CONFIGURATION=Release
RUN dotnet publish "./Codebreaker.GameAPIs.NativeAOT.csproj" -c
$BUILD_CONFIGURATION -o /app/publish /p:UseAppHost=true

FROM mcr.microsoft.com/dotnet/runtime-deps:8.0 AS final
WORKDIR /app
EXPOSE 8080
COPY --from=publish /app/publish .
ENTRYPOINT ["./Codebreaker.GameAPIs.NativeAOT"]
```

To compile the application to native code, using the same SDK-included base image, the `clang` and `zlib1g-dev` dependencies need to be installed with the Linux environment. This is done as the first step before copying the project files. For production, a different base image is used: `dotnet/runtime-deps`. This is the new base image containing native dependencies needed by .NET. This image does not include the .NET runtime; instead, it can be used for self-contained applications.

Building the Docker image can be done using the following command:

```
docker build . -f Codebreaker.GameAPIs.NativeAOT\Dockerfile -t
codebreaker/gamesapi-aot:latest -t codebreaker/gamesapi-aot:3.5.6
```

Running the solution with the native AOT container

After building the image, you can start the Docker container and use different clients (for example, the bot service, the HTTP files, and the client from the previous chapter) to test the service. With this, you can also do some performance comparisons – but remember that some features have been removed to be native AOT compatible.

> **Note**
>
> With .NET 8, native AOT is in its early stages. I expect many libraries to be updated to support native AOT in time. With a microservices architecture, for services that can improve from fast startup times, it can be useful to already use native AOT. A native AOT service can make use of gRPC (covered in *Chapter 14, gRPC for Binary Communication*), and the service accessible via gRPC can access the database. In any case, non-AOT services can also get improvements from features you've seen here, such as the slim builder or the JSON serializer source generator.

Summary

In this chapter, you learned the foundation of Docker, pulling, creating, and running Docker images. You used containers that store state with volumes running databases in Docker containers, passed environmental variables and secrets to running Docker containers, and used .NET Aspire to run multiple containers at once.

Using .NET Aspire, you configured orchestration for multiple services – including the configuration of a SQL Server Docker container. Comparing this to the work needed with Docker, this was an easy task – but it's still useful to understand the foundations.

With native AOT, you reduced startup times and the memory footprint, which you might be able to use with some of your services.

Before moving on to the next chapter, using the bot, you can now easily play thousands of games. The bot uses a simple algorithm to set random moves from a list of possible moves. Using the games query, check the number of moves the bot needs to find the result. Try to play a game using the client you created in the previous chapter to access the Docker container from this chapter. Can you solve games in fewer moves?

As you've seen in this chapter, you can run databases in Docker containers. With this, you still need to manage your database in the same way as you manage your natively installed on-premises database. Another option you'll see in the next chapter is to use **platform-as-a-service** (**PaaS**) cloud services such as Azure Cosmos DB. In the next chapter, we'll create Azure resources and publish Docker images we created in this chapter to **Azure Container Registry** (**ACR**) and Azure Container Apps.

Further reading

To learn more about the topics discussed in this chapter, you can refer to the following links:

- *Get started with Docker remote containers on WSL 2*: https://learn.microsoft.com/windows/wsl/tutorials/wsl-containers

- *How does Docker generate container names?*: https://frightanic.com/computers/docker-default-container-names/

- *Configure and customize SQL Server Docker containers*: https://learn.microsoft.com/sql/linux/sql-server-linux-docker-container-configure

- Azure SQL Edge Docker image: https://hub.docker.com/_/microsoft-azure-sql-edge

- Run the Azure Cosmos DB emulator on Docker for Linux: https://learn.microsoft.com/azure/cosmos-db/docker-emulator-linux

- Dockerfile instructions: https://docs.docker.com/engine/reference/builder/

- *Containerize a .NET app with dotnet publish*: https://learn.microsoft.com/dotnet/core/docker/publish-as-container

- .NET service discovery: https://learn.microsoft.com/en-us/dotnet/core/extensions/service-discovery

- *.NET Aspire documentation*: https://learn.microsoft.com/en-us/dotnet/aspire

- *Native AOT deployment*: https://learn.microsoft.com/dotnet/core/deploying/native-aot/

6

Microsoft Azure for Hosting Applications

After creating Docker images with the previous chapters, and running the complete application using Docker containers locally, let's move over to run the solution with Microsoft Azure.

In this chapter, you'll learn how to push Docker images to an Azure container registry, run Docker containers with Azure Container Apps, access a database using Azure Cosmos DB, and configure environment variables and secrets with Azure Container Apps.

Using Bicep scripts, you learn how to create multiple Azure resources at once.

In this chapter, you'll learn about the following topics:

- Experiencing Microsoft Azure

- Creating Azure resources

- Creating an Azure Cosmos database

- Pushing images to the **Azure Container Registry** (**ACR**) instance

- Creating Azure container apps

- Creating Azure resources using .NET Aspire and the **Azure Developer CLI** (**azd**)

Technical requirements

For this chapter, you need to have Docker Desktop installed. You also need a Microsoft Azure subscription. You can activate Microsoft Azure for free at `https://azure.microsoft.com/free`, which gives you an amount of about USD 200, Azure credits that are available for the first 30 days, and several services that can be used for free for the time after.

What many developers miss: if you have a Visual Studio Professional or Enterprise subscription, you also have a free amount of Azure resources every month. You just need to activate this with your Visual Studio subscription: `https://visualstudio.microsoft.com/subscriptions/`.

To work through the samples of this chapter, besides Docker Desktop, the Azure CLI and `azd` are needed.

To create and manage resources, install the Azure CLI and `azd`:

```
winget install microsoft.azureCLI
winget install microsoft.azd
```

These tools are available on Mac and Linux as well. To install the Azure CLI on different platforms, see `https://learn.microsoft.com/cli/azure/install-azure-cli`, and for `azd`, see `https://learn.microsoft.com/azure/developer/azure-developer-cli/install-azd`.

An easy way to use the Azure Cloud Shell is from a web browser. As you log in to the Azure portal at `https://portal.azure.com` using your Microsoft Azure account, on the top button bar, you'll see an icon for *Cloud Shell*. Clicking on this button, a terminal opens. Here, the Azure CLI is already installed – along with many other tools such as `wget` to download files, `git` to work with repositories, `docker`, the .NET CLI, and more. You can also use a Visual Studio Code editor (just run `code` from the terminal) to edit files. All the files you create and change are persisted within an Azure Storage account that is automatically created when you start the Cloud Shell. For a fullscreen Cloud Shell, you can open `https://shell.azure.com`.

The code for this chapter can be found in the following GitHub repository: `https://github.com/PacktPublishing/Pragmatic-Microservices-with-CSharp-and-Azure`.

In the `ch06` folder, these are the important projects:

- `Codebreaker.GameAPIs` – The `gamesAPI` project we used in the previous chapter. There's one change: instead of including the projects with the database access code and the models, NuGet packages are referenced.

- `Codebreaker.Bot` – The bot service calling the game APIs.

- `Codebreaker.AppHost` – This project contains important changes in this chapter to define the app model with Azure resources.

- `Codebreaker.ServiceDefaults` – This project is unchanged in this chapter.

You can start with the results from the previous chapter to work on your own through this chapter.

Experiencing Microsoft Azure

Microsoft Azure offers cloud services from many different categories. You can create **virtual machines (VMs)**, which belong to the **Infrastructure as a Service (IaaS)** category, where you are in control of the machines but also need to manage them as you do in on-premises environments, up to ready-to-use software such as Office 365 from the **Software as a Service (SaaS)** category. Something in between is **Platform as a Service (PaaS)**, where you don't have full control over the VMs but instead get many functionalities out of the box.

The focus here is on PaaS services. With the PaaS category, there's also a category named **serverless**. This category allows for easy scaling, starting from zero, where no or low costs are associated, up to a maximum amount of automatic scaling based on the needs. Many Azure services have offerings in this category.

Cost

When creating resources in a cloud environment, there's always a question about the cost. Many are afraid of the need to pay unexpected amounts, but this fear is not necessary. Some subscriptions (such as a Visual Studio subscription) are limited to the amount available every month. If this amount is reached, resources are automatically stopped (unless you explicitly allow the cost to go above the limits), so no additional cost applies.

With subscriptions and also just with resource groups, you can specify a budget to specify the amount that's planned to be spent. To do this, open the Azure portal and select a resource group. Within a resource group, you'll see the **Cost Management** category with the **Budgets** option. By creating a budget (see *Figure 6.1*), you can define limits month by month. Before this limit is reached, you specify alerts where you can be informed. With an alert, you can specify to receive a notification via email, SMS, push, or voice notification, and in addition to that, you can define an action that should be invoked to call an Azure function, a logic app, an Automation runbook, or other Azure resources where custom functionality can be implemented. Based on usage and requirements, stopping services could be an option:

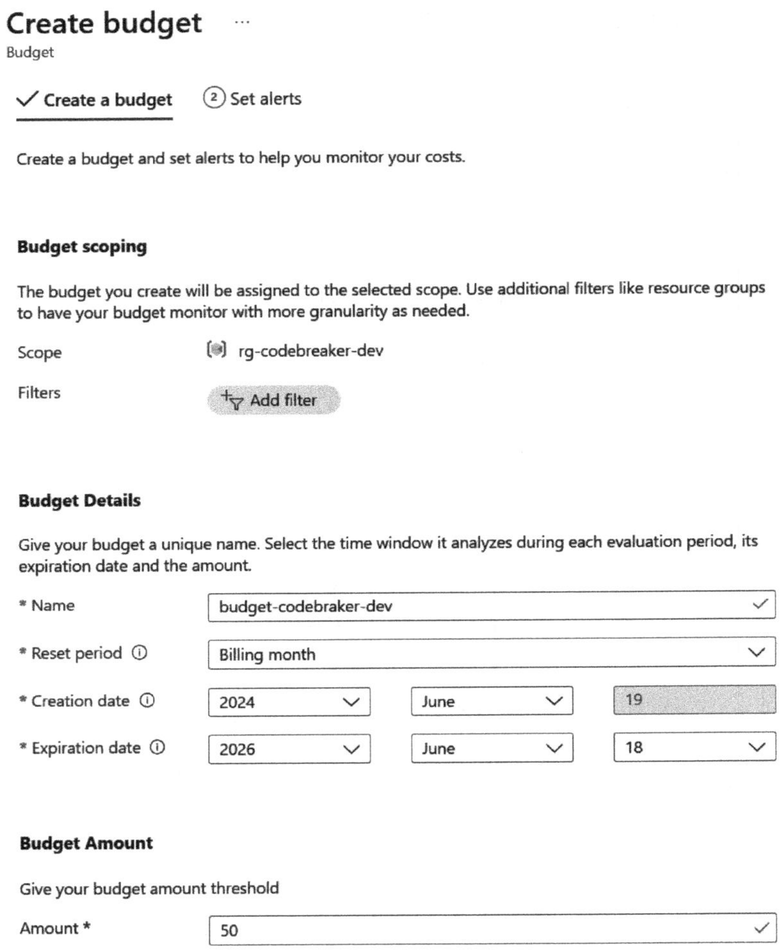

Create budget   ···
Budget

✓ **Create a budget**   ② Set alerts

Create a budget and set alerts to help you monitor your costs.

Budget scoping

The budget you create will be assigned to the selected scope. Use additional filters like resource groups to have your budget monitor with more granularity as needed.

Scope   [⬡] rg-codebreaker-dev

Filters   ⁺▽ Add filter

Budget Details

Give your budget a unique name. Select the time window it analyzes during each evaluation period, its expiration date and the amount.

* Name | budget-codebraker-dev   ✓ |

* Reset period ⓘ | Billing month   ∨ |

* Creation date ⓘ | 2024   ∨ | | June   ∨ | | 19 |

* Expiration date ⓘ | 2026   ∨ | | June   ∨ | | 18   ∨ |

Budget Amount

Give your budget amount threshold

Amount * | 50   ✓ |

Figure 6.1 – Specifying budgets

To get price information about services, at `https://azure.microsoft.com`, you can select **Azure pricing**, search for products, or select a product from a category to get details of different offerings available. You will also see a pricing calculator where you select multiple products and get complete price information based on selections made.

Naming conventions and more

When creating resources within Azure, we should think about some important foundations to easily find resources based on the needs of IT but also based on the needs of business organizations. Which resources are for production, and which are for testing environments? Which resources are used by different organizations in the company? Which resources are used by one product? What are the resources potentially being impacted by a technical issue? For all these scenarios, these features help:

- Every resource needs to be put into one **resource group**. For the `codebreaker` solution, resource groups will be created for the test and production environments.

- It should also be easy to find multiple resources across resource groups. **Resource tags** can be used here.

- Define a convention for how you name your resources. The number of resources will grow over time! You might create multiple instances for scaling, run the same services around the globe in different regions for better latency, run services in different environments... there are many reasons the number of resources grows. To deal with this, a good naming strategy used from the start can help a lot!

With the `codebreaker` application, we can use `rg-codebreaker-dev`, `rg-codebreaker-test`, and `rg-codebreaker-prod` resource groups for *development*, *test*, and *production* environments.

> **Note**
> Besides separating the environments with different resource groups, it's a good practice to separate the development and production environments into different subscriptions. As there's an Azure subscription with some free amount available with a Visual Studio subscription, this subscription can be used with the development environment.

Some resources are used across different resource groups. For example, you might use a central Azure DNS resource. You might also share resources across different applications. You can share an Azure app service that hosts many small websites. With every Azure resource, you can add custom tags and search for resources using different tags and their values. For example, you can specify a tag named `cc` (for cost center), and the value specifies the cost center.

To define a naming convention for resources, Microsoft not only has a guideline (available at `https://learn.microsoft.com/azure/cloud-adoption-framework/ready/azure-best-practices/resource-naming`), but also an Excel template you can use (available at `https://raw.githubusercontent.com/microsoft/CloudAdoptionFramework/master/ready/naming-and-tagging-conventions-tracking-template.xlsx`), and even a Blazor application that you can host on-premises (or in the cloud) for your administrators to manage naming conventions with a simple user interface: `https://github.com/mspnp/AzureNamingTool`.

Components that can be part of the name of resources include the following:

- The *resource type*. Microsoft has a list of proposed abbreviations; for example, `rg` for resource group, `cosmos` for an Azure Cosmos DB database, `cr` for ACR, `ca` for container apps, and `cae` for Container Apps environments.

- The project, application, or service name. We'll use `codebreaker` for the application name.

- The environment where the resource is used; for example, `prod` for production, `dev` for development, and `test` for testing.

- The location of the Azure resource; for example, `eastus2` for the second East US region, and `westeu` for West Europe. Creating resources in multiple regions can be useful for failover scenarios, for better performance for customers around the globe, and because of data regulations.

Now, we are ready to create Azure resources.

Creating Azure resources

Using Microsoft Azure, there are different ways to create and manage Azure resources. Azure resources are accessible via a REST API. You can send `GET` requests to read information about resources and `POST` requests to create new resources, but of course, there's an easier way to do it. The Azure portal (`https://portal.azure.com`) is a great way to learn and see the different options you have. To automatically create Azure resources, you can use the Azure CLI, PowerShell scripts, and many more options to use. In this book, we'll use the Azure portal, Bicep scripts, .NET Aspire, and `azd`. Bicep scripts give you a simple syntax from Microsoft to easily recreate Azure resources. .NET Aspire offers to define Azure resources using .NET code and directly create the resources.

Within company environments, there are different ways Azure resources are created and how teams are organized. .NET Aspire, together with `azd`, offers great functionality for creating Azure resources, but this might not (yet?) fit into your environment. You can also decide to use parts of .NET Aspire that fit into your company environment, or use all that .NET Aspire and `azd` offer. The second option is the easiest one. To better understand the options, and for you to map it into your environment, we'll start using the Azure CLI and the Azure portal. With this, you can easily see what options a resource offers for configuration. Later in this chapter, we'll use .NET Aspire and `azd`. Specifying Azure resources with .NET code just needs a few statements to create all the resources needed with the solution.

What are the resources we create? In this section, we'll do the following:

1. Create a resource group that groups all Azure resources together.

2. Create an Azure Cosmos DB database that is added to the previously created resource group and used by the EF Core context we created in *Chapter 3*.

3. Create an Azure container registry to publish the Docker images we created in *Chapter 5*.

4. Create two Azure container apps to run the `gamesAPI` service and the bot service.

Creating a resource group

Resource groups are used to manage Azure resources together. With a resource group, you can specify permissions for who is allowed to create or manage resources within the resource group. From a price standpoint, you easily can see the cost of the complete resource group and which resources of this resource group were responsible for which cost. You can also delete a resource group, which deletes all resources within the group.

To create a resource group, let's use the Azure CLI.

To log in to Azure, use the following command:

```
az login
```

This command opens the default browser to authenticate the user.

If you have multiple Azure subscriptions, you can check these with `az account list`. The current active subscription where you create resources is shown with `az account show`.

To create a resource group, use the `az group` command:

```
az group create -l westeurope -n rg-codebreaker-test
```

The `create` subcommand creates a resource group. With `-l`, we specify the location of this Azure resource. Here, I'm using `westeurope` because this region is near my location. With the `-n` value, the name of the resource group is set.

The location of the resource group is independent of the location of the resources within the resource group. Resources within the resource group can have other regions. A resource group is just metadata. The location for the resource group specifies the primary location for the resource group. In a fatal case where the location is not available, you cannot make changes to the resource group.

To get the regions available with your subscription, you can use `az account list-locations -o table`.

After the resource group is created, we can create resources within this resource group.

Creating an Azure Cosmos DB account

In *Chapter 3*, we used the Azure Cosmos DB emulator to store games and moves. Now, let's change this to the real database in the Azure cloud. First, we'll use the Azure portal to create an Azure Cosmos DB account.

Within the Azure portal, by clicking on **Create a Resource**, you can select **Databases** from a list of categories, or just enter a search term, `Azure Cosmos DB`. When you click **Create** with the Azure Cosmos DB resource, this does not immediately create the resource. Instead, you need to do some configuration beforehand.

With the Azure Cosmos DB resource, you first need to select one of the APIs available. Read back to *Chapter 3* for the different APIs available and what they offer. Now, select **Azure Cosmos DB for NoSQL**, and click the **Create** button. This opens the configuration, as shown in *Figure 6.2*:

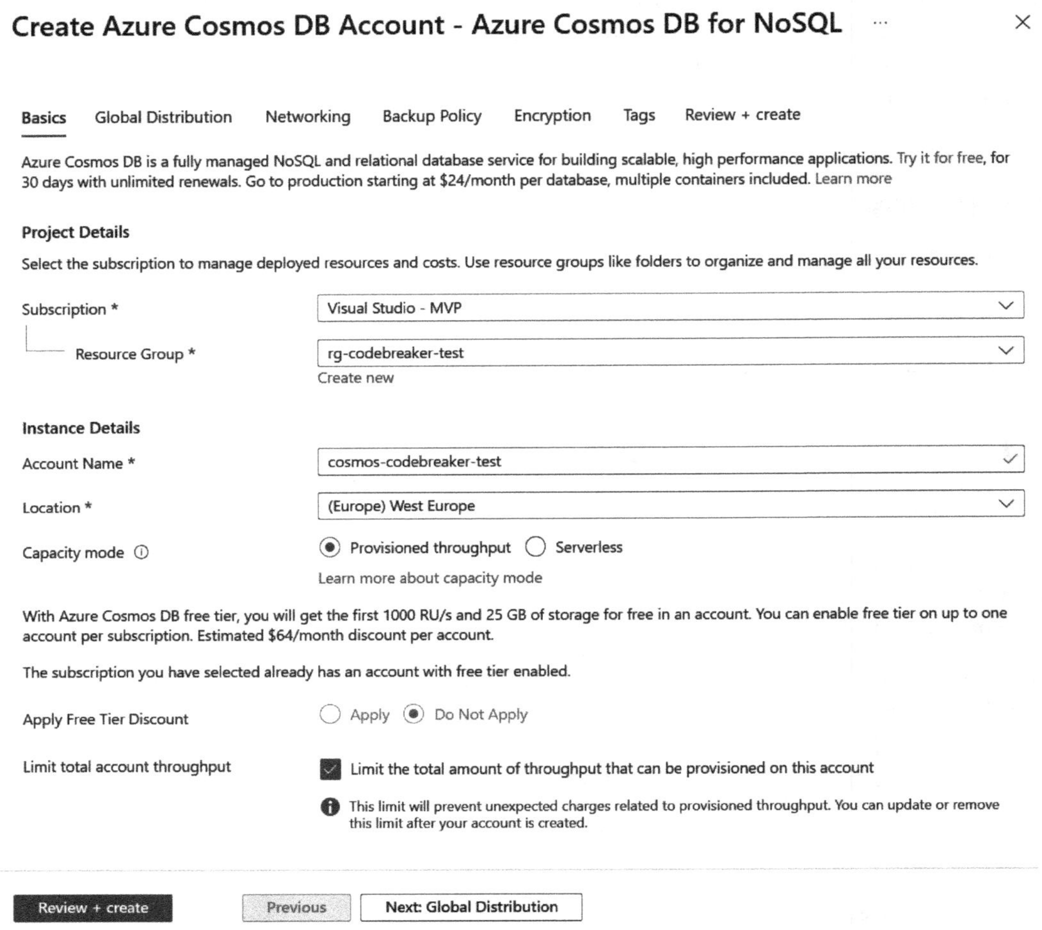

Create Azure Cosmos DB Account - Azure Cosmos DB for NoSQL   ⋯   ✕

Basics   Global Distribution   Networking   Backup Policy   Encryption   Tags   Review + create

Azure Cosmos DB is a fully managed NoSQL and relational database service for building scalable, high performance applications. Try it for free, for 30 days with unlimited renewals. Go to production starting at $24/month per database, multiple containers included. Learn more

Project Details

Select the subscription to manage deployed resources and costs. Use resource groups like folders to organize and manage all your resources.

Subscription * | Visual Studio - MVP   ⌄ |

    └─── Resource Group * | rg-codebreaker-test   ⌄ |
                                   Create new

Instance Details

Account Name * | cosmos-codebreaker-test   ✓ |

Location * | (Europe) West Europe   ⌄ |

Capacity mode ⓘ   ⦿ Provisioned throughput   ◯ Serverless
                                   Learn more about capacity mode

With Azure Cosmos DB free tier, you will get the first 1000 RU/s and 25 GB of storage for free in an account. You can enable free tier on up to one account per subscription. Estimated $64/month discount per account.

The subscription you have selected already has an account with free tier enabled.

Apply Free Tier Discount   ◯ Apply   ⦿ Do Not Apply

Limit total account throughput   ☑ Limit the total amount of throughput that can be provisioned on this account

                                   ⓘ This limit will prevent unexpected charges related to provisioned throughput. You can update or remove
                                      this limit after your account is created.

[Review + create]   [Previous]   [Next: Global Distribution]

Figure 6.2 – Creating an Azure Cosmos DB account

You have some pages for configuration before you can click the **Create** button one last time. With the basic configurations, you need to specify the subscription in which the resource will be created, the resource group name, the name of the account (I used `cosmos-codebreaker-test` for the test environment, but be aware this name needs to be globally unique), and the capacity. Azure Cosmos DB offers one free tier with a subscription. In case you haven't used this yet with your subscription, you can choose this option. This gives you 1,000 **request units per second** (**RU/s**) and 25 GB of storage for free. Using provisioned throughput, you define a RU/s limit by database or by database container, at least 400 RU/s. The serverless option starts with a higher minimum limit but automatically scales to the RU/s needed. With serverless, you need to be aware of some limits. With serverless, the maximum database container size is 1 TB; there's no limit with the provisioned configuration. Serverless also doesn't support geo-distribution, which is available with the provisioned setting.

> **Note**
>
> Creating an Azure Cosmos DB account registers a DNS name, thus the name needs to be globally unique. For your account, you can add a number following the account name, and check by clicking **Review + create** if your selected name is available.

With the next configurations, you can configure the global distribution of a database, networking, a policy to automatically create backups, encryption with a service-managed key or a customer-managed key, and tags (which are available with every resource). You can use the default values with all the settings other than the basic configuration. Upon clicking on the **Review + create** button, final checks are made, and you can click the final **Create** button. Now, you just need to wait for a few minutes until the database account is created.

Using the Azure CLI, you can use the `az cosmosdb create` command.

The database account is created! Next, we'll create an Azure container registry.

> **Note**
>
> In *Chapter 3*, we not only created a library to write to Azure Cosmos DB but also to SQL Server. With Microsoft Azure, you can also configure Azure SQL Database. Just be aware of the low cost in the development environment; select a **Database Transaction Unit** (**DTU**) tier instead of the vCore tier. With 5 Basic DTUs, there's just a cost lower than USD 5, for a month for 2 GB storage (at the time of this writing) compared to USD 400,- where a VM with 2 vCores is allocated.

Creating an Azure container registry

In the previous chapter, we created Docker images and used them locally. You can publish Docker images to the Docker hub, or any container registry. ACR offers a registry for Docker images that greatly integrates with Microsoft Azure.

While creating an ACR instance, three different tiers are available:

- **Basic**: For the purpose of the `codebreaker` application, the Basic tier (SKU) fits the purpose and is a lot cheaper than the other options. You just need to be aware of the limits.

- **Standard**: The Standard tier offers more storage (the Basic tier is limited to 10 GB storage) and image throughput.

- **Premium**: The Premium tier adds some features, such as geo-replication replicates images across different regions and private access points.

Figure 6.3 shows how to create an ACR instance via the portal. Clicking on **Create a resource**, **Container Registry** is available in the **Containers** category. Instead, you can also enter `Container Registry` in the search box. Selecting the **Azure services only** checkbox doesn't show the many third-party offerings:

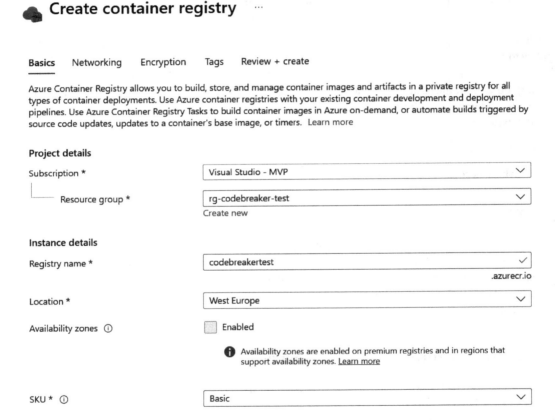

Figure 6.3 – Creating a container registry

With the configuration, we just need the name of the resource group, the name of the registry, the location, and the SKU. Availability zones, where the images are stored in different data centers in the same region, are only available with the Premium tier. Other configurations to change the networking and encryption require the Premium tier as well.

Filling out this form, after clicking **Review + create**, you still can verify all the options before clicking **Create** to create the resources.

> **Note**
>
> The name of the registry is a globally available DNS name (with the `azurecr.io` extension) and thus needs to be unique. Choose your own name where creating the resource succeeds with an available name.

We now have a database and a registry for container images running. With the first resources created, all we need to run the application in the cloud is a compute service where we can run the Docker images. We will use Azure Container Apps.

Creating an Azure Container Apps environment

Microsoft Azure has several compute offerings to run Docker containers. You can publish a Docker image to an **Azure app service**, and use a Windows or Linux server to run your APIs. Another option is to use **Azure Container Instances** (**ACI**), which allows you to host a group of Docker containers, including one frontend container (the API service) and multiple backend containers. While Azure app services offer automatic scaling to create multiple instances based on rules, this feature is not available with ACI. ACI is great with a fast startup – you spin up a VM where just the smaller Docker image needs to be uploaded, but it doesn't offer orchestration and scaling features.

For full-blown orchestration and scaling of Docker containers, Azure offers **Azure Kubernetes Service** (**AKS**), where we can deploy and manage Docker images using the well-known Kubernetes tool, `kubectl`. To remove the complexity of Kubernetes, defining an Ingress controller is just a matter of changing some settings; **Azure Container Apps** instances are available. This service makes use of Kubernetes behind the scenes but removes a lot of its complexity.

Let's get into creating an Azure container app.

Creating a Log Analytics workspace

When creating an Azure container app, having space for logging is a good idea. With previous versions of Azure Container Apps, it was a requirement to have a **Log Analytics workspace**. This is no longer a requirement, as you can also use Azure Monitor to log to an Azure storage account, an Azure event hub, or a third-party monitoring solution. Azure Monitor can also be configured to route logs to Log Analytics.

A Log Analytics workspace is a storage unit for log data to analyze data and metrics. In *Chapter 10*, *Logging*, we'll dive into logging and metrics with microservices, and make use of Log Analytics, Azure Monitor, and Application Insights to get information about running services.

To create a Log Analytics workspace, we will use the Azure CLI:

```
az monitor log-analytics workspace create -g rg-codebreaker-test -n
logs-codebreaker-test-westeu
```

Log Analytics belongs to Azure Monitor, thus the `az monitor log-analytics` command is used to create and manage Log Analytics. With the `workspace create` subcommand, a Log Analytics workspace is created. This command requires the resource group and the name of the workspace. If the location is not supplied with the command, the workspace uses the same location as the resource group.

Creating a container app environment

Creating a container app environment uses a Kubernetes cluster behind the scenes. You can create this environment to create a Log Analytics workspace automatically. Using an existing workspace (we created one in the previous step), we need the customer ID and a key from the workspace. Get the customer ID using the following command:

```
az monitor log-analytics workspace show -g rg-codebreaker-test -n
logs-codebreaker-test-westeu --query customerId
```

Without supplying `--query customerId`, you get more complete information about the workspace, including the `customerId` value. Using the `--query` command, we can supply a **JMESPath** query. Check `https://jmespath.org` for more information on this query syntax. Using `customerId` with the query, just the unique identifier of this id (a GUID) is returned. Copy this GUID as well as the key from the next command as we'll need these values when creating the environment.

This command returns keys to connect to the log workspace:

```
az monitor log-analytics workspace get-shared-keys -g rg-codebreaker-
test -n logs-codebreaker-test-westeu
```

The output returns primary and secondary shared keys. Copy the primary shared key.

Using the customer ID as well as the key from the Log Analytics workspace, we can create a container app environment:

```
az containerapp env create -g rg-codebreaker-test -n cae-codebreaker-
test-westeu --logs-workspace-id <customer-id> --logs-workspace-key
<logs-key> --location westeurope
```

To create the environment, you need to specify the resource group, the name of the environment, information to connect log analytics, as well as the location of the newly created resource. This command does not use the location of the resource group if the location is not supplied. Be aware that this command might take several minutes. But think about how many minutes you would need to create a Kubernetes cluster manually.

Creating a hello container app

After creating the environment, let's create our first app within this environment:

```
az containerapp create -n ca-hello-westeu -g rg-codebreaker-test
--environment cae-codebreaker-test-westeu --image mcr.microsoft.com/
azuredocs/containerapps-helloworld:latest --ingress external --target-
port 80 --min-replicas 0 --max-replicas 2 --cpu 0.5 --memory 1.0Gi
```

Using the create command creates a new app. The name of this app is specified with the -n parameter. The environment is specified with the resource group (-g) and the --environment parameter. The image referenced with the --image parameter is a sample Docker image from Microsoft that hosts a web server with a static page. To access the web server running on port 80 within the container, the Ingress service is configured with the --ingress and --target-port parameters. Using the --min-replicas and --max-replicas parameters, scaling is defined to scale from 0 up to 2 instances. With 0 instances, the first user accessing the service needs to wait until the container is started. With the supplied configuration, the application scales up to 2 running containers. One container allocates 0.5 CPUs and 1.0 Gi memory.

> **Note**
>
> *Chapters 9* to *11* will give you information about scaling services. In *Chapter 9*, you'll create load tests to stress-test services, in *Chapter 10*, we'll use these load tests to monitor metrics information, and in *Chapter 11*, we'll configure scaling with information learned in the previous two chapters.

When the app is created, a link for the app service is shown. You can also get the URL using this command:

```
az containerapp show -n ca-hello-westeu -g rg-codebreaker-test --query
properties.configuration.ingress.fqdn
```

The containerapp show command shows properties of the Azure container app. Using the properties.configuration.ingress.fqdn JMESPath query returns the **fully qualified domain name** (**FQDN**) of the Ingress service. Using the returned domain name with a prefixed https:// instance shows the running application (see *Figure 6.4*):

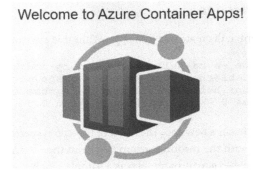

Figure 6.4 – Accessing the hello Azure container app

Now, opening the `rg-codebreaker-test` resource group with the Azure portal, we can see the Azure Cosmos DB, ACR, and Azure Container Apps environments, the Log Analytics workspace, and the container app, as shown in *Figure 6.5*. Just check the options you have with the categories on the left side of the **Resource group** view. The **Overview** view shows the resources, as shown here. Clicking on **Access control**, you can configure who has access to the resources of this group. The *activity log* shows who created, updated, and deleted resources within this group. The *resource visualizer* gives a graphical view of resources and how they relate to each other. The **Cost Management** category might also be of interest.

You might need to wait for a day before seeing the detailed cost of each resource. With the tiers we used, the cost will be within a few cents. But you can also click on **Recommendations** to see what should be changed and configured with a production environment. Some of these recommendations require different tiers where you need to check into the cost changes. In case your company already experienced hacking into the company's site, the cost of turning security features on with Microsoft Azure is really low compared to the cost of a hacking attack:

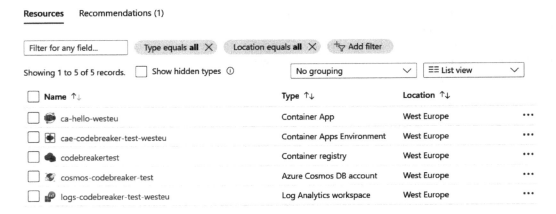

Figure 6.5 – Resource group with Azure resources

Now, as the Azure resources are created, let's publish the `codebreaker` services to Microsoft Azure.

Let's start with creating an Azure Cosmos database within an Azure Cosmos DB account.

Creating an Azure Cosmos database

From the Azure portal, you can open the page for your Azure Cosmos DB account, open **Data Explorer**, and from there, click on **New Database** to create a new database, and **New Container** to create a container within the database. Here, we'll use the Azure CLI instead:

```
az cosmosdb sql database create --account-name <your cosmos account
name> -n codebreaker -g rg-codebreaker-test --throughput 400
```

This command creates a database named `codebreaker` in the existing account. Setting the throughput option with this command defines the scale of the database. Here, all containers within this database share the 400 RU/s throughput. 400 is the smallest value that can be set. Instead of supplying this value when creating the database, scaling can also be configured with every container. In case some containers should not take away scaling from other containers, configure the RU/s with every container – but here, the minimum value to be used with each container is 400 as well.

After creating the database, let's create a container:

```
az cosmosdb sql container create -g rg-codebreaker-test -a <your
cosmos account name> -d codebreaker -n GamesV3, --partition-key-path
"/PartitionKey"
```

The implementation of the `gamesAPI` service uses a container named `GamesV3`. This container is created within the previously created database, using the `/PartitionKey` partition key, as was specified with the EF Core context in *Chapter 3*.

After this command is completed, check **Data Explorer** in the Azure portal, as shown in *Figure 6.6*:

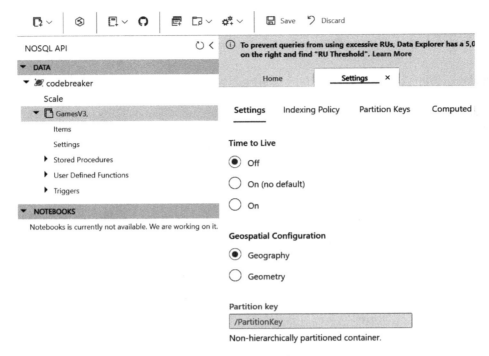

Figure 6.6 – Data Explorer

You can see the database, the container, and, with the container, the configured partition key.

Configuring replication with Azure Cosmos DB

A great feature of Azure Cosmos DB is global data replication. Within the Azure portal, in the **Settings** category, click on **Replicate data globally**. *Figure 6.7* shows the replication view:

Figure 6.7 – Replication with Azure Cosmos DB

You just need to click on the Azure regions that are available with your subscription to replicate data within the selected regions. You can also configure it to write to multiple regions.

With the `codebreaker` application where users around the world can play, for faster performance for users in the US, Europe, Asia, and Africa, writing to multiple regions can be configured. For this option to be available, automatic scaling cannot be configured. For the best scalability across the globe, we also need to think about the partition key. By using different partition key values for every game that's stored, games can be stored within different partitions.

Configuring consistency

With the **Settings** category in the Azure portal of Azure Cosmos DB, we can configure the default consistency level. The outcomes are shown using music notes, reading, and writing from multiple regions, as shown in *Figure 6.8*:

Figure 6.8 – Outcome shown using music notes

The default setting is **Session consistency** – the data is consistent within the same session. With this setting, write latencies, availability, and read throughput are comparable to **Eventual consistency**. Using the Azure Cosmos DB API, a session can be created and distributed within the application.

The **Strong consistency** option is not available if multiple regions are configured. With multiple regions, **Bounded staleness** can be configured, which specifies a maximum lag time and a number of maximum lag operations before the data is consistently replicated.

The database is now ready to use, so let's publish Docker images to the registry!

Pushing images to the ACR instance

The ACR instance is ready, and we created Docker images in the previous chapter – now, let's publish the images to this registry.

After you've logged in to Microsoft Azure (using az login), to log in to the ACR instance, you can use az acr login. Make sure to use the name you defined with the ACR instance:

```
az login
az acr login -n <the name of your azure container registry>
```

This command needs to have Docker Desktop installed and running.

> **Note**
> Referencing the ACR instance using the Azure CLI, just the name of the registry is needed (such as codebreakertest). The docker and dotnet commands support different registries, thus with these commands, the complete domain name is needed, such as codebreakertest.azurecr.io.

Next, let's build the images. With the game APIs, we created a Dockerfile in the previous chapter. With the Windows terminal, make sure to set the current directory to the ch06 folder and build the game image locally:

```
docker build -f Codebreaker.GameAPIs\Dockerfile . -t codebreaker/
gamesapi:3.5.1
```

This command – as in the previous chapter – builds the Docker image locally, referencing the Dockerfile, setting the context for docker build, and setting the tag.

To publish the image to ACR, we need to tag the local images:

```
docker tag codebreaker/gamesapi:3.5.1 <full DNS name of your ACR>/
codebreaker/gamesapi:3.5.1
docker tag codebreaker/gamesapi:3.5.1 <full DNS name of your ACR>/
codebreaker/gamesapi:latest
```

The images are tagged with a link to the ACR instance. The same image is tagged with the version number as well as the latest tag. The latest tag is a convention where the newest version is stored and is always overridden in the repository.

Next, push the image to the registry with docker push:

```
docker push <full DNS name of your ACR>/codebreaker/gamesapi:3.5.1
docker push <full DNS name of your ACR>/codebreaker/gamesapi:latest
```

Make sure you are already logged in to the ACR instance; otherwise, the push will fail.

With a successful push, you can see the images in the Azure portal within the **Repositories** menu, as shown in *Figure 6.9*:

Figure 6.9 – Repositories in ACR

With the bot service, we didn't create a Dockerfile in the previous chapter and used the dotnet CLI instead. Using `dotnet publish`, we just need to add this `PropertyGroup` instance to the project file:

Codebreaker.Bot/Codebreaker.Bot.csproj project file

```
<PropertyGroup>
  <ContainerRegistry>add your registry
  </ContainerRegistry>
  <ContainerRepository>codebreaker/bot
  </ContainerRepository>
  <ContainerImageTags>3.5.3;latest</ContainerImageTags>
</PropertyGroup>
```

The `dotnet publish` command uses `ContainerRegistry`, `ContainerRepository`, and `ContainerImageTags` elements to create the image and publish it to the registry. Pay attention to configuring your own registry with the `ContainerRegistry` element.

All that needs to be done with the current directory is to set the directory of the project file of the bot and run `dotnet publish`:

```
cd Codebreaker.Bot
dotnet publish --os linux --arch x64 /t:PublishContainer -c Release
```

This command builds the image and directly publishes it to the registry, as specified with the `ContainerRegistry` element. Just make sure to enter the link to your registry, and be logged in using `docker login`.

As the images are ready, let's continue using them with Azure container apps!

Creating Azure container apps

Let's create a `gamesAPI` service running with an Azure container app. This one needs a configuration containing a secret to the Azure Cosmos database.

With the Azure Cosmos database, in the Azure portal, go to the **Settings** category and open **Keys**. From this page, copy the primary or the secondary connection string.

With such keys, it's useful to regenerate them from time to time – that's why pairs are available. When you use the primary key from apps, regenerate the secondary key. After the regeneration, use the secondary key from within the apps, and regenerate the primary key. With this, you have some time to configure all the apps for the new key.

When creating an Azure container app for game APIs, there are quite some values to configure. While you can pass all the configuration values to the `az containerapp create` command, let's start creating this with the Azure portal. Opening the resource of the Azure container app environment, click on **Apps**, and create a new app. *Figure 6.10* shows the basic settings:

Create Container App   ...

Basics   Container   Ingress   Tags   Review + create

Azure Container Apps are containerized apps that scale on demand without requiring you to manage cloud infrastructure. You'll need a container and an environment for your first app. Select existing resources, or create them now. Learn more

Project details

Select a subscription to manage deployed resources and costs. Use resource groups like folders to organize and manage all your resources.

Subscription *	Visual Studio - MVP ˅
Resource group *	rg-codebreaker-test ˅
	Create new
Container app name *	cae-codebreaker-gamesapi-3

Container Apps Environment

The environment is a secure boundary around one or more container apps that can communicate with each other and share a virtual network, logging, and Dapr configuration. Container Apps Pricing

Show environments in all regions ⓘ	☐
Region *	West Europe ˅
Container Apps Environment *	cae-codebreaker-test-westeu (rg-codebreaker-test) ˅
	Create new

Figure 6.10 – Basic settings for the Azure container app

With the basic settings, these values need to be configured:

- The subscription for the resource.

- The resource group (`rg-codebreaker-test`).

- The name of the container app. We use `cae-codebreaker-gamesapi-3`. The suffix 3 names version 3 of this API. You can run different versions of this app in parallel.

- The region – select the region that best fits your location.

- The container app environment. Select the environment created earlier.

The configuration screen for the container app is shown in *Figure 6.11*:

Basics   **Container**   Ingress   Tags   Review + create

Select a quickstart image for your container, or deselect quickstart image to use an existing container.

Use quickstart image   ☐

Container details

You can change these settings after creating the Container App.

Name *   cae-codebreaker-gamesapi-3

Image source   ⦿ Azure Container Registry

                                  ◯ Docker Hub or other registries

Registry *   codebreakertest.azurecr.io   ⌄

Image *   codebreaker/gamesapi   ⌄

Image tag *   latest   ⌄

Command override ⓘ   *Example: /bin/bash, -c, echo hello; sleep 100000*

Container resource allocation

CPU and Memory *   0.25 CPU cores, 0.5 Gi memory   ⌄

Environment variables

Name	Value	Delete
DataStorage	Cosmos	🗑
Enter name	*Enter value*	

Figure 6.11 – Container settings for the Azure container app

Here, we select the image that will be published by selecting the ACR instance and the image name and tag, the CPU and memory resources that should be allocated for one running instance, and environment variables. Setting the `DataStorage` environment variable to `Cosmos` overrides the values defined in the `appsettings.json` file.

Figure 6.12 shows the **Ingress** configuration:

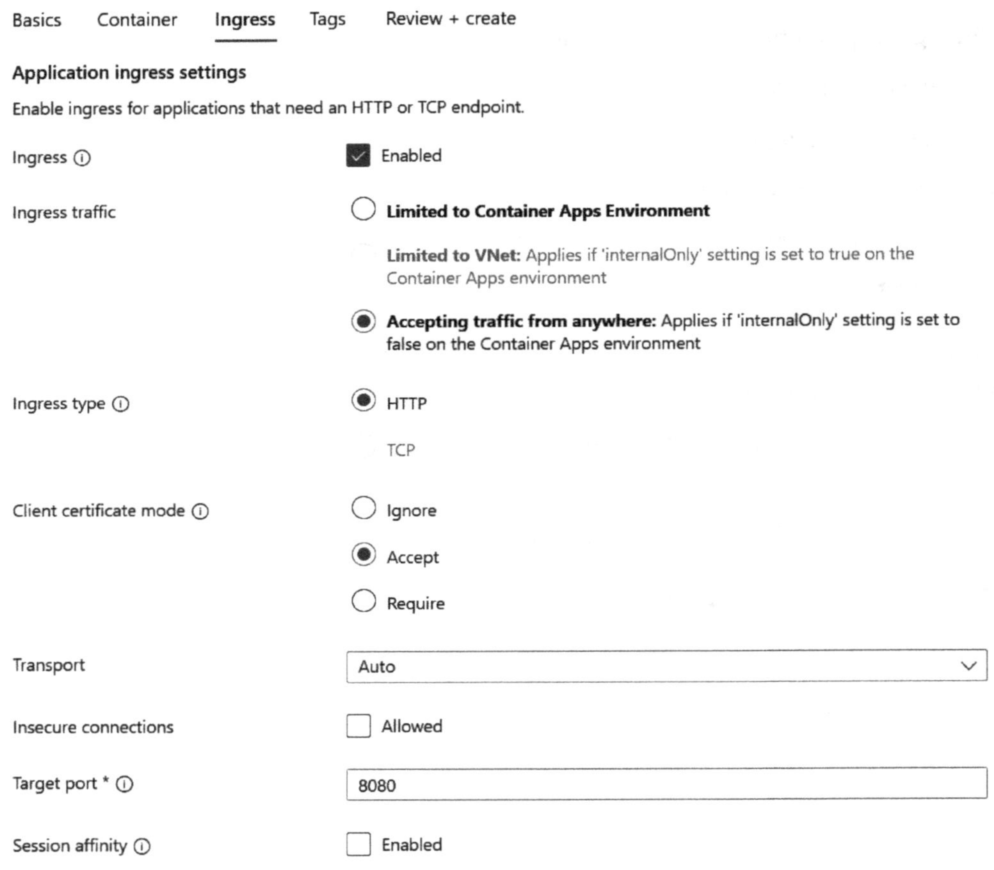

Basics   Container   **Ingress**   Tags   Review + create

Application ingress settings

Enable ingress for applications that need an HTTP or TCP endpoint.

Ingress ⓘ	☑ Enabled
Ingress traffic	◯ **Limited to Container Apps Environment**
	Limited to VNet: Applies if 'internalOnly' setting is set to true on the Container Apps environment
	◉ **Accepting traffic from anywhere:** Applies if 'internalOnly' setting is set to false on the Container Apps environment
Ingress type ⓘ	◉ HTTP
	TCP
Client certificate mode ⓘ	◯ Ignore
	◉ Accept
	◯ Require
Transport	Auto ⌄
Insecure connections	☐ Allowed
Target port * ⓘ	8080
Session affinity ⓘ	☐ Enabled

Figure 6.12 – Ingress settings for the Azure container app

We need to enable **Ingress** and accept traffic from anywhere to make the service available from the outside. The target port of the Docker container is `8080`, as defined by the .NET 8 images.

Clicking on **Create** creates the Azure container app by getting the image from the ACR instance. Be aware that starting the app will fail, as the connection string to the Cosmos database still needs to be configured. We will do this after creating the app for the bot service.

To create the app for the bot service, open the newly created container app in the Azure portal, and copy the *application URL* from the **Overview** view. This URL is needed for the configuration of the bot.

When creating an app for the bot service, you can configure it similarly to the game APIs. The **Ingress** configuration needs to be the same as before to have the bot accessible via its REST interface. With the container configuration, select the `codebreaker/bot` image. Create an environment variable with the name `ApiBase`, and the value with the application URL from the game APIs.

We still need to add some configuration values, which we'll do next.

Configuring secrets and environment variables

Defining secrets for the application is not directly possible with the portal when creating the app. This would be possible directly using the `az containerapp create` command.

With the portal, the secret can be configured afterward. When opening the container app within the Azure portal, in the **Settings** category, you can click on **Secrets**. One secret that was already stored on creating the app is the password for the ACR instance. For a successful connection to the Azure Cosmos DB instance, create a new secret named `cosmosconnectionstring`, as shown in *Figure 6.13*, and copy the connection string you copied from Azure Cosmos DB to the value:

> **Note**
> The screenshot in *Figure 6.13* shows another option where secrets can be stored: a Key Vault reference. In *Chapter 7*, we'll discuss using other options to use configurations, which include **Azure Key Vault**.

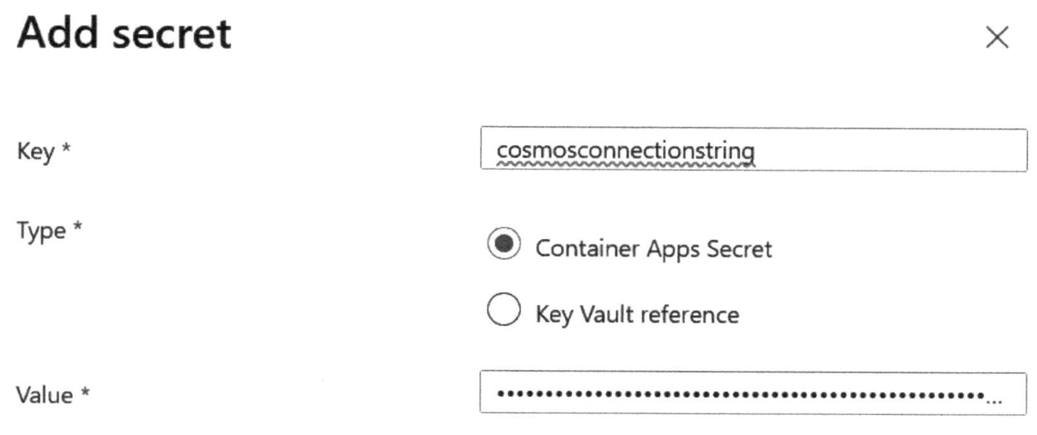

Add secret   ✕

Key *   cosmosconnectionstring

Type *   ⦿ Container Apps Secret

                             ◯ Key Vault reference

Value *   •••...

Figure 6.13 – Secret configuration with the Azure container app

To create an environment variable that references the secret, we can use the Azure CLI:

```
az containerapp update -n cae-codebreaker-gamesapi-3 -g
rg-codebreaker-test --set-env-vars ConnectionStrings__
GamesCosmosConnection=secretref:cosmosconnectionstring
```

Using the `az containerapp update` command, we need to reference the container app and the resource group and set environment variables using `--set-env-vars`. Contrary to passing hierarchical configuration values to the command line where `:` is used as a separator, such as `ConnectionStrings:GamesCosmosConnection`, with environment variables, using `:` is not possible. Instead, here, `__` maps values. The key that's specified for the connection to the Azure Cosmos DB instance is `ConnectionStrings__GamesCosmosConnection`. The value for this is stored in a secret. A secret is referenced with `secretref`, followed by the secret key.

The application should be running now, but let's make sure to configure scaling.

Configuring scaling with Azure Container Apps

The default scaling configured with Azure Container Apps is scaling from 0 to 10. If no load is on the app, it scales down to 0 where CPU and memory costs are reduced to zero. However, scaling to 0 also means that the first user accessing the service needs to wait for a few seconds before the service returns results. With the bot service that's running in the background and doesn't need some user interaction after the first invocation, this can be fast enough. With app jobs that are triggered from messages or events, this is also OK. However, with the `gamesAPI` service, this should be responsive for the first user accessing the service.

Configuring the minimum scale to 1 has a reduced price for the CPU if there's no load. With idle pricing, the memory doesn't have a price difference, but the CPU is about 10% of the cost compared to a running service.

Let's configure the `gamesAPI` service to scale from 1 to 3 replicas, and the bot service to scale from 0 to 3. Within the Azure portal, select the container app, the **Applications** category, and the **Scale and replicas** menu. Click the **Edit and deploy** menu, select **Scale**, and change the replicas to 1 to 3 and to 0 to 3 depending on the app. In case the UI elements are not easily movable to change the values accordingly, you can use the arrow keys to change the values one by one. The maximum scale count (at the time of this writing) is 300.

Clicking on **Create** does a redeploy and creates a new revision of the app. By default, only one revision is active at a time. As soon as the new revision successfully starts up, the load balancer moves 100% of the traffic to the new revision. With the **Applications** | **Revisions** menu, you can see active and inactive revisions. There, you also can configure the revision mode. The default revision mode is **Single**, where just one revision is active. You can change this to **Multiple**, where several revisions are running concurrently, and you can configure how much percentage of the traffic should be distributed to which revision. This can be useful to test different versions running on a user load.

With this in place, let's try to run the application. You can open the Swagger page of the bot to let the bot play some games. You can also use a client you created in *Chapter 4*, configure the address with the Container App game API's URL, and play a game. Check into the **Data Explorer** section of Cosmos DB to see the games stored.

As you now know all the Azure services used in the first place, you can delete Azure resources and recreate them easily in the next section.

Creating Azure resources with .NET Aspire and azd

Here, we'll look into how to easily create Azure resources from the development system. First, we use some resources from the Azure cloud, while most of the projects are running locally on the development system, before we publish the complete solution to Azure.

Provisioning Azure resources while debugging

When creating API services and using databases, you might not need any Azure resources when debugging the application locally. The API can run locally; even building Docker images is not required here. To run the database, a Docker image can be used easily, as you've already seen in *Chapter 5*. However, for some Azure resources that you might also use during development, creating and connecting Azure resources is required. One example is Azure Application Insights (which is covered in detail in *Chapter 8*).

To use Azure resources with the application map in the AppHost project, you need to add the Aspire. Hosting.Azure.* packages. To use Azure resources to define the app model, packages such as Aspire.Hosting.Azure and Aspire.Hosting.Azure.cosmosDB are available.

Provisioning of Azure resources happens automatically when the Azure resources are specified with the app model:

Codebreaker.AppHost/Program.cs

```
var builder = DistributedApplication.CreateBuilder(args);

string dataStore = builder.Configuration["DataStore"] ?? "InMemory";

var cosmos = builder.AddAzureCosmosDB("codebreakercosmos")
  .AddDatabase("codebreaker");
// code removed for brevity
```

The `AddAzureProvisioning` method creates Azure resources or retrieves the connection string when starting the application. Before this can successfully run, you need to specify your subscription ID and the location where the resources are created:

```
{
  "Azure": {
    "SubscriptionId": "your subscription id",
    "Location": "westeurope",
    "CredentialSource":"AzureCli"
  }
}
```

This information should not be part of the source code repository, thus add it to the user secrets. The `SubscriptionId` and `Location` keys need to be specified within the `Azure` category. Adding a *CredentialSource* is optional. The user to create the resources is selected using the DefaultAzureCredential (see *Chapter 7* for details). In case this is not working in your environment, you can configure AzureCli which uses the account you are logged in with the Azure CLI.

To get the subscription ID, you can use the following command:

```
az account show --query id
```

You need to be logged in to your subscription not only to see the subscription ID but also to deploy the resources automatically. After running the application, you can see the deployed resources are written to the user secrets as well.

This app model defines provisioning and using the Azure Cosmos DB database without using the emulator:

Codebreaker.AppHost/Program.cs

```csharp
var builder = DistributedApplication.CreateBuilder(args);

string dataStore = builder.Configuration["DataStore"] ?? "InMemory";

var cosmos = builder.AddAzureCosmosDB("codebreakercosmos")
  .AddDatabase("codebreaker");

var gameAPIs = builder.AddProject<Projects.Codebreaker_
GameAPIs>("gameapis")
  .WithExternalHttpEndpoints()
  .WithReference(cosmos)
  .WithEnvironment("DataStore", dataStore);

builder.AddProject<Projects.CodeBreaker_Bot>("bot")
.WithExternalHttpEndpoints()
```

```
    .WithReference(gameAPIs);

builder.AddProject<Projects.Codebreaker_CosmosCreate>("cosmoscreate")
    .WithReference(cosmos);

builder.Build().Run();
```

To avoid the need to install and run the local Azure Cosmos DB emulator, and to get rid of some issues using Azure Cosmos DB in the Docker image, we can use Azure Cosmos DB in the cloud. Not using the `RunAsEmulator` method with `AddAzureCosmosDB`, we use the resource running in Azure. The `AddDatabase` method adds the `codebreaker` database to the account. The `Codebreaker.CosmosCreate` project is used to run once, invoking the `EnsureCreatedAsync` method of the EF Core context to create a container with the partition key. The `WithReference` method used both with the `gamesAPI` service and the `CosmosCreate` project passes the newly created Azure Cosmos DB connection string to these resources. The `WithExternalEndpoints` method configures the Ingress controller of the Azure App Configuration to make this service external available.

Figure 6.14 shows the .NET Aspire dashboard with the application running:

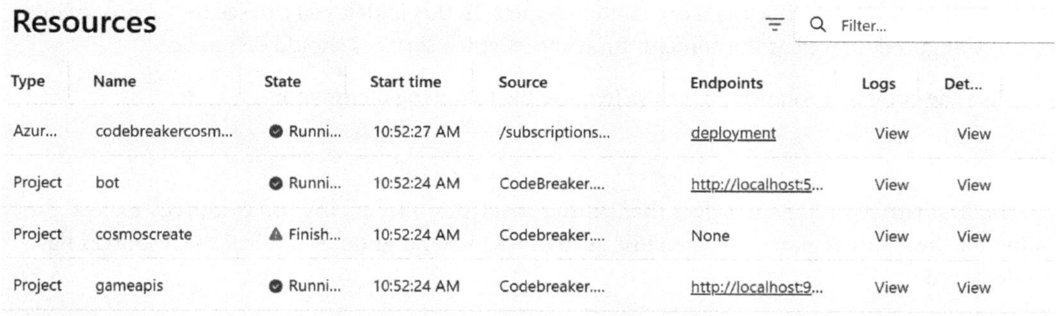

Resources

Type	Name	State	Start time	Source	Endpoints	Logs	Det...
Azur...	codebreakercosm...	● Runni...	10:52:27 AM	/subscriptions...	deployment	View	View
Project	bot	● Runni...	10:52:24 AM	CodeBreaker....	http://localhost:5...	View	View
Project	cosmoscreate	⚠ Finish...	10:52:24 AM	Codebreaker....	None	View	View
Project	gameapis	● Runni...	10:52:24 AM	Codebreaker....	http://localhost:9...	View	View

Figure 6.14 – .NET Aspire dashboard with deployed Azure resources

The `codebreakercosmos` resource shows an endpoint with the `deployment` link text. This is a resource deployed to Azure. Clicking this link, you are directly navigated to this cloud resource and can check the database and container name are created. The `cosmoscreate` reference is in the **Finished** state, thus the creation of the container is done.

Now let's start the bot and let it run some games, then open **Data Explorer** with Azure Cosmos DB, and you'll see the games created. You can debug the solution locally while using some of the resources in the cloud, just by adding one API method.

The resource group created here uses the name `rg-aspire-{yourhost}-codebreaker.apphost`. If multiple developers use the same Azure subscription, resources are created independently of each other to not get into conflicts. Make sure to delete resources when not needed.

Next, let's create the complete solution to run with Azure.

Provisioning the complete solution with azd up

For this, we use `azd`. First, in the directory of the solution, use the following command:

```
azd init
```

This initializes an application to be used with `azd`. You can use a template to create a new solution or analyze an existing application. As we already have a running application, select `Use code in the current directory` to analyze the application. The application needs to be stopped because `azd` also starts the compilation. With a successful scan, `azd` informs to host the app using **Azure Container Apps**. Confirm this to continue the initialization. Next, projects are listed where you need to select which of these should be accessible on the internet (the Ingress service will be configured accordingly). Select the `bot` and `gamesAPI` services. Then, define an environment (for example, `codebreaker-06`) using a chapter suffix.

What happened? This command created a `.azure` folder and `azure.yaml` and `next-steps.md` files. `next-steps.md` gives information about what you can do next. `azure.yaml` is a short file containing information that references the `AppHost` project running with a `containerapp`. The most interesting generated information can be found in the `.azure` folder. This folder is excluded from the source code repository as it can contain secrets. In this folder, you can see the environments that are configured, as well as the configuration about which services should be public.

To publish the complete solution to Azure, just use the following command:

```
azd up
```

With the first run, you need to select the Azure subscription to deploy the resources to, and the location for the Azure region. Next, you just need to wait several minutes until all the resources have been deployed.

In the provisioning phase, these resources are deployed:

- A resource group
- A container registry
- A key vault
- A Log Analytics workspace
- A Container Apps environment

After the provisioning phase, the deployment phase starts with these actions:

- Pushing Docker images to the ACR instance
- Creating container apps in the Container Apps environment using the images from the ACR instance

As you make any changes to the source code or the configuration, you just need to use `azd up` again to deploy the updates. As the created environment is no longer needed, use `azd down` to delete all the resources again. Make sure to wait until you are asked for verification as to whether the number of resources should really be deleted.

Checking the resource group within the Azure portal, you can see all resources created, as shown in *Figure 6.15*:

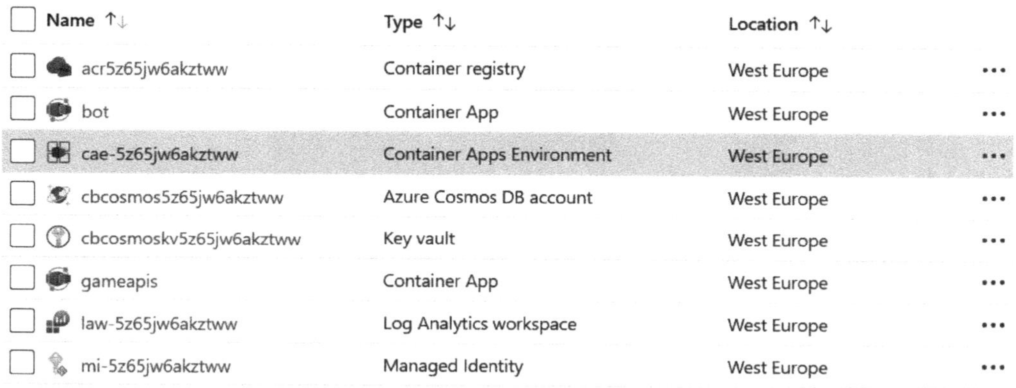

Name ↑↓	Type ↑↓	Location ↑↓	
acr5z65jw6akztww	Container registry	West Europe	...
bot	Container App	West Europe	...
cae-5z65jw6akztww	Container Apps Environment	West Europe	...
cbcosmos5z65jw6akztww	Azure Cosmos DB account	West Europe	...
cbcosmoskv5z65jw6akztww	Key vault	West Europe	...
gameapis	Container App	West Europe	...
law-5z65jw6akztww	Log Analytics workspace	West Europe	...
mi-5z65jw6akztww	Managed Identity	West Europe	...

Figure 6.15 – Resources created from azd up

Now, you can check the resources deployed, the images published to the container registry, the applications published to the container apps environment, the key vault containing a secret, and the Azure Cosmos DB account with the database and the configured container. Let the bot play games and verify if everything is running.

Next, let's get into the details of what happened with `azd up`.

Diving into azd up stages

Running the `AppHost` project, command-line arguments can be passed to create a manifest file describing all the resources:

```
dotnet run --project Codebreaker.AppHost/Codebreaker.AppHost.csproj --
--publisher manifest --output-path aspire-manifest.json
```

When using `dotnet run`, command-line arguments can be passed to the application by using `--` to differentiate from the arguments of `dotnet run`. Using the `--publisher manifest` option creates an Aspire manifest describing the app model of the application. This manifest specifies all resources with the resource type, bindings, environment variables, and paths to the projects. This information is used by `azd` to create Azure resources and can be used from other tools to, for example, deploy the solution to Kubernetes.

Next, `azd provision` is used. In case you just want to provision Azure resources without pushing Docker images and deploying the Azure container apps, use the following command:

```
azd provision
```

`azd provision` uses the manifest file to create Bicep files in memory and creates Azure resources.

You can use this command when any Azure resources are added to the app model, then just these resources are created.

The next step is the following:

```
azd deploy
```

`azd deploy` pushes the container images to the ACR instance using `dotnet publish`, and then creates or updates Azure resources using these images.

`azd up` creates Bicep scripts in memory. It's also possible to create Bicep scripts on disk to use them to create Azure resources, as we'll do next.

Creating Bicep files using azd

Bicep is a domain-specific language using declarative syntax. Before Bicep was available, we created **Azure Resource Manager** (**ARM**) templates to create Azure resources. ARM templates are defined using JSON. Bicep is simpler to write than ARM templates. During deployment, Bicep files are converted to ARM templates.

Here, we use `azd infra` to create Bicep files for the solution.

> **Note**
> At the time of this writing, `azd infra` is in an early stage. Check the README file of this chapter for updates.

Start this command from the directory of the solution:

```
azd infra synth
```

This command creates an `infra` folder with these files:

- `main.bicep` – The main bicep file that creates a resource group and references modules to create more resources.
- `main.parameters.json` – A parameters file that is used to pass parameters such as the environment name and the location to the `main.bicep` file.

- `resources.bicep` – This file is referenced by `main.bicep` and contains resources such as the ACR instance, the Log Analytics workspace, the Container Apps environment, and the Azure key vault that are created.

- `codebreakercosmos/codebreakercosmos.bicep` – This file is referenced from `main.bicep` as well and contains resource information for Azure Cosmos DB, as well as an Azure Key Vault secret that's written to the Azure key vault. The secret itself is not part of this file; the secret is retrieved dynamically from the Azure Cosmos DB account when creating this resource.

In case you customize these generated Bicep files, the customized files are used by `azd up` or `azd provision` on creating the Azure resources.

With the `bot` and `gamesAPI` projects, `azd infra` also creates an infra folder with template files in the AppHost project; for example `gameapis.tmpl.yaml`. With these files, the Azure Container Apps instances can be customized; for example, by changing CPU and memory sizes or changing the number of replicas that should be used. Changing these values, `azd up` or `azd deploy` makes use of these files.

When you open the resource group you previously created with `azd up`, open **Deployments** in the **Settings** category. This shows deployments of the resource group, as shown in *Figure 6.16*:

	Deployment name	Status	Last modified	Duration	Related events
☐	cbcosmos	✅ Succeeded	3/15/2024, 5:29:45 PM	2 minutes, 3 seconds, 583 millise...	Related events
☐	resources	✅ Succeeded	3/15/2024, 5:27:21 PM	47 seconds, 767 milliseconds	Related events

Figure 6.16 – Deployments

The deployments match the Bicep files used. When you open **Related events**, you can see all the steps that have been done with these deployments.

After you don't need the resources anymore, use this command to delete all the resources again:

```
azd down
```

This tool retrieves the number of Azure resources to delete and asks if this should be done – thus, make sure to wait until you can answer yes. After the deletion of the resources is complete, which usually takes more than 10 minutes, another question is asked if the data from the key vault should be purged. If you don't answer yes to this question, this data can be recovered for 90 days, and during that time you cannot create the resource with the same name again until the end of this recovery time.

Summary

In this chapter, you learned to create Microsoft Azure resources using the Azure CLI, the Azure portal, and .NET Aspire with `azd`. The `gamesAPI` service is now running with Microsoft Azure resources using ACR, Azure Container Apps, and the Azure Cosmos DB database. When using `azd` together with .NET Aspire, just one command was needed to deploy all the services.

Before moving on to the next chapter, let's configure the client application you used in previous chapters to now use the URL from the Azure container app instead of the local services, and play some games.

In this chapter, the Azure key vault was already created. In the next chapter, we look into the configuration of the backend services, which includes Azure Key Vault, and use Azure App Configuration as a central place for the configuration of all `codebreaker` services.

Further reading

To learn more about the topics discussed in this chapter, you can refer to the following links:

- Azure round-trip latency statistics with Azure regions: `https://learn.microsoft.com/azure/networking/azure-network-latency`

- Azure Cosmos DB – databases, containers, and items: `https://learn.microsoft.com/azure/cosmos-db/resource-model`

- Stored procedures, triggers, and user-defined functions: `https://learn.microsoft.com/azure/cosmos-db/nosql/how-to-write-stored-procedures-triggers-udfs`

- Azure Container Apps: `https://learn.microsoft.com/azure/container-apps/`

- Bicep: `https://learn.microsoft.com/azure/azure-resource-manager/bicep`

- `azd` reference: `https://learn.microsoft.com/en-gb/azure/developer/azure-developer-cli/reference`

7

Flexible Configurations

.NET offers flexible configurations using a provider-based model to read configurations from different sources. In the last chapter, we configured environment variables with Azure Container Apps to override the JSON file configuration.

In this chapter, you will learn how to use the app configuration with .NET and how to add a configuration provider to use a central configuration store: **Azure App Configuration**. For secrets, we have another Azure service available: **Azure Key Vault**. In this chapter, you'll also learn how to combine Azure Key Vault with Azure App Configuration and reduce the number of secrets you need to store by using **Azure managed identities**.

In this chapter, you'll learn how to do the following:

- Explore the functionality of .NET configurations
- Store configurations with Azure App Configuration
- Store secrets with Azure Key Vault
- Reduce the number of secrets needed using managed identities
- Use environments with Azure App Configuration

Technical requirements

Similar to the previous chapter, an Azure subscription, the Azure CLI, Azure Developer CLI, and Docker Desktop are required.

The code for this chapter can be found in the following GitHub repository: `https://github.com/PacktPublishing/Pragmatic-Microservices-with-CSharp-and-Azure`.

In the ch07 folder, you'll see these projects with the final result of this chapter:

- ConfigurationPrototype – This is a new project that shows some concepts with configuration before implementing this with the games API and the bot service.

- Codebreaker.InitializeAppConfig – This is a new project to initialize values with Azure App Configuration.

- Codebreaker.AppHost – The app model defined with this project is enhanced to include the ConfigurationPrototype and Codebreaker.InitializeAppConfig projects and add App Configuration and Azure Key Vault resources to the app model.

- Codebreaker.GameAPIs – The games API project we used in the previous chapter is enhanced using App Configuration.

- Codebreaker.Bot – This is the implementation of the bot service that plays games. This project is enhanced with App Configuration as well.

You can start with the results from the previous chapter to work on your own through this chapter.

To publish the solution to Azure (which is needed later in this chapter when we use managed identities), use the Azure Developer CLI with the current directory set to the solution folder:

```
azd init
azd up
```

With azd init, select to analyze the code in the folder, accept to deploy Azure Container Apps, specify an environment such as codebreaker-07, and select the game APIs, the bot, and the configuration prototype to be accessible from the Ingress controller. With azd up, the resources are deployed to your configured environment.

Check the README file of the ch07 folder of the repository for the latest updates.

Experiencing .NET configurations

In this chapter, we will create a new Web API project to try out .NET configuration features before adding configuration features to the game APIs and the bot service:

```
dotnet new webapi -o ConfigurationPrototype
```

.NET is flexible in how to read configuration values. Configuration values can be retrieved from different sources such as JSON files, environment variables, and command-line arguments. Depending on the environment (for example, production and development), different configuration values are also retrieved. Using this core .NET feature, it's easily possible to add other configuration sources and customize environments.

Behind the scenes, the `ConfigurationManager` class is used to configure sources for the application configuration. This configuration is done at application startup when invoking `WebApplication.CreateBuilder`.

> **Note**
>
> With .NET 8, other builder methods, such as `CreateSlimBuilder` and `CreateEmptyBuilder`, are available. With these builders, the number of services registered is reduced to increase performance.

With the default configuration as done by `WebApplicationBulder.CreateBuilder`, a list of configuration providers has already been added:

- **Memory configuration provider**: The memory configuration provider is great for values that can be retrieved programmatically during runtime. For example, the `webroot` key is set to the path of the web directory. Instead of using other APIs to get this information, you can retrieve it using configuration keys.

- **Environment variable configuration provider**: For accessing environment variables, multiple providers are configured. Built-in provider configurations add environment variables with the `ASPNETCORE_` and `DOTNETCORE_` prefixes to have them available early in the process, which allows overriding the values by all providers following. Another environment variable configuration provider adds all other environment variables. The `ASPNETCORE_HTTP_PORTS` and `ASPNETCORE_HTTPS_PORTS` environment variables are new since .NET 8 to easily change the listening ports of the Kestrel server. .NET Aspire passes environment variables to the configured projects.

- **JSON configuration provider**: The JSON configuration provider is one of the file providers. XML and INI providers are available as well to read configuration values from files with the defined syntax. The files that are referenced with the built-in configuration are `appsettings.json` and `appsettings.{environmentName}.json`. In case the environment name is `Development`, the values from `appsettings.Development.json` are retrieved. This overwrites settings from the previously loaded `appsettings.json` file.

  In your environment, you can use multiple JSON files (for example, `connectionstrings.json`) in case you prefer to have all the connection strings separated:

  ```
  builder.Configuration.AddJsonFile("connectionstrings.json",
  optional: true);
  ```

  The `AddJsonFile` extension method adds the filename as another JSON configuration provider. If the `optional` parameter is not configured to be `true`, an exception is thrown in case the file cannot be found.

- **Command-line configuration provider**: The command-line provider allows overriding all the settings (because it is last in the providers' list). Starting the application, you can pass configuration values to overwrite other settings.

  Imagine a case where a hierarchical setting is specified with JSON, such as this connection string:

  ```
  {
     "ConnectionStrings": {
        "GamesSqlServerConnection": "server=(localdb)\\
  mssqllocaldb;database=CodebreakerGames;trusted_connection=true"
     }
  }
  ```

  In such a case, you can pass the value using command-line arguments with a `:` separator:

  ```
  ConnectionStrings:GamesSqlServerConnection = "the new connection
  string"
  ```

  Using `:` is not possible with environment variables. As you saw in the previous chapter, when passing environment variables for hierarchical configurations, two underscores (__) are used as separators.

- **User secrets configuration provider**: The user secrets provider is only used during development time. Don't add secrets to the source code repository. With this, configuration values are stored within the user profile and thus not stored in the directory of the source code. But be aware that every developer running the application locally needs to configure the needed user secrets.

  The user secrets provider is only added by the default builder if the application is running in debug mode and a value for `UserSecretsId` is set with the project file:

  ```
  cd ConfigurationPrototype
  dotnet user-secrets init
  ```

  This command adds `UserSecretsId` to the project file and uses a unique identifier to reference the corresponding secrets from the user profile.

  To add a secret, use this command:

  ```
  dotnet user-secrets set SecretKey1 "This is a secret"
  ```

  Use `dotnet user-secrets -h` to see the other commands available.

Note

Running a .NET Aspire solution on the development system, the app model with its dependencies is used to create environment variables containing referenced information with the processes running the services. When deploying the solution to Microsoft Azure, with Azure Container Apps, environment variables and secrets are created. Because environment variables by default are configured as configuration providers, nothing special needs to be done running the services.

Retrieving configuration values

How can we access configuration values? To get custom configuration values, let's enhance the `appsettings.json` file:

ConfigurationPrototype/appsettings.json

```json
{
  "Logging": {
    "LogLevel": {
      "Default": "Information",
      "Microsoft.AspNetCore": "Warning"
    }
  },
  "AllowedHosts": "*",
  "Config1": "config 1 value",
  "Service1": {
    "Config1": "config 1 value",
    "Config2": "config 2 value"
  }
}
```

The `Config1` key is added to the root elements of the file. With `Service1`, we use a parent-child relationship and define multiple child elements, `Config1` and `Config2`.

To retrieve configuration values, we just need to inject the `IConfiguration` interface, as shown in the following code snippet. You need to add this code snippet before the `app.Run` method:

ConfigurationPrototype/Program.cs

```csharp
app.MapGet("/readconfig", (IConfiguration config) =>
{
  string? config1 = config["Config1"];
  return $"config1: {config1}";
});
```

The `IConfiguration` interface is injected in the GET request of the API implementation. Using a C# indexer, we retrieve the value for the `Config1` key. To retrieve child elements, we can use the `GetSection` method and use the indexer from the returned section. `GetSection` returns an object implementing the `IConfigurationSection` interface. This interface itself derives from `IConfiguration`, thus the members of the `IConfiguration` interface are available.

Try it out: start the `ConfigurationPrototype` app, and with the OpenAPI test page, test the `/readconfig` endpoint.

To retrieve the children, we'll use a different approach with options next.

Using options

When configuration values are needed, many .NET services make use of the **options pattern**. This adds some flexibility in where to get these values – this can be the configuration, but these service configuration values can also be assigned programmatically.

Strongly typed configuration values are another feature of this pattern. Add this class to map the configuration values:

ConfigurationPrototype/Program.cs

```
internal class Service1Options
{
  public required string Config1 { get; set; }
  public string? Config2 { get; set; }
}
```

The class to map the configuration values needs a parameter-less constructor and properties that match the configuration values.

To fill the values, the `Service1Options` class is configured with the **dependency injection container** (DIC). Add this code before the `builder.Build` method:

ConfigurationPrototype/Program.cs

```
builder.Services.Configure<Service1Options>(
  builder.Configuration.GetSection("Service1"));
```

The `IServiceCollection Configure` extension method offers two overloads. With one overload, a delegate can be assigned to fill the `Service1Options` instance programmatically. The second overload – which is used here – receives an `IConfiguration` parameter. Remember – in the configuration file created earlier, a `Service1` parent element was defined. The `GetSection` method retrieves the values within this section. Because the configuration keys map to the class, the values are filled.

> **Note**
> A new .NET 8 feature with a binding configuration is a source generator. Using native AOT (see *Chapter 5*), this source generator is enabled by default. With non-AOT projects, `EnableConfigurationBindingGenerator` can be added to the project file to turn off this source generator.

With this configuration in place, let's retrieve these configuration values. Add this code before app. Run to configure the endpoint:

ConfigurationPrototype/Program.cs

```
app.MapGet("/readoptions", (IOptions<Service1Options> options) =>
{
  return $"options - config1: {options.Value.Config1}; config 2:
    {options.Value.Config2}";
});
```

The IOptions interface with the Service1Options generic parameter is injected, and with this, the configured values can be used.

After you have made these code changes, run the ConfigurationPrototype project again. Use the /readoptions endpoint to retrieve the configured values.

Using environments

As the application runs in different environments (for example, production, staging, and development), there's a need for different configuration values. For example, in the development environment, you don't want to use the production database. The .NET configuration easily supports different environments.

With the default configuration, an appsettings.{environment}.json file is loaded to specify environment-specific configuration values – for example, appsettings.staging.json in the staging environment.

Aside from using different filenames to load environment-specific configuration values, we can programmatically verify the current environment.

The template-generated code contains this code:

ConfigurationPrototype/Program.cs

```
if (app.Environment.IsDevelopment())
{
  // code removed for brevity
}
```

The IsDevelopment extension method compares the environment with the Development string. Environment is a property of the WebApplication class. Other methods available are IsProduction, IsStaging, and IsEnvironment. Invoking the IsEnvironment method, any string can be passed to check if the application is running in the specified environment. Instead of using the IsEnvironment method, you can also create a custom extension method extending the IHostEnvironment type to compare with the environment.

What environment the application is running in is defined by the `ASPNETCORE_ENVIRONMENT` environment variable, as mentioned earlier. While debugging locally, the `launchsettings.json` file (in the `Properties` folder) defines the environment to the `Development` value. If the environment variable is not set, the default environment is `Production`. For all other environments, you need to set this environment variable.

Using configurations with Azure Container Apps

Azure Container Apps supports specifying environment variables, and secrets. In *Chapter 6*, when we created the container app, we configured environment variables and secrets. Environment variables of a container app can be configured on creating the application or afterward when updating the application – for example, using `az containerapp update`.

Environment variables may be visible in log files. For secrets, this can be a security issue. Security sniffers can catch secrets that are configured in environment variables and alert system administrators when these are found. With container apps, secrets are stored within the scope of an application but independent of revisions of the application.

To get even better security for secrets, container app secrets can be connected to secrets with the Azure Key Vault service. The Key Vault service and additional features we get for secrets are discussed later in this chapter.

When you use multiple Azure services (for example, Azure App Service, Azure Functions, Azure Container Apps…), how configuration is managed is different from service to service. Also if you just run a large list of services within only container apps, you might prefer a central place where all the configuration is managed. Azure App Configuration offers this functionality without the need to create a custom configuration service.

Using configurations with Azure App Configuration

In this chapter, we add Azure App Configuration and Azure Key Vault to the solution, as shown in *Figure 7.1*:

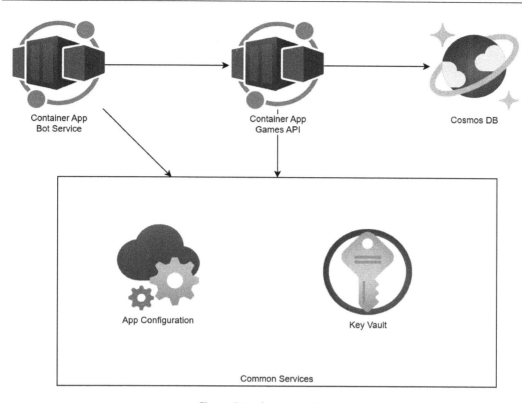

Figure 7.1 – Azure services

Both of these services can be used with any service that needs configuration values. The Key Vault service is used to store secrets and adds great features for this.

Let's create an Azure App Configuration resource.

Creating an Azure App Configuration service

We use .NET Aspire to create an Azure App Configuration service. To use the .NET Aspire `AppHost` configuration from the `ConfigurationPrototype` project, add **.NET Aspire Orchestrator Support** to this project (which adds a project reference to the `AppHost` project and references the project with the app model definition):

Codebreaker.AppHost/Program.cs

```
var builder = DistributedApplication.CreateBuilder(args);
var appConfig = builder.AddAzureAppConfiguration("codebreakerconfig")
  .WithParameter("sku", "Standard");
```

```
builder.AddProject<Projects.
ConfigurationPrototype>("configurationprototype")
  .WithReference(appConfig);
// code removed for brevity
```

For using the Azure App Configuration resource with the `AppHost` project, we also need to add the `Aspire.Hosting.Azure.AppConfiguration` NuGet package. Calling the `AddAzureAppConfiguration` method adds the resource to the app model. In case you don't use any App Configuration features with your Azure subscription yet, you can set the `sku` value to `Free` to use a free version of the App Configuration service. The free version does not offer any SLAs and is limited to 1,000 invocations per day, but for development, this limit can be fine. The App Configuration service is referenced from the `ConfigurationPrototype` project with the `WithReference` method.

Starting the `AppHost` project, the resources are provisioned. Remember to have user secrets configured with the `AppHost` project:

```
{
  "Azure": {
    "SubscriptionId": "<enter your subscription id>",
    "Location": "westeurope"
    "CredentialSource": "AzureCli"
  }
}
```

Change the subscription ID to your subscription ID and change the location to your chosen Azure region. It can also be helpful to specify the source of credentials used to create Azure resources. Setting the value to `AzureCli`, the same account is used that you used to log in with the Azure CLI.

Because user secrets store the configuration inside the user profile, when using the same `UserSecretsId` value with multiple projects, this information might already show up. .NET Aspire also adds information about resources created to user secrets.

When you start the application, an additional Azure resource will be created. After this is completed successfully, as you can see with the .NET Aspire dashboard, let's add some configuration values.

Configuring values with Azure App Configuration

After the creation process of the App Configuration service is completed, we can define configuration values using the Configuration explorer in the Azure portal (see *Figure 7.2*):

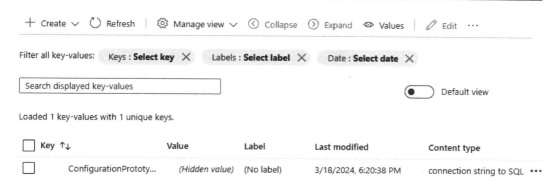

Figure 7.2 – Configuration explorer

With App Configuration, key-value pairs are stored. Creating a `ConfigurationPrototype:-ConnectionStrings:SqlServer` key, we define a string value for a database connection. As the configuration values for all `Codebreaker` services can be configured in one place, it's good practice to use the first part of the key with the name of the service – this way, we know which configuration values belong to which service. It's also possible to use JSON content with the value, as we'll do later with the games API. This reduces the number of requests for this service and can simplify the configuration.

Next, let's get the configuration from the `ConfigurationPrototype` project.

Initializing app configuration values

We can also add configuration values programmatically when the application is deployed. To do this, let's create a background service that runs once.

Create a new background worker service:

```
dotnet new worker -o Codebreaker.InitializeAppConfig
```

To publish a worker project as a Docker image, you also need to enable SDK container support:

Codebreaker.InitializeAppConfig/Codebreaker.InitializeAppConfig.csproj

```
<PropertyGroup>
  <IsPublishable>true</IsPublishable>
  <EnableSdkContainerSupport>true</EnableSdkContainerSupport>
</PropertyGroup>
```

When creating worker projects, without this setting, a Docker image cannot be created using `dotnet publish`.

Add this project to the .NET Aspire orchestration (use .NET Aspire Orchestrator Support, or add a reference to the ServiceDefaults project and add a project reference from the AppHost project to this project). Rename the Worker class that's created from this template to AppConfigInitializer.

Add the Azure.Data.AppConfiguration and Microsoft.Extensions.Azure NuGet packages. The Azure.Data.AppConfiguration package offers functionality to access the App Configuration API to create, read, and update settings. Microsoft.Extensions.Azure provides integration with the **dependency injection (DI)** system.

To write configuration settings, add this code to the AppConfigInitializer class:

Codebreaker.InitializeAppConfig/AppConfigInitializer.cs

```
public class AppConfigInitializer(ConfigurationClient
configurationClient, IHostApplicationLifetime hostApplicationLifetime,
ILogger<AppConfigInitializer> logger) : BackgroundService
{
  private Dictionary<string, string> s_6x4Colors = new()
  {
    { "color1", "Red" },
    { "color2", "Green" },
    { "color3", "Blue" },
    { "color4", "Yellow" },
    { "color5", "Orange" },
    { "color6", "Purple" }
  };
  protected override async Task ExecuteAsync(CancellationToken
    stoppingToken)
  {
    foreach ((string key, string color) in s_6x4Colors)
    {
      ConfigurationSetting setting = new($"GameAPIs.Game6x4.{key}",
        color);
      await configurationClient.AddConfigurationSettingAsync(setting);
      logger.LogInformation("added setting for key {key}", key);
    }
  }
}
```

With the constructor of the AppConfigInitializer class, the ConfigurationClient class and the IHostApplicationLifetime interface are injected. ConfigurationClient is the class to communicate with App Configuration. We add settings by invoking the AddConfigurationSettingAsync method. IHostApplicationLifetime is the interface for background services to be informed about start and stop events and is used to stop the service at the end. After the settings are written, the application ends, invoking the StopApplication method.

Now, we can configure the `AppConfigInitializer` class with the DIC configuration:

Codebreaker.InitializeAppConfig/Program.cs

```
using Codebreaker.InitalizeAppConfig;
using Microsoft.Extensions.Azure;

var builder = Host.CreateApplicationBuilder(args);

builder.AddServiceDefaults();
builder.Services.AddHostedService<AppConfigInitializer>();
builder.Services.AddAzureClients(clients =>
{
  string appConfigUrl = builder.Configuration.
  GetConnectionString("codebreakerconfig") ??
    throw new InvalidOperationException("codebreakerconfig not
    configured");
  clients.AddConfigurationClient(new Uri(appConfigUrl));
});

var host = builder.Build();
host.Run();
```

The `AddHostedService` method requires an object to implement the `IHostedService` interface. This interface is implemented with the base class of the `AppConfigInitializer` class, `BackgroundService`. When the service is started, the `StartAsync` method of `BackgroundService` is invoked, which in turn invokes the `ExecuteAsync` method of `AppConfigInitializer` where the configuration values are set.

`AddAzureClients` is an extension method that allows configuring clients to access many of the Azure services. Here, we use the `AddConfigurationClient` extension method, passing the URL of the App Configuration resource.

Starting this initializer project now adds configuration settings to the App Configuration service. The game APIs service can now be changed to read colors for games from the configuration, which allows easy changes of colors without recompiling.

> **Note**
>
> Before .NET Aspire was available, I configured non-secret configuration values such as URLs to different Azure resources with Azure App Configuration. As the orchestration of .NET Aspire covers this aspect and makes it easy to run the solution with different environments and automatically configures these dependencies, App Configuration is now mainly used for other application-specific configuration values.

With this initialization in place, let's continue to read configuration values from the application.

Using Azure App Configuration from the application

To use the Azure App Configuration service from the .NET application, we need to add the `Microsoft.Azure.AppConfiguration.AspNetCore` NuGet package. This NuGet package offers a configuration provider.

This provider is configured with the following code snippet:

ConfigurationPrototype/Program.cs

```
var builder = WebApplication.CreateBuilder(args);

builder.Configuration.AddAzureAppConfiguration(appConfigOptions =>
{
  DefaultAzureCredential cred = new();
  string appConfigUrl = builder.Configuration.
    GetConnectionString("codebreakerconfig") ??
    throw new InvalidOperationException("could not read
    codebreakerconfig");
  appConfigOptions.Connect(new Uri(appConfigUrl), cred);
});
// the code from the repository also includes the Key Vault
configuration added later
```

The `AddAzureAppConfiguration` extension method adds the App Configuration service to the configuration providers. One overload uses a string parameter to pass a connection string including a secret. The default orchestration configuration of .NET Aspire just passes the URL from the App Configuration service without the secret. **Role-based access control (RBAC)** is configured, which doesn't need a secret. When an App Configuration resource is created with Aspire provisioning, your identity is added to role-based access to allow access to the configuration values. During development time, your user credentials can be used as well, programmatically accessing the API. This is the role of the `DefaultAzureCredential` class. This class uses a defined order to try different credentials, including *Visual Studio credentials*, *Azure CLI credentials*, and *Azure Developer CLI credentials*. The first credentials that are successfully retrieved are used to access the configuration service. The URL to the App Configuration service is forwarded from the .NET Aspire orchestrator and retrieved with the configuration API. After this, invoking the `Connect` method of the `AzureAppConfigurationOptions` class, the URL of the configuration service as well as the credentials are used to connect. After adding this configuration provider, App Configuration can be used like any other configuration provider.

> **Note**
>
> When a solution is deployed to Azure, the local credentials cannot be used. A managed identity is used when the solution is running within Azure. This is covered later in this chapter.

Now, all that needs to be done is to retrieve the configuration values. There's no difference where the configuration is coming from:

ConfigurationPrototype/Program.cs

```
app.MapGet("/azureconfig", (IConfiguration config) =>
{
  string? connectionString = config.
    GetSection("ConfigurationPrototype")
    .GetConnectionString("SqlServer");
  return $"Configuration value from Azure App Configuration:
    {connectionString}";
});
```

Again, the IConfiguration interface is injected. The key configured with App Configuration has a hierarchical name: ConfigurationPrototype:ConnectionStrings:SqlServer. The first hierarchy is accessed using the GetSection method. Next, the GetConnectionString method is used. This accesses the section named ConnectionString and then uses the SqlServer key to get its value.

With this last change, you can run the application and retrieve the configuration value from the App Configuration service!

Using the environment on your local system uses **user secrets**. In the production environment, you already know from the previous chapter how to configure secrets with Azure Container Apps to add a connection string to App Configuration in a secure manner with secrets of container apps. The Azure Key Vault service covered next offers an even more secure environment.

Storing secrets with Azure Key Vault

To get secret configuration values, the Azure Key Vault service can be used. The Key Vault service can be used to store secrets such as **passwords**, **certificates**, and **keys**. This service adds hardware-level encryption, automatic certificate renewals, and granular access control. With predefined roles, the service decides who is allowed to read secrets (*Key Vault Secrets User*, the application), who is allowed to create and update secrets but not read secrets (*Key Vault Contributor*), and who is allowed to monitor which users use secrets but not to create and read secrets (*Key Vault Secrets Officer*).

With .NET applications, the Key Vault service can be added as a configuration provider, as with Azure App Configuration. Another way to use this service is to link secrets stored with Key Vault to an Azure App Configuration instance. We will use the second option.

When you add a key to App Configuration, instead of just supplying a key and value, the key can be linked to a secret stored within the Key Vault service. While the same API as used with App Configuration can be used with secrets, the user running the service needs access to the Key Vault service.

Let's create a key vault using the .NET Aspire app model:

Codebreaker.AppHost/Program.cs

```
var appConfig = builder.AddAzureAppConfiguration("codebreakerconfig");
var keyVault = builder.AddAzureKeyVault("codebreakervault");

builder.AddProject<Projects.
ConfigurationPrototype>("configurationprototype")
   .WithReference(appConfig)
   .WithReference(keyVault);
```

The AddAzureKeyVault method adds a Key Vault resource to the app model. This resource is referenced from the following project configuration to pass the URL. As with the App Configuration before, secret information is not part of this URL passed.

Run the application to create the resource. Then, you can verify the **Access configuration** page (in the **Settings** section) to verify the permission model, as shown in *Figure 7.3*:

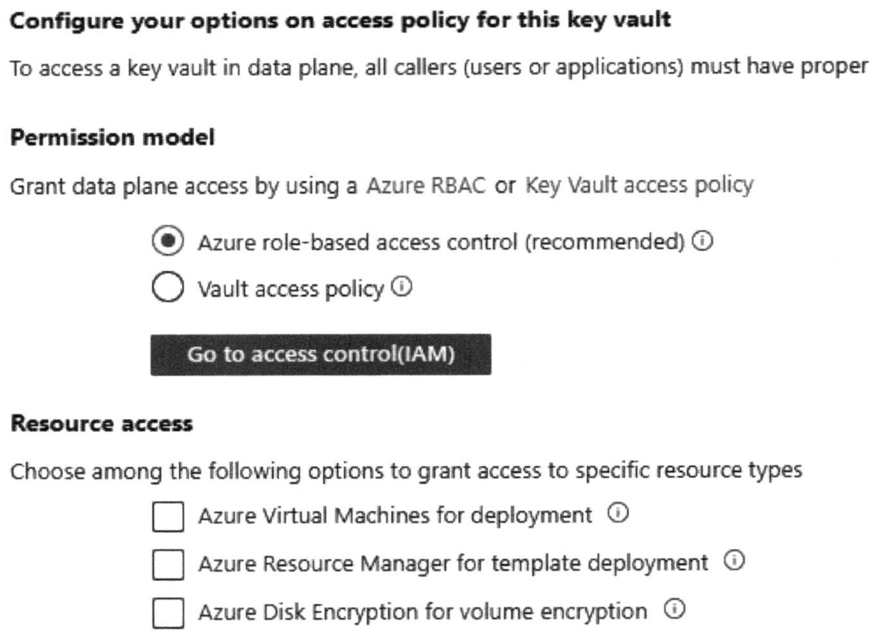

Configure your options on access policy for this key vault

To access a key vault in data plane, all callers (users or applications) must have proper

Permission model

Grant data plane access by using a Azure RBAC or Key Vault access policy

◉ Azure role-based access control (recommended) ⓘ

◯ Vault access policy ⓘ

Go to access control(IAM)

Resource access

Choose among the following options to grant access to specific resource types

☐ Azure Virtual Machines for deployment ⓘ

☐ Azure Resource Manager for template deployment ⓘ

☐ Azure Disk Encryption for volume encryption ⓘ

Figure 7.3 – Key Vault access configuration

The Azure Key Vault service supports two access permission models: **vault access policy** is the older (legacy) option. **Azure role-based access control** is the preferred configuration. User roles are defined to allow read or write access to different Key Vault objects such as keys, secrets, and certificates. Another setting in this category is for allowing **Azure Resource Manager** (**ARM**)-based deployment (which includes Bicep); for this specific resource, access needs to be granted.

After the key vault creation succeeds, you can create and import **secrets**, **keys**, and **certificates**. In this chapter, we just use the secrets from the Key Vault service. Create a secret, as shown in *Figure 7.4*:

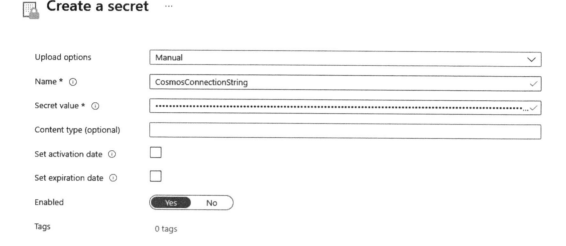

Figure 7.4 – Creating a secret

In addition to the name and the secret value, you can set activation and expiration dates.

After creating the secret, we can switch back to the Azure App Configuration service. Create a Key Vault reference to map a configuration value to a value from the Azure Key Vault service (as shown in *Figure 7.5*):

Create   ×

Create a new Key Vault reference

Key *

| ConfigurationPrototype:ConnectionStrings:Cosmos   ✓ |

Label

| 🔍 (No label) |

Browse   Input
_____

Subscription

| Visual Studio - MVP   ⌄ |

Resource group

| rg-aspire-empyrean-codebreaker.apphost   ⌄ |

Key Vault *

| codebreakervaultolttwmle   ⌄ |

Secret *

| CosmosConnectionString   ⌄ |

Secret version

| Latest version   ⌄ |

Figure 7.5 – Mapping Key Vault secrets with App Configuration

Adding a Key Vault reference from the Configuration explorer, key values can be specified that correspond to the configuration keys, but for the value, a Key Vault resource and a secret are referenced.

To connect the App Configuration service to the Key Vault service, the App Configuration service needs to be updated:

ConfigurationPrototype/Program.cs

```
builder.Configuration.AddAzureAppConfiguration(appConfigOptions =>
{
  DefaultAzureCredentialOptions credentialOptions = new();
```

```
DefaultAzureCredential cred = new();
string appConfigUrl = builder.Configuration.
  GetConnectionString("codebreakerconfig") ?? throw new
  InvalidOperationException("could not read codebreakerconfig");
appConfigOptions.Connect(new Uri(appConfigUrl), cred)
  .ConfigureKeyVault(keyVaultOptions =>
  {
    keyVaultOptions.SetCredential(cred);
  });
});
```

The Connect method of the AzureAppConfigurationOptions class is a fluent API that returns the same options type. With this, the ConfigureKeyVault method is now invoked to connect the Key Vault service to the same App Configuration resource. The SetCredential method defines the credentials that should be used to access the secrets. Here, we use the same credentials as used with the App Configuration service, but it's also possible to use different credentials.

With this configuration in place, secrets can be accessed in the same way as other configuration values:

ConfigurationPrototype/Program.cs

```
app.MapGet("/secret", (IConfiguration config) =>
{
  string? connectionString = config.
    GetSection("ConfigurationPrototype").GetConnectionString("Cosmos");
  return $"Configuration value from Azure Key Vault via App
    Configuration: {connectionString}";
});
```

Having the Key Vault service connected to the App Configuration service, we can use the same configuration API we used previously. Behind the scenes, different access mechanisms are used.

Run the application and check how secrets can be retrieved successfully using the DefaultAzureCredential type.

Before we integrate the App Configuration and Key Vault services with our game APIs and bot services, we can get rid of some needed secrets with configuration values using managed identities.

Reducing the need for secrets with managed identities

Managed identities (now known by the full name **Microsoft Entra managed identities for Azure resources**) remove the hassles we had with service principals. Managed identities abstract service principals, creating and deleting them automatically.

Using an Azure service (such as Azure Container Apps), the identity of the service can be configured to run with a managed identity. Services that are accessed (such as Azure App Configuration) use role management, whereby you configure who has access to this resource – which includes a simple option to select a managed identity.

The kinds of managed identities that are available are **system-assigned managed identities** and **user-assigned managed identities**:

- A system-assigned managed identity is directly associated with the Azure resource. If the Azure resource is deleted, the managed identity and its role-based access are removed as well.
- A user-assigned managed identity is created independent of an Azure service. As with other Azure resources, user-assigned managed identities are resources within a resource group.

Each of these two options has advantages but also disadvantages.

Properties and advantages of system-assigned managed identities include the following:

- They have the same lifetime as the service
- Deleting the service also deletes the managed identity and its role assignments

Advantages of user-assigned managed identities include the following:

- One user-assigned managed identity can be used by multiple services.
- Deleting the service does not delete the managed identity – it can be used from other services.
- Multiple services can use the same managed identity. If multiple services need the same permissions, you only need to specify this once with the shared managed identity.

One service can use multiple user-assigned managed identities. This also includes a disadvantage: using a user-assigned managed identity requires you to configure the principal ID to specify which managed identity to use.

Figure 7.6 shows a user-assigned managed identity that is used with the bot service and the games API to access the App Configuration and Key Vault services:

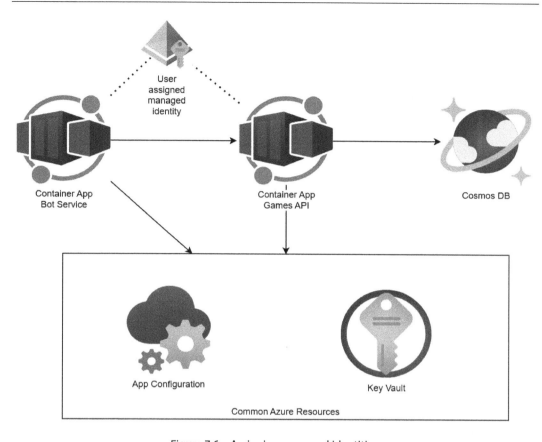

Figure 7.6 – Assigning managed identities

Let's create a managed identity and assign permissions to this managed identity.

Creating a managed identity and assigning roles

Running the application from the local system, managed identities are not used. To use managed identities within Azure, let's deploy the solution to Azure, as described in the *Technical requirements* section.

After the resources are successfully deployed, open the Azure Container App service for the game APIs with the Azure portal and select **Identity** in the **Settings** section, as shown in *Figure 7.7*:

System assigned   **User assigned**

User assigned managed identities enable Azure resources to authenticate to cloud services
type of managed identities are created as standalone Azure resources, and have their own I
multiple user assigned managed identities. Similarly, a single user assigned managed identi
Machine).

$+$ Add   🗑 Remove   ⟳ Refresh | ᷓ Got feedback?

Name		Resource group
☐ mi-533gstq35zse4		rg-codebreaker-07

Figure 7.7 – Managed identity

The system-assigned identity is turned off, but a user-assigned managed identity is created. If you
open the identity configuration with the other container apps, you can see that the same managed
identity is assigned to all these apps, making it easy to define permissions.

Clicking on this managed identity, select **Add role assignment**, as shown in *Figure 7.8*:

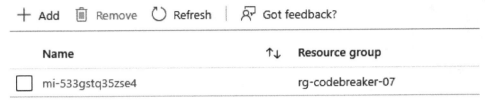

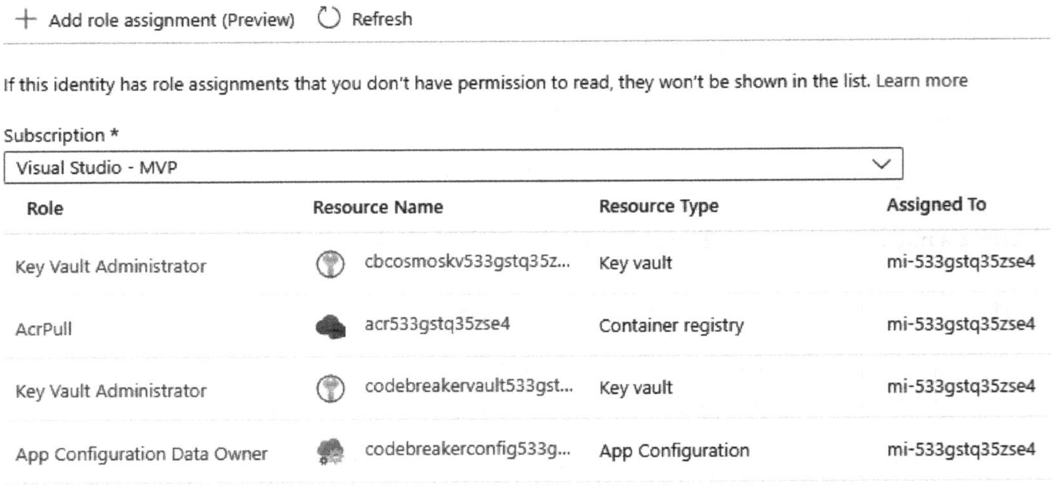

$+$ Add role assignment (Preview)   ⟳ Refresh

If this identity has role assignments that you don't have permission to read, they won't be shown in the list. Learn more

Subscription *

Visual Studio - MVP   ⌄

Role	Resource Name	Resource Type	Assigned To
Key Vault Administrator	🔵 cbcosmoskv533gstq35z...	Key vault	mi-533gstq35zse4
AcrPull	☁ acr533gstq35zse4	Container registry	mi-533gstq35zse4
Key Vault Administrator	🔵 codebreakervault533gst...	Key vault	mi-533gstq35zse4
App Configuration Data Owner	🔵 codebreakerconfig533g...	App Configuration	mi-533gstq35zse4

Figure 7.8 – Role-based access for the managed identity

Here, you see that this managed identity already has several roles assigned – it can pull Docker images
from the Azure Container Registry service, which is needed on deployment of the Azure Container
App service, it can access the Azure Key Vault, and it has access to the Azure App Configuration service
with the **App Configuration Data Owner** role. This allows setting configuration values that could be
changed to read access if the configuration values are not set by an application running with this identity.

> **Note**
>
> You might wonder why the managed identity has the administrator role assigned to the Key Vault service and the **App Configuration Data Owner** role with the App Configuration service. This managed identity is also used during deployment. When the Azure Container Apps service is deployed, a secret containing the connection string of the Azure Cosmos DB database is added to the Key Vault service. Specifying configuration values can be supplied to the App Configuration service.
>
> To support the **principle of least privilege** (**PoLP**), only the necessary permissions should be applied. You can create multiple managed identities for different container apps or use system-assigned managed identities, whereby every container app has a different identity, and specify the required roles for every identity. A container app can have just one system-assigned managed identity but multiple user-assigned managed identities. Different identities can be used for deployment and while running the application.

Let's get back to some C# code – the games API, the bot service, and the configuration prototype startup code to configure Azure App Configuration with a managed identity.

Configuring the Azure App Configuration provider with managed identities

Previously, with the `ConfigurationPrototype` project, we already used the `AddAzureAppConfiguration` method overload, which doesn't require a connection string containing a secret. Invoking the `Connect` method, we supplied a `DefaultAzureCredential` instance. Using a user-assigned managed identity, a change is needed here. An application can have just one system-assigned managed identity assigned but multiple user-assigned managed identities. The one we use needs to be specified.

Let's check the configuration that has been applied to Azure Container Apps. Open the configuration prototype container app in the Azure portal, and within the **Settings** category, open **Secrets**, as shown in *Figure 7.9*:

+ Add   ◯ Refresh   ⋈ Send us your feedback

Secrets are key/value pairs that can be used to protect sensitive data like passwords and connection strings. Secrets that you store here will be valid across all your revisions. Note that changing secrets will not create a new revision.

Key ↓	Value
connectionstrings--codebreakervault	👁 Click to show value
connectionstrings--codebreakerconfig	⟳ https://codebreakerconfig533gstq.azconfig.io

Figure 7.9 – Secrets with Azure Container Apps

The connection strings to the Azure App Configuration and Azure Key Vault services are stored with the **Secrets** configuration. This wouldn't really be required as secret keys are not part of these links – but it would help if the configuration were changed to not only the endpoint link but also the connection string containing the endpoint and the secret.

Because the secret is not part of this secret configuration, check the environment variables configured. This setting is available in the **Application** category. Click on **Containers** and select **Environment variables**, as shown in *Figure 7.10*:

Properties	**Environment variables**	Health probes	Volume mounts	Bindings

Search		
Name ↑	**Source**	**Value**
AZURE_CLIENT_ID	Manual entry	dc57230e-813a-43e2-925c-d144cac8d4e4
ASPNETCORE_FORW...	Manual entry	true
OTEL_DOTNET_EXPER...	Manual entry	true
OTEL_DOTNET_EXPER...	Manual entry	true
ConnectionStrings__c....	Reference a secret	connectionstrings--codebreakerconfig
ConnectionStrings__c...	Reference a secret	connectionstrings--codebreakervault

Figure 7.10 – Environment variables with Azure Container Apps

The identifier of the user-assigned managed ID is passed as an environment variable named AZURE_CLIENT_ID. This environment variable can be used to select the managed identity. Let's use this to configure the DefaultAzureCredential object. We used this class earlier, but now it's time to investigate the different options offered. DefaultAzureCredential uses accounts in this order:

- EnvironmentCredential – This authentication needs environmental variables containing client IDs, tenant IDs, and secrets to be set. We don't use this here.

- WorkloadIdentityCredential – When running on **Azure Kubernetes Service** (**AKS**), Microsoft Entra workload identities can be enabled.

- ManagedIdentityCredential – This is the authentication used when the application runs with managed identities configured within Microsoft Azure.

- SharedTokenCacheCredential – This is a legacy mechanism that has been replaced by VisualStudioCredential.

- `VisualStudioCredential` – Using Visual Studio, in the options dialog, you can configure the account to be used with Azure Service authentication. This is the account used with `VisualStudioCredential`. Just make sure within Visual Studio that you don't need to re-authenticate – otherwise, authentication via `DefaultAzureCredential` might not succeed.

- `VisualStudioCodeCredential` – This is a similar mechanism to `VisualStudioCredential` for Visual Studio Code but doesn't work with the current version of the *Azure Account extension*. A new authentication mechanism is going to be built for Visual Studio Code, but this requires some time before it's ready. Using Visual Studio Code, use the next option.

- `AzureCliCredential` – This is the account used by the Azure CLI. With the `az account list` command, you can see the Azure accounts and subscriptions you are logged in to. `az account show` gives you the default account and subscription that will be used. If this is not the correct one, use `az account set` to set the current active subscription.

> **Note**
>
> In case you have issues using the `DefaultAzureCredential` class in your development environment, you can enable diagnostic information and also enable or disable specific accounts explicitly to find the issues. Check this troubleshooting guide in case you have any errors: `https://github.com/Azure/azure-sdk-for-net/blob/main/sdk/identity/Azure.Identity/TROUBLESHOOTING.md`.

Let's update the configuration to use Azure App Configuration while the application is running within Azure:

ConfigurationPrototype/Program.cs

```
builder.Configuration.AddAzureAppConfiguration(appConfigOptions =>
{
#if DEBUG
  DefaultAzureCredential credential = new();
#else
  string managedIdentityClientId = builder.Configuration["AZURE_
    CLIENT_ID"] ?? string.Empty;
  DefaultAzureCredentialOptions credentialOptions = new()
  {
    ManagedIdentityClientId = managedIdentityClientId,
    ExcludeEnvironmentCredential = true,
    ExcludeWorkloadIdentityCredential = true
  };
  DefaultAzureCredential credential = new(credentialOptions);
```

```
#endif
  string appConfigUrl = builder.Configuration.
    GetConnectionString("codebreakerconfig") ??
    throw new InvalidOperationException("could not read
    codebreakerconfig");
  appConfigOptions.Connect(new Uri(appConfigUrl), credential)
    .Select("ConfigurationPrototype*")
    .ConfigureKeyVault(keyVaultOptions =>
    {
      keyVaultOptions.SetCredential(cred);
    });
});
```

DefaultAzureCredential not only works in the development environment but also when the application runs within Azure. Using system-assigned managed identities, a change would not be required. With user-assigned managed identities, the ManagedIdentityClientId property needs to be set to the ID of the managed identity. We do this by reading the AZURE_CLIENT_ID environment variable and passing the value to this setting.

Using the Select method with the AzureAppConfigurationOptions class returned from the Connect method filters the configuration value. Because configuration values are specified for all services of the solution, we just need the ones that start with the ConfigurationPrototype key. With the bot and the game API services, the filtering is done with the bot and gameapis keys.

Run the configuration prototype with these changes, and then let's continue using .NET environments with App Configuration.

Using environments with Azure App Configuration

The Azure container apps are deployed and running, using all the Azure services we have created so far. What's missing with App Configuration are the different environments that are supported with the .NET configuration. Is the application running in the local development environment, in the Azure test environment, or on the production server? Running in the test environment, the production database should not be used.

.NET configuration supports different environments – depending on the environment, either appsettings.development.json or appsettings.production.json is loaded. Similar functionality is possible with Azure App Configuration using **labels**. We can specify development, production, and testing labels to differentiate environment configurations. This can be mapped to .NET environments.

> **Note**
>
> It's a good practice to separate the production and the development environments across different Azure subscriptions, probably also using different Azure Active Directory services. Here, you also use separate Azure App Configuration services. Some environments can use the same subscription; for example, the production and the staging environments can be configured to run in the same subscription. In such cases, labels can be used to map different configuration values to environments.

Using App Configuration labels to map .NET environments

In the Azure portal, open the Azure App Configuration service again. Create a new key-value pair and use the `BotService` key again, but this time, set the label to `Development`. The default setting of this key should contain the `ApiBase` configuration to the games API running in the container app, whereas the `Development` label should reference `localhost`.

With the startup code of the bot service, we can now change the filtering code:

Codebreaker.Bot/Program.cs

```
builder.Configuration.AddAzureAppConfiguration(options =>
{
  options.Connect(new Uri(endpoint), credential)
    .Select("BotService*", labelFilter: LabelFilter.Null)
    .Select("BotService*", builder.Environment.EnvironmentName);
});
```

Invoking the `Select` method multiple times works the same way as you saw at the beginning of this chapter with multiple configuration providers. If a setting is configured multiple times, the last one wins. The first `Select` method loads all configuration values where the key starts with `BotService`, and no label filter is applied. Next, all configuration values where, again, the key starts with the name `BotService` are loaded, but this time, only values where a label with the same name as the current environment name are loaded. All configuration values that are not overwritten from the specific environment label are unchanged – the value is the active one. With all keys with a matching label, the new value is now active.

This is all that needs to be done to map different environment configuration values with the Azure App Configuration service.

> **Note**
>
> If you don't need the Azure resources for some time, delete the resource group. In the next chapter, we'll re-create the services again. `azd up` makes this easy!

Summary

This was quite a journey around using Azure services for common needs such as Azure App Configuration and Azure Key Vault in relation to configuration with .NET. You learned how .NET configuration offers features to attach different providers and used Azure App Configuration for storing configuration values for a large list of services. The Azure Key Vault service was used to store secrets. In addition to this, you learned about using managed identities, which can help get rid of a lot of secrets.

In this chapter, we used the Azure Developer CLI to create Docker images, publish them to the Azure Container Registry service, and create new replicas of Azure container apps using the new image. While `azd up` makes this easy, this can be automated. This is of special interest in testing, staging, and production environments. In the next chapter, we will automate these activities using GitHub Actions. There's also more that can be done with Azure App Configuration – using feature flags with modern deployment patterns. This is covered in *Chapter 8* as well.

Further reading

To learn more about the topics discussed in this chapter, you can refer to the following links:

- Configuration providers in .NET: `https://learn.microsoft.com/en-us/dotnet/core/extensions/configuration-providers`
- Options pattern in .NET: `https://learn.microsoft.com/dotnet/core/extensions/options`
- Microsoft Entra managed identities: `https://learn.microsoft.com/azure/active-directory/managed-identities-azure-resources/`
- Troubleshooting with `AzureDefaultCredential`: `https://github.com/Azure/azure-sdk-for-net/blob/main/sdk/identity/Azure.Identity/TROUBLESHOOTING.md`
- Built-in roles: `https://learn.microsoft.com/azure/role-based-access-control/built-in-roles`
- Azure Key Vault keys, secrets, and certificates: `https://learn.microsoft.com/en-us/azure/key-vault/general/about-keys-secrets-certificates`
- PoLP: `https://learn.microsoft.com/en-us/entra/identity-platform/secure-least-privileged-access`
- Microsoft Entra Workload ID: `https://learn.microsoft.com/en-us/azure/aks/workload-identity-overview`

8

CI/CD – Publishing with GitHub Actions

One of the features of microservices is their ability to continuously build and deploy services. In the previous chapters, we automatically created the infrastructure that's used by our service solution.

In this chapter, we'll continue to automatically build and update services and use protection rules before deploying applications to staging and production environments. While doing this, you'll learn how to use feature flags with Azure App Configuration.

In this chapter, you'll learn how to do the following:

- Use GitHub Actions
- Build and test the application automatically after a pull request
- Deploy the application to test environments
- Use deployment protection rules before deploying the application to production environments
- Publish NuGet packages
- Use feature flags with modern deployment patterns

Technical requirements

In this chapter, similar to the previous chapter, you'll need an Azure subscription, the Azure CLI, the Azure Developer CLI, and .NET Aspire. You'll also need your own GitHub repository so that you can store secrets, create environments, and run GitHub actions. These features are available in public repositories. If you create a private repository, the GitHub Team feature is required for creating environments (see https://github.com/pricing).

The source code for this chapter can be found in this book's GitHub repository: https://github.com/PacktPublishing/Pragmatic-Microservices-with-CSharp-and-Azure.

The ch08 folder contains the following projects, along with the output for this chapter:

- Codebreaker.GameAPIs: The game-apis project we used in the previous chapter has been enhanced using feature flags.

- Codebreaker.Bot: This is the implementation of bot-service, which plays games.

- Codebreaker.GameAPIs.KiotaClient: This is the client library we created in *Chapter 4* to be used by clients.

- Workflows: This folder is new. Here, you will find all the GitHub Actions workflows. However, these don't become active until you copy them to the .github/workflows folder in your repository.

To work through the code with this chapter, you can use the service and bot projects from the previous chapter, as well as the Kiota library from *Chapter 4*.

For this chapter, you'll need GitHub rights to run GitHub workflows, as well as to create and use GitHub environments with protection rules. The easiest way to do this is to create a public repository and copy just the code from this chapter into it. Create the src folder in this new repository and copy the source code to this folder.

Check out the README file in the ch08 folder of this book's GitHub repository for the latest updates.

Preparing the solution using the Azure Developer CLI

First, let's prepare the solution using the Azure Developer CLI. When initializing the solution, set the current folder to the root folder of the repository (not the folder of the solution file, as we did previously):

```
azd init
```

Select **Use code in the current directory**, confirm that you wish to use **Azure Container Apps**, select **Continue initializing my app**, select bot and game-apis as projects to be exposed to the internet, and enter a new environment name – for example, codebreaker-08-dev. The generated azure. yaml file, which contains a link to the AppHost project file, needs to be committed to the source code repository. The generated .azure folder can contain secrets and has been – because of the generated .gitignore file – excluded from the source code repository.

> **Note**
>
> The reason to use the root directory of the azd pipeline command used later; At the time of writing, this command requires the .github/workflows directory to be in the same folder. Some changes are planned for a later release, so please check the README file for this chapter for updates.

Now, let's deploy the resources to Azure:

```
azd auth login
azd up
```

With `azd up`, the resources are deployed to your configured environment. Select the Azure subscription you wish to use and the Azure region where you want to deploy the resources.

The generated file, `azure.yaml`, references the AppHost project. The generated folder, `.azure` (which has been excluded from the source code repository because of possible secrets being stored), contains the current environment and a folder that has the same name as the environment. This folder contains the `config.json` file, which lists the publicly accessible service configuration, and the `.env` file, which contains variables referencing the created Azure resources.

Now, we are ready to use GitHub Actions. You can remove the Azure resources with `azd down` again since the complete infrastructure should have been deployed via GitHub Actions:

```
azd down
```

Answer y to delete the resources, and then y again to permanently delete the resources that have soft delete enabled.

If you want to permanently delete resources, open the Azure portal (`https://portal.azure.com`), go to **Key Vault**, and click on **Manage deleted vaults**. Key vaults that are deleted need to be purged so that you can create a resource with the same name again. Purge the key vaults. Similarly, check for Azure App Configuration services that need to be purged.

Exploring GitHub Actions

GitHub Actions is a feature of GitHub that you can use to automatically build, test, and deploy source code. GitHub Actions is a product that consists of *workflows*, *events*, *jobs*, *actions*, and *runners*:

- A **workflow** is a YAML file stored in the `.github/workflows` folder of a repository. A workflow contains events and jobs.

- An **event** specifies what triggers a workflow. When should the workflow be started?

- A **job** consists of steps that are executed on a **runner** machine.

- A **step** can run a script or an action.

- An **action** is a reusable GitHub extension that reduces the need to write scripts. Many of these reusable extensions can be used to build and deploy applications.

Now that we've set the foundation with these terms, let's get into the details by creating a workflow using the Azure portal.

Creating a GitHub Actions workflow

There are several options to automatically create GitHub Actions workflows to deploy services to Microsoft Azure. Using the Azure portal, upon opening **Container App**, you can select **Continuous deployment** under **Settings**, as shown in *Figure 8.1*:

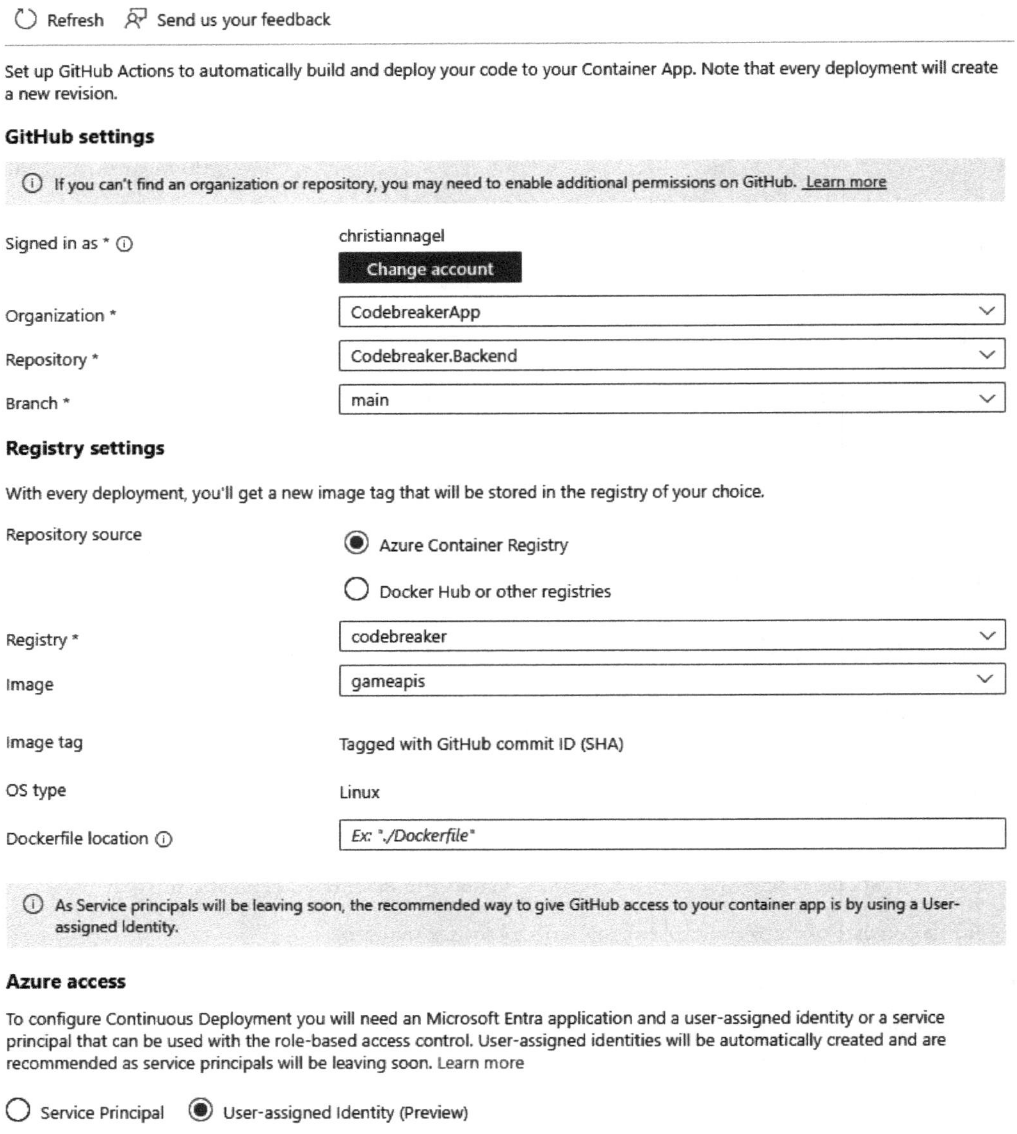

○ Refresh   ⊠ Send us your feedback

Set up GitHub Actions to automatically build and deploy your code to your Container App. Note that every deployment will create a new revision.

GitHub settings

ⓘ If you can't find an organization or repository, you may need to enable additional permissions on GitHub. Learn more

Signed in as * ⓘ   christiannagel

                               Change account

Organization *   CodebreakerApp   ⌄

Repository *   Codebreaker.Backend   ⌄

Branch *   main   ⌄

Registry settings

With every deployment, you'll get a new image tag that will be stored in the registry of your choice.

Repository source   ◉ Azure Container Registry

                            ○ Docker Hub or other registries

Registry *   codebreaker   ⌄

Image   gameapis   ⌄

Image tag   Tagged with GitHub commit ID (SHA)

OS type   Linux

Dockerfile location ⓘ   Ex: "./Dockerfile"

ⓘ As Service principals will be leaving soon, the recommended way to give GitHub access to your container app is by using a User-assigned Identity.

Azure access

To configure Continuous Deployment you will need an Microsoft Entra application and a user-assigned identity or a service principal that can be used with the role-based access control. User-assigned identities will be automatically created and are recommended as service principals will be leaving soon. Learn more

○ Service Principal   ◉ User-assigned Identity (Preview)

Figure 8.1 – Creating a GitHub Actions workflow from the Azure portal

Using the Azure portal, you can select the GitHub repository, configure the Azure Container Registry you wish to use, and specify a **Service Principal** or a **User-assigned Identity** value to be used to publish the project.

Another option is to use Visual Studio. With Visual Studio, you can select a project (for example, `game-apis`) and select **Publish...** from the context menu. Upon adding a new publish profile, which you can do by selecting **Azure | Azure Container Apps (Linux)**, then selecting **Container App**, then Container Registry, the following dialogue appears:

Figure 8.2 – Creating a GitHub Actions workflow via Visual Studio

From this dialogue, you can directly publish to the Azure Container App or create a GitHub Actions workflow.

What's common with these options is that you can publish service by service. Here, you used `azd up` to deploy the complete solution. Let's have a look at what the Azure Developer CLI has to offer to create GitHub Actions workflows.

First, you need to create a `.github` folder in the root directory of the repository. Files that are used by specific GitHub functionality are stored in this folder. To this folder, add a `workflows` folder (`.github/workflows`). All the GitHub Actions workflows need to be stored within this folder.

Next, create the `codebreaker-deploy.yml` file. Now, copy the content of the `azure-deploy.yaml` file to this file. This file is from the *Azure-Samples* repository: `https://github.com/Azure-Samples/azd-starter-bicep/blob/main/.github/workflows/azure-dev.yml`.

Now that we've created this workflow file, we can take a closer look at it.

Workflow file with YAML syntax

The syntax of workflow files makes use of **YAML Ain't Markup Language** (**YAML**, a recursive acronym) syntax. YAML is a data-oriented human-readable serialization language that uses indentation to specify what belongs together.

See `https://yaml.org/` for the YAML spec and links to libraries. You can check out the following cheat sheet for the syntax: `https://yaml.org/refcard.html`.

Let's take a closer look at the workflow file while making some small changes.

Triggers

A workflow file starts with a name followed by a trigger:

workflows/codebreaker-deploy.yml

```yaml
name: Codebreaker backend workflow

on:
  workflow-dispatch:
  push:
    branches:
      - main
    paths:
    - 'src/**'
```

The name of a workflow is shown in the list of workflows. The `on` keyword specifies the events that trigger the workflow. GitHub offers many events that can be used with workflows. In this YAML file, the workflow is triggered with a `workflow_dispatch` event. This allows you to manually trigger the workflow. The second event, `push`, is triggered when changes are pushed to the repository. Because of the filtering that follows as part of `push`, the trigger is only done with a push to the `main` branch with changes specified by the files specified with `path`. If we don't specify branches and path filters, the workflow will be triggered with every change in this repository.

Permissions for secretless Azure federated credentials

The `permissions` section is a new construct that's used with secretless Azure federated credentials to deploy to Azure:

workflows/codebreaker-deploy.yml

```yaml
permissions:
  id-token: write
  contents: read
```

Permissions are used to access the identity token and the content. With `contents read`, the workflow has read access to the content of the repository. `id-token write` grants write access to the identity token. This token is used to authenticate GitHub with Azure.

Jobs and runners

In the workflow file, after the trigger is defined, the `jobs` keyword can be used to list one or more jobs that should run:

workflows/codebreaker-deploy.yml

```
jobs:
  build-and-deploy:
    runs-on: ubuntu-latest
    env:
      AZURE_CLIENT_ID: ${{ vars.AZURE_CLIENT_ID }}
      AZURE_TENANT_ID: ${{ vars.AZURE_TENANT_ID }}
      AZURE_SUBSCRIPTION_ID: ${{ vars.AZURE_SUBSCRIPTION_ID }}
      AZURE_ENV_NAME: ${{ vars.AZURE_ENV_NAME }}
      AZURE_LOCATION: ${{ vars.AZURE_LOCATION }}
```

`build-and-deploy` is the name of the job. A job needs a runner. GitHub offers hosted runners to run jobs on Linux, Windows, and Mac. You can find out what runners are available, as well as their versions, at: `https://docs.github.com/en/actions/using-github-hosted-runners/about-github-hosted-runners/about-github-hosted-runners#supportedrunners-and-hardware-resources`. If other hardware or operating system versions are needed, a custom runner can be used.

Using the `env` keyword, environment variables are defined that can be used with the steps in this runner. The values for these variables come from the GitHub project variables using the `vars` object. The `${{ }}` expressions are evaluated during the execution of the workflow, and the values that are retrieved are added to the workflow at runtime. We will specify these values later using `azd pipeline config`.

Steps and actions

A job consists of steps and actions:

workflows/codebreaker-deploy.yml

```
jobs:
  build-and-deploy:
    runs-on: ubuntu-latest
```

```
    steps:
      - name: Checkout
        uses: actions/checkout@v4

      - name: Install azd
        uses: Azure/setup-azd@v1.0.0

      - name: Install .NET Aspire workload
        run: dotnet workload install aspire
# Code removed for brevity
```

The first step consists of an action, `actions/checkout@v4`. This action checks out the source code to ensure it's available alongside the runner. `@v4` defines the version number to be used for this GitHub action. Actions are available via GitHub Marketplace: `https://github.com/marketplace?category=&query=&type=actions`. Every action has documentation that you can read to learn which parameters are available. `actions/checkout`, for example, allows you to include submodules at checkout and also allows you to check out source code from other repositories.

The next action installs the Azure Developer CLI using `Azure/setup-azd`.

.NET is installed with the hosted runner, but we need to make sure we install the .NET Aspire workload. This can be done using a one-line script specified with the `run` field – that is, `dotnet workload install aspire`.

Next, the `azd auth` command is used:

---

workflows/codebreaker-deploy.yml

```
      - name: Log in with Azure (Federated Credentials)
        if: ${{ env.AZURE_CLIENT_ID != '' }}
        run: |
          azd auth login `
            --client-id "$Env:AZURE_CLIENT_ID" `
            --federated-credential-provider "github" `
            --tenant-id «$Env:AZURE_TENANT_ID»
        shell: pwsh
```

Here, we have the step where the permissions we defined earlier are needed: authentication with Federated Identity. `if` specifies that this step is conditional – only if `AZURE_CLIENT_ID` is not empty. We'll have another option for authentication if `AZURE_CLIENT_ID` is empty with the following step. This step is not using a GitHub action; instead, the `run` field defines that it will invoke a multi-line script. Using multiple lines is specified by `|` at the end of the line. PowerShell (which is specified with the `shell` field) uses the backtick (`` ` ``) as a line continuation character.

The command that's running with PowerShell is `azd auth login`, passing a few parameters. `--client-id` uses the identifier of the service principal, which has the necessary Azure permissions. `--federated-credential-provider` uses GitHub for federated authentication. Federated authentication allows you to use GitHub identities to access resources on Azure. `--tenant-id` specifies the Azure directory identifier that is used with Azure authentication.

The next step uses another conditional script if the AZURE_CREDENTIALS variable is set:

workflows/codebreaker-deploy.yml

```
- name: Log in with Azure (Client Credentials)
  if: ${{ env.AZURE_CREDENTIALS != '' }}
  run: |
    $info = $Env:AZURE_CREDENTIALS | ConvertFrom-Json -AsHashtable;
    Write-Host "::add-mask::$($info.clientSecret)"

    azd auth login `
      --client-id "$($info.clientId)" `
      --client-secret "$($info.clientSecret)" `
      --tenant-id "$($info.tenantId)"
  shell: pwsh
  env:
    AZURE_CREDENTIALS: ${{ secrets.AZURE_CREDENTIALS }}
```

The AZURE_CREDENTIALS environment variable is stored as a JSON script. This is converted into a PowerShell hash table variable named `info` and allows us to access each part of the JSON content, such as the client ID, the client secret, and the tenant ID. These parts are then passed to the `azd auth login` command. AZURE_CREDENTIALS itself is retrieved using `secrets.AZURE_CREDENTIALS`. Secrets are stored encrypted alongside the GitHub project and are not part of the source code of the repository. We'll configure these secrets in the next section.

Finally, two one-line commands are invoked:

workflows/codebreaker-deploy.yml

```
    - name: Provision Infrastructure
      run: azd provision --no-prompt

    - name: Deploy Application
      run: azd deploy --no-prompt
```

The first command, `azd provision`, creates the Azure infrastructure, as specified by the app-model definition of the AppHost project. If the infrastructure already exists, it is checked if a change is needed, and only updates are applied. `azd deploy` then deploys the services to Azure, thus building the Docker images, publishing them to Azure Container Registry, and creating various Azure Container Apps with the created images. The `--no-prompt` option doesn't wait for the user to interact with this command and just uses defaults.

The `azd up` command we used previously used `azd provision` and `azd deploy`.

Create the `.github/workflows` folder in your repository and copy the `codebreaker-deploy.yml` workflow file to this folder.

GitHub variables and secrets

Secrets shouldn't be part of the source code, and the workflow file is stored with the source code. GitHub has a vault where you can store secrets outside of the source code repository. Using `azd`, secrets and variables can automatically be configured with GitHub.

The `azd pipeline` command supports this. Start by having the current directory set to the root folder of the repository. `azd pipeline` needs the `.github/workflows` folder to be in the same directory where you run `azd pipeline`; you must also have the .NET Aspire application initialized in this directory. This requirement might change – check the README file in this chapter's GitHub repository for more information.

Run the following command to configure the pipeline:

```
azd auth login
azd pipeline config --auth-type federated --principal-name github-
codebreaker-dev
```

`azd pipeline config` uses the Azure subscription you configured earlier to create GitHub variables and secrets, as well as create an Azure app registration, which allows GitHub to deploy to Azure. By default, the principal name created starts with `az-dev-` and contains the date and time it was created. Here, we specify the principal name as `github-codebreaker-dev`. Repository variables for `AZURE_ENV_NAME`, `AZURE_LOCATION`, `AZURE_SUBSCRIPTION_ID`, `AZURE_TENANT_ID`, and `AZURE_CLIENT_ID` are created.

Open the GitHub repository in your browser. In the portal, click **Settings**. With the **Security** category open in the left pane, you'll see **Secrets and variables**. In this sub-category, when you open **Actions**, you'll see the **Actions secrets and variables** page, which contains **Repository secrets**. This is shown in *Figure 8.3*:

Figure 8.3 – Repository secrets

The AZD_INITIAL_ENVIRONMENT_CONFIG secret contains the content of the .azure/[environment]/config.json file. This file contains a list of publicly accessible services and is read by azd deploy to configure the Ingress controller. The needed environment name, location, subscription ID, and other details are stored within repository variables.

Because azd pipeline config creates Federated Identity credentials for GitHub, secrets to access Azure are not required with the default (federated) configuration. Instead of using a value of federated, you can pass client-credentials, which configures credentials to be stored within a repository secret.

> **Note**
>
> In *Chapter 6*, you understood how to separate secrets and variables with Azure Container Apps, and then with Azure App Configuration and Key Vault in *Chapter 7*. The reason for this separation is similar here using GitHub Actions.

GitHub allows you to specify different levels where you can store secrets and variables. The organization level can be used when secrets should be shared across different repositories within the organization. Repository secrets are stored within the scope of the repository and are not available from other repositories. Environment secrets are scoped within deployment environments. These will be covered later in the *Using deployment environments* section.

Run the GitHub Actions workflow now – either by pushing a source code update to the GitHub repository or by running the workflow explicitly from the GitHub portal. You'll see that the workflow is in progress before it completes:

Figure 8.4 – Workflow in progress

At this stage, you might need to wait until an agent is available. When it's in progress, you can click on it to see progress information about what's going on. *Figure 8.5* shows the steps that appear when the workflow is completed successfully:

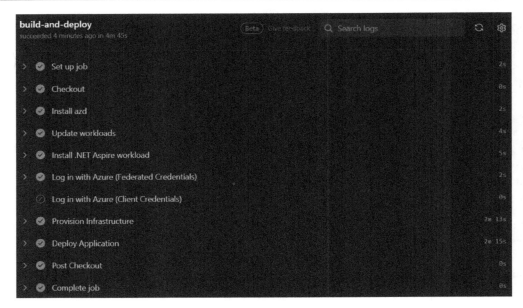

Figure 8.5 – Workflow steps

Upon checking the logs, you'll see all the steps that have been completed. In the preceding figure, you can see that federated credentials have been used, and not the client credentials. You can click on each of these steps to find more details.

When creating a GitHub Actions workflow with azd, just a few statements are required to deploy the complete solution. With every source code change that is not pushed to the main branch, the deployment is updated.

Getting more with GitHub Actions

It's great to have an easy way to create a GitHub action using integration from Visual Studio, the Azure portal, or via the azd pipeline command. azd pipeline is great for deploying a complete solution, but because of its early development stages, some features are missing. With our solution, some more features are required; we'll customize these manually.

Let's have a look at some of the goals we have:

- All the services should be built, tested, and deployed
- NuGet packages should be published to GitHub Packages and made available there
- We don't want to repeat code, so we'll create reusable workflows
- Deploying should be done to multiple environments, such as development, staging, and production

Let's get into the details.

Enhancing GitHub Actions workflows

To build our services, we must create reusable workflows. First, let's configure the variables and secrets that are needed by these workflows.

Configuring variables and secrets

We've configured variables and secrets using `az pipeline config` previously. if you need more customization, you might need to set these values yourself. You've already seen how to access the repository secrets and variables with the GitHub portal. Now, let's add these to secrets:

- `AZURE_TENANT_ID`
- `AZURE_SUBSCRIPTION_ID`
- `AZURE_CLIENT_ID`

To get the tenant ID, use the Azure CLI:

```
az account show --query tenantId -o tsv
```

`az account show` returns JSON information about the logged-in Azure account. With the JSONPath `--query tenantId` query, the Microsoft Entra tenant ID is returned. `-o tsv` returns the result in tab-separated values. Set the returned value with the `AZURE_TENANT_ID` repository secret.

The subscription ID can also be listed with `az account show`:

```
az account show --query id -o tsv
```

Here, `id` contains the subscription ID. Set this value with the `AZURE_SUBSCRIPTION_ID` repository secret.

Earlier in this chapter, we used the `azd pipeline` command to create an account for federated authentication. Let's check out this account within the Microsoft Entra portal: `https://entra.microsoft.com`. After logging in, from the left bar, within the **Identity** category, open **Applications** and click **App registrations**. Select the **All applications** tag. Look for an app registration with a display name starting with `github-codebreaker-dev`. If you didn't supply a name, `azd` creates an account starting with `az-dev`. Open this account and, within the **Manage** category, click **Certificates & secrets**. Open **Federated credentials**. You will see credentials named based on the GitHub organization and the repository, with entity types of **Pull request** and **Branch**. A predefined federated credential scenario for GitHub actions deploying Azure resources will be available.

> **Note**
>
> To create a new app registration with federated credentials using the Azure portal, the Azure CLI, or Azure PowerShell, check out the following documentation: `https://learn.microsoft.com/en-us/azure/developer/github/connect-from-azure?tabs=azure-portal%2Clinux#add-federated-credentials`.

Copy the value for **Application (client) ID** and set this identifier with the `AZURE_CLIENT_ID` repository secret.

Now that we've specified the necessary secrets and variables, let's get back to creating workflows.

Running unit tests

When triggering the workflow by updating the source code of a service, the first step should be to run unit tests. Let's create a reusable workflow:

workflows/shared-test.yml

```yaml
name: Shared workflow to build and test a .NET project
on:
  workflow_call:
    inputs:
      project-name:
        description: 'The name of the project'
        required: true
        type: string
      solution-path:
        description: 'The solution file of the project to build and
run tests'
        required: true
        type: string
      dotnet-version:
        description: 'The version of .NET to use'
        required: false
        type: string
        default: '8.0.x'
```

A reusable workflow is triggered by calling this workflow. The trigger specified by `on` uses the `workflow_call` keyword. At this point, the input values that are needed are also defined. With this workflow, `project-name` and `solution-path` are required input values. The `dotnet-version` input value has a default value assigned and is not required.

After the trigger and the input values, a job with a runner is defined, followed by the steps to be invoked:

work flows/shared-test.yml

```
jobs:
  run-test:
    runs-on: ubuntu-latest

    steps:
      - name: Checkout to the branch
        uses: actions/checkout@v4

      - name: Setup .NET
        uses: actions/setup-dotnet@v4
        with:
          dotnet-version: ${{ inputs.dotnet-version }}

      - name: Install .NET Aspire workload
        run: dotnet workload install aspire

      - name: Restore NuGet Packages
        run: dotnet restore ${{ inputs.solution-path }}

      - name: Run unit tests
        run: dotnet test --logger trx --results-directory
"TestResults-${{ inputs.project-name}}" --no-restore ${{ inputs.
solution-path }}

      - name: Upload the test results
        uses: actions/upload-artifact@v4
        with:
          name: test-results-${{ inputs.project-name}}
          path: TestResults-${{ inputs.project-name}}
        if: always()
```

After checking out the source code with the actions/checkout action, the .NET SDK is installed using actions/setup-dotnet. Here, the .NET version is retrieved from the input values. As the .NET SDK is now installed with this runner, the .NET CLI can be used. In the next step, instead of invoking an action, the run keyword is used to execute the dotnet restore command. dotnet restore retrieves the NuGet packages of the referenced solution. If this fails, there's no need to continue with the next step. The next step runs the unit tests using dotnet test. The --logger option specifies to write log output with the TRX logger format – a Visual Studio **Test Results File**

(**TRX**). This result file is then uploaded with the `actions/upload-artifact` action. Artifacts can be used to share data between runners and also downloaded with workflow runs. By default, a step only runs if the previous step succeeds. In this case, we want to download the test result from the artifacts if the test fails – that's why `if: always()` was added on uploading the artifact.

This shared workflow is started from the `codebreaker-test.yml` workflow:

workflows/codebreaker-test.yml

```
# code removed for brevity
jobs:
  build-and-test:
    uses: ./.github/workflows/shared-test.yml
    with:
      project-name: 'Codebreaker-Backend'
      solution-path: 'src/Chapter08.sln'
```

The job that's been defined uses the name `build-and-test`, references the shared workflow file with the `uses` keyword, and sets the input values using the `with` keyword.

This workflow is triggered when a change is made in the specified files and folders in the main branch, explicitly. *Figure 8.6* shows the result of running the workflow:

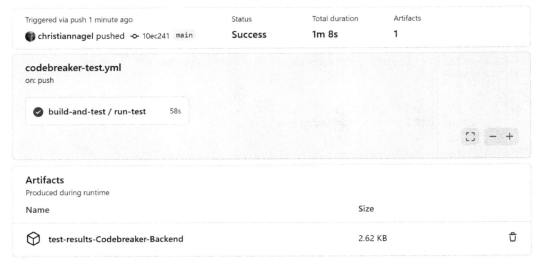

Figure 8.6 – Running the workflow

With this result, you can see the downloadable artifacts for viewing the test results.

Now that we've run the unit tests, let's combine this with the previously created build and deploy job.

Running multiple jobs

To run multiple jobs from one workflow, we'll need to create a shared workflow from the deployment project:

workflows/shared-deploy.yml

```
name: Shared workflow to deploy a .NET Aspire project

on:
  workflow_call:
    inputs:
# code removed for brevity
    secrets:
      AZURE_CLIENT_ID:
        required: true
      AZURE_TENANT_ID:
        required: true
      AZURE_SUBSCRIPTION_ID:
        required: true
```

Because this is a shared workflow that's triggered from other workflows, on specifies `workflow_call`. This workflow is very similar to the previously created deployment workflow, so the code for this hasn't been repeated here. Check out the source code repository for the complete workflow. What's important here is that not only inputs are passed from the calling workflow, but also secret information. These secrets are referenced using the `$ {{ secrets.<secret> }}` expression.

The `codebreaker-testanddeploy.yml` workflow invokes both shared workflows:

workflows/codebreaker-testanddeploy.yml

```
# code removed for brevity
jobs:
  build-and-test:
    uses: ./.github/workflows/shared-test.yml
    with:
      project-name: Codebreaker-Backend
      solution-path: src/Chapter08.sln

  build-and-deploy:
    needs: build-and-test
    uses: ./.github/workflows/shared-deploy.yml
    with:
      environment-name: ${{ vars.AZURE_ENV_NAME }}
      location: ${{ vars.AZURE_LOCATION }}
```

```
secrets:
  AZURE_CLIENT_ID: ${{ secrets.AZURE_CLIENT_ID }}
  AZURE_TENANT_ID: ${{ secrets.AZURE_TENANT_ID }}
  AZURE_SUBSCRIPTION_ID: ${{ secrets.AZURE_SUBSCRIPTION_ID }}
```

Using the `needs` keyword, the `build-and-deploy` job is defined to require the `build-and-test` job to run beforehand. If the `build-and-test` job does not succeed, `build-and-deploy` will not run. Secrets need to be forwarded to the shared workflow. With secrets, you can specify every secret to pass with the calling workflow, or to share all the secrets available from the calling workflow with the called workflow. Inheriting these secrets is required when using environments (as shown in a later section).

When you run the workflow at this stage, you'll see a graphical view of how the two jobs are connected, as shown in *Figure 8.7*:

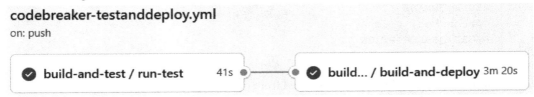

Figure 8.7 – Running multiple jobs

Both jobs were completed successfully.

Next, we'll dive into environments that can be used from multiple jobs to, for example, deploy the solution to staging and production environments.

Using deployment environments

When running the solution locally on the developer system, projects can be built and debugged locally. Just a few services, such as App Insights and Key Vault, need to be run in the Azure cloud environment. This is done automatically by .NET Aspire, which provisions `app-model` in the AppHost project. You just need to make sure you configure `Azure:SubscriptionId` with the user secrets. To run and test the application while it's running within Azure, and to try out different Azure offerings, every developer of the team can use `azd init` and `azd up` to have all the services running in the personal Azure subscription that's part of the Visual Studio Professional and Enterprise offerings.

It's also useful to use a shared environment where the services of the solution running in Microsoft Azure are used together by the developer team. One example is for client application developers to use a new daily build to test the client applications accessing the services in the cloud. This is the *development environment*.

To run load tests, it's useful to have *test environments*. Such environments can be created on demand before running the load test. They can be deleted again after the load tests are finished and the results have been documented. See *Chapter 10* for more details on running tests.

Before moving into production, *staging environments*, which mirror the *production environments*, are used to make final tests if the application is behaving as expected.

We can deploy the solution to all these environments by using GitHub Actions. However, some of these environments are more restrictive, which means that deployments can only be done when it's been verified that the solution runs successfully with the defined constraints.

Let's take a closer look.

Create environments with the Azure Developer CLI

To create environments with the Azure Developer CLI, you can use the `azd env new` command:

```
azd env new codebreaker-08-prod
```

This not only creates a new environment named `codebreaker-08-prod` but also sets the current environment to this new one. To show all environments that have been configured, run the following command:

```
azd env list
```

This shows all the environments that have been configured, as well as the currently selected environment. To change the current one, run the following command:

```
azd env select codebreaker-dev
```

Creating an environment with `azd` creates the `.azure` subdirectory. Upon opening this folder, you will see the `config.json` file. This shows the currently selected environment.

With every environment that's created, a subdirectory containing the name of the environment is created that contains values for the resource group, the Azure region, and the Azure subscription ID. When you're creating a new environment, you can change the subscription with the `--subscription` option. To change the location of the resources, use `--location`.

To see the configuration values for an environment, run the following command:

```
azd env get-values
```

To change the Azure region afterward, you can use `azd env set`:

```
azd env set AZURE_LOCATION eastus3
```

While the Azure Developer CLI supports using multiple environments, using this in combination with GitHub environments is (not yet) directly available but can be easily customized. At the time of writing, the `azd pipeline config` command only supports one environment per repository. However, this is expected to change, and integration with GitHub environments is already being discussed. Check the README file in this chapter's repository for updates.

You can still use `azd pipeline` to create federated accounts for every environment:

```
azd pipeline config --auth-type federated --principal-name github-
codebreaker-prod
```

This creates the account that we'll use with the `codebreaker-08-prod` environment.

At this point, we need to learn how to use GitHub environments. So, we'll start by creating GitHub environments.

Creating GitHub environments

Before using GitHub environments, you need to be aware that this GitHub feature is only available for free with public repositories. With private repositories, a Team license is required (see `https://github.com/pricing`).

Open your GitHub repository in your browser and click **Settings**. In the left pane, under the **Code and automation** category, click **Environments**. *Figure 8.8* shows the environments for development, testing, staging, and production:

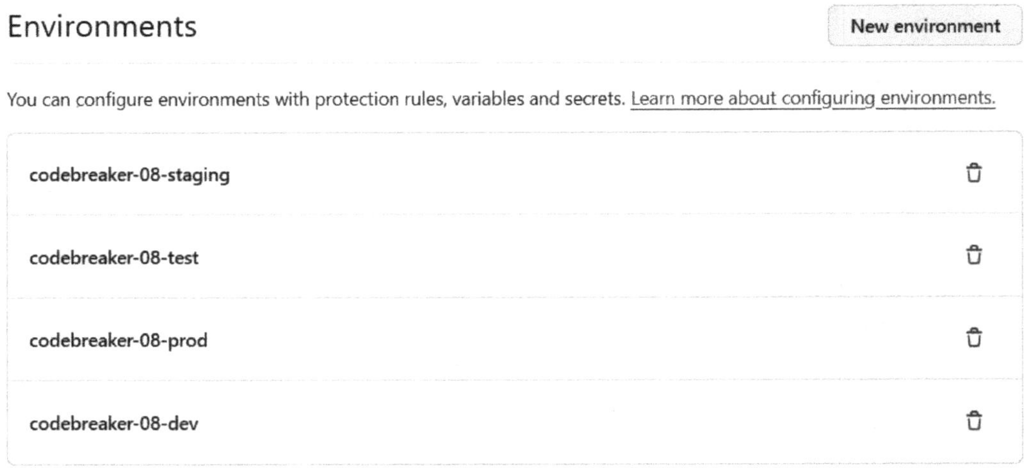

Environments   [New environment]

You can configure environments with protection rules, variables and secrets. Learn more about configuring environments.

codebreaker-08-staging	🗑
codebreaker-08-test	🗑
codebreaker-08-prod	🗑
codebreaker-08-dev	🗑

Figure 8.8 – GitHub environments

You can create these environments using your browser by accessing your repository. As the environments are being created, protection rules can be applied.

Defining deployment protection rules

Before publishing to another environment, you can enforce **deployment protection rules**. Publishing to the production environment might only be allowed from **protected branches**, specific branches that fulfill a naming convention, and only with commits from specific tag names. Up to six reviewers can be specified to approve the deployment. There's also the option to implement custom protection rules, which, for example, might check the results of different test runs (tests will be covered in *Chapter 10*) or check for issues within the GitHub repository. Third-party protection rules are also available.

> **Note**
>
> With the first few versions of the application, where you'll start with deployments across different environments, it's good practice to add reviewers that do some manual checks. Before the solution is deployed to the **production** environment, it needs to be deployed to the **staging** environment. In the staging environment, manual checks are used. On the road to improving the CI/CD process, you might add more and more automatic checks. Automated tests, code analysis, checking for issues, and more can be done before moving on to the next stage. You just need to be able to trust how you set up the environment and have your tests running.

Within the production environment, add yourself as a required reviewer with deployment protection, as shown in *Figure 8.9*:

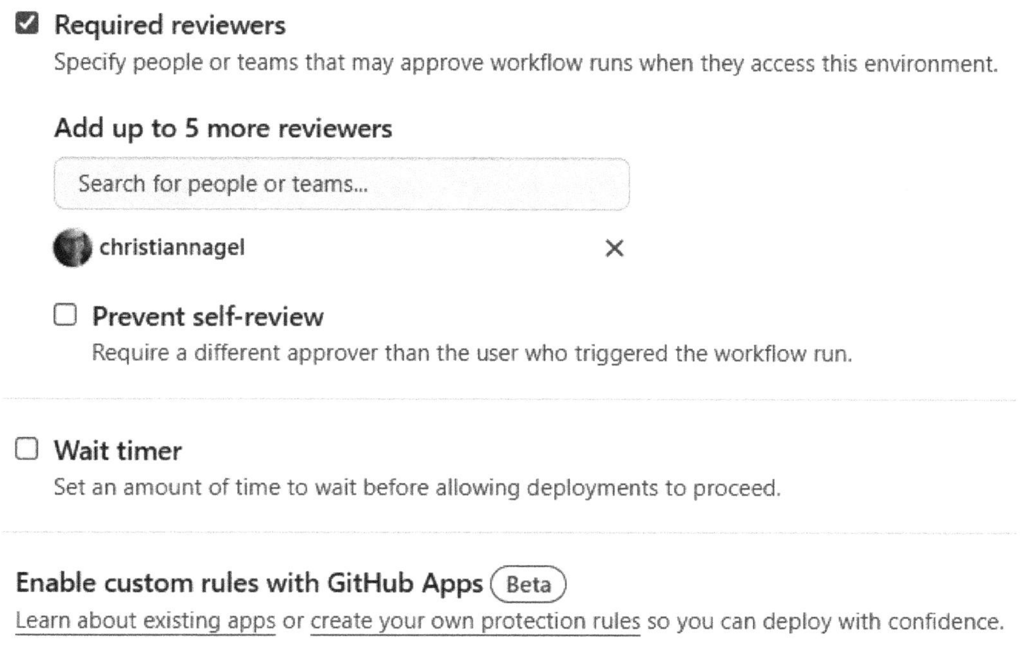

☑ **Required reviewers**
Specify people or teams that may approve workflow runs when they access this environment.

Add up to 5 more reviewers

    Search for people or teams...

  🌑 christiannagel   ✕

☐ **Prevent self-review**
Require a different approver than the user who triggered the workflow run.

☐ **Wait timer**
Set an amount of time to wait before allowing deployments to proceed.

Enable custom rules with GitHub Apps (Beta)

Learn about existing apps or create your own protection rules so you can deploy with confidence.

Figure 8.9 – Required reviewers with GitHub environments

Other than requiring reviewers, you can use rules defined by existing apps from GitHub partner applications to require some source code or issue checks, and also implement custom protection rules

> **Note**
>
> When using deployment protection rules for branches and tags, you should specify that not everyone is allowed to create branches and tags that are used with the rules. See *Configuring tag protection rules* at `https://docs.github.com/en/repositories/managing-your-repositorys-settings-and-features/managing-repository-settings/configuring-tag-protection-rules` for more details.

Next, we'll configure secrets and variables with environments.

Setting environment secrets and variables

With environments, you can also specify variables and secrets that are only available within these environments. We need the tenant ID, the subscription ID , and the account ID of the federated account we created earlier. This information was configured in the *Enhancing GitHub Actions workflows* section.

As a reminder, to get the tenant ID, use `az account show -query tenantId -o tsv`). To get the subscription ID, use `az account show --query id -o tsv`. With the account, to use an environment, an additional credential is required.

Open the Entra portal (`https://entra.microsoft.com`) and select **App registrations** in the left pane. Then, select the **All applications** tab and look for the previously created app registration – that is, `az-dev<date>`. Select **Certificates & secrets** from the left pane and select the **Federated credentials** tab. Previously, two credentials with the `repo:<github org/repo>:pull_request` and `repo:<github org/repo:refs/heads/main` subject identifiers were added. Add a new credential and select **GitHub Actions deploying Azure resources**, as shown in *Figure 8.10*:

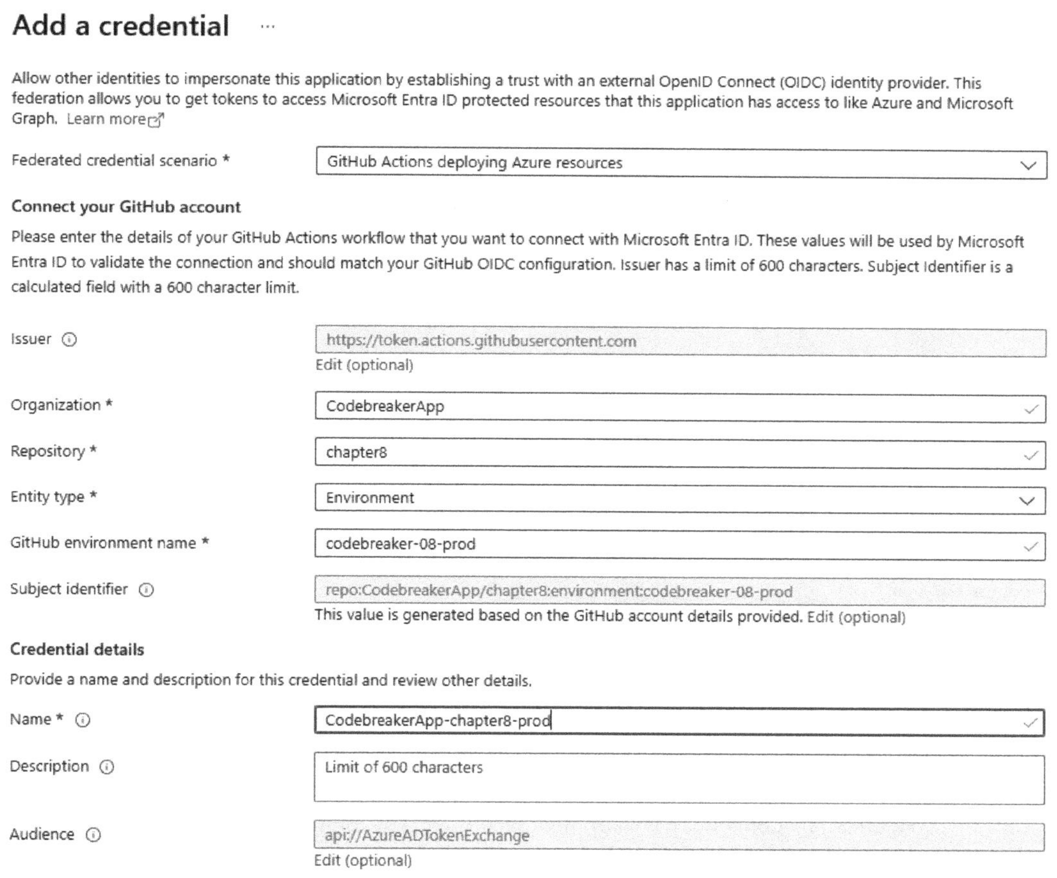

Add a credential   ···

Allow other identities to impersonate this application by establishing a trust with an external OpenID Connect (OIDC) identity provider. This federation allows you to get tokens to access Microsoft Entra ID protected resources that this application has access to like Azure and Microsoft Graph. Learn more ☝

Federated credential scenario * | GitHub Actions deploying Azure resources ∨ |

Connect your GitHub account

Please enter the details of your GitHub Actions workflow that you want to connect with Microsoft Entra ID. These values will be used by Microsoft Entra ID to validate the connection and should match your GitHub OIDC configuration. Issuer has a limit of 600 characters. Subject Identifier is a calculated field with a 600 character limit.

Issuer ⓘ | https://token.actions.githubusercontent.com |
Edit (optional)

Organization * | CodebreakerApp ✓ |

Repository * | chapter8 ✓ |

Entity type * | Environment ∨ |

GitHub environment name * | codebreaker-08-prod ✓ |

Subject identifier ⓘ | repo:CodebreakerApp/chapter8:environment:codebreaker-08-prod |
This value is generated based on the GitHub account details provided. Edit (optional)

Credential details

Provide a name and description for this credential and review other details.

Name * ⓘ | CodebreakerApp-chapter8-prod| ✓ |

Description ⓘ | Limit of 600 characters |

Audience ⓘ | api://AzureADTokenExchange |
Edit (optional)

Figure 8.10 – Environment credentials

Within this dialogue, add your GitHub **Organization** and **Repository**, select **Environment** for **Entity type**, enter the **GitHub environment name** value that matches your GitHub environment, and provide **Credential details**.

To configure the secret, copy the **Application (client) ID** value of this app registration.

Once you have these values, open the environment in the GitHub portal and add **Environment secrets** and **Environment variables**, as shown in *Figure 8.11*:

Environment secrets

Secrets are encrypted environment variables. They are accessible only by GitHub Actions in the context of this environment.

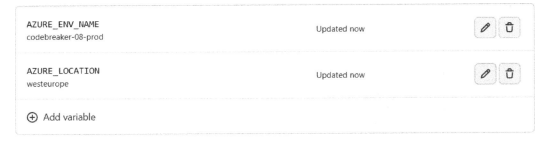

🔒 AZURE_CLIENT_ID	Updated 3 minutes ago	✏️ 🗑️
🔒 AZURE_SUBSCRIPTION_ID	Updated 2 minutes ago	✏️ 🗑️
🔒 AZURE_TENANT_ID	Updated 1 minute ago	✏️ 🗑️
⊕ Add secret		

Environment variables

Variables are used for non-sensitive configuration data. They are accessible only by GitHub Actions in the context of this environment. They are accessible using the vars context.

AZURE_ENV_NAME codebreaker-08-prod	Updated now	✏️ 🗑️
AZURE_LOCATION westeurope	Updated now	✏️ 🗑️
⊕ Add variable		

Figure 8.11 – Configuring Environment secrets and Environment variables

The following variables are required:

- `AZURE_ENV_NAME`: The resource group name that should be used without the `rg-` prefix – for example, `codebreaker-08-prod`
- `AZURE_LOCATION`: Your preferred Azure region

You will need the following secrets:

- `AZURE_SUBSCRIPTION_ID`
- `AZURE_TENANT_ID`
- `AZURE_CLIENT_ID`

With this configuration in place, let's update the workflows.

Using environments with workflows

To use an environment from a workflow, all you need to do is reference the environment name. Copy the shared workflow, `shared-deploy.yml`, to `shared-deploy-withenvironment.yml` and enhance it with an environment configuration:

workflows/shared-deploy-withenvironment.yml

```
# code removed for brevity
  workflow_call:
    inputs:
      environment-name:
        description: 'The environment to deploy to'
        required: true
        type: string
jobs:
  build-and-publish:
    runs-on: ubuntu-latest
    environment: ${{ inputs.environment-name }}
```

When configuring the job, the `environment` keyword is used to reference an environment name. With this implementation, a required input parameter is used to pass the name of the environment. No changes need to be made regarding the secrets and variables. When running in the environment, these values are retrieved from the environment configuration.

The `codebreaker-production.yml` workflow, which uses the various workflows to create and push the Docker image and publish the Container App, is different from the development environment, as shown here:

workflows/codebreaker- produnction.yml

```
# code removed for brevity
jobs:
  build-and-deploy:
    uses: ./.github/workflows/shared-deploy-withenvironment.yml
    secrets: inherit
    with:
      environment-name: codebreaker-08-prod
```

The environment parameter is now set to `codebreaker-08-prod`. This time, the secrets are not explicitly declared, but all secrets this workflow has access to are given to the called workflow. Because of the environment specified by the called workflow, secrets and variables are referenced from the GitHub environment.

Now, you can try triggering the workflow. The first stage runs, but the second stage must be reviewed, as shown in *Figure 8.12*:

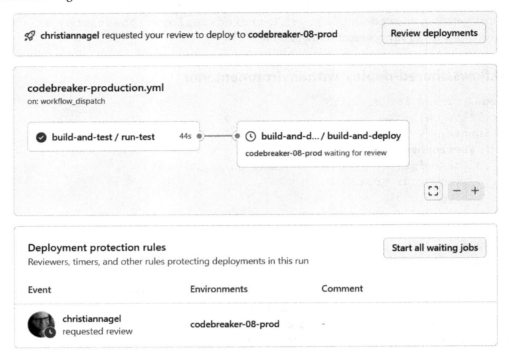

Figure 8.12 – Workflow review requested

Review the results of the workflow and approve it, as shown in *Figure 8.13*:

Figure 8.13 – Approve and deploy

At this point, you need to wait a few minutes until all the resources are deployed to the production environment. Verify that the deployment succeeded. After a successful deployment, you can use a client, update the link to the new environment, and play a game.

For the client programmers, we'll create a NuGet package.

Publishing NuGet packages

With our solution, we also have libraries that are used by client applications. Having NuGet packages helps with using these libraries. By creating a GitHub action, we can automatically build and publish the NuGet package. If you want to make a package publicly available, you can publish it to the NuGet server (you've already used packages that are available for this book). To make packages private with authentication, GitHub offers **GitHub Packages**.

Preparing the library project

Adding some metadata, such as a README Markdown file to the project describing the package, and a custom icon to replace the default icon, enhances usability.

The `Codebreaker.GameAPIs.KiotaClient` project contains a `readme.md` file and a JPG file for an icon. These additions need to be uploaded within the project file:

Codebreaker.GameAPIs.KioataClient/Codebreaker.GameAPIs.KiotaClient.csproj

```
<ItemGroup>
  <None Include="package/readme.md" Pack="true" PackagePath="\" />
  <None Include="package/codebreaker.jpeg" Pack="true"
  PackagePath="\" />
</ItemGroup>
```

The README file and the icon don't need to be built into the library, which is why `None` is used within `ItemGroup` to exclude them from the build result of the library. Adding these items to the NuGet package is specified by the `Pack` attribute. `PackagePath` specifies the folder within the package where these items can be found.

The following `PropertyGroup` definitions specify the use of the README file and the package icon and add some metadata:

Codebreaker.GameAPIs.KioataClient/Codebreaker.GameAPIs.KiotaClient.csproj

```
<PropertyGroup>
  <PackageId>
```

```
      CNinnovation.Codebreaker.KiotaClient
    </PackageId>
    <PackageTags>
      Codebreaker;CNinnovation;Kiota
    </PackageTags>
    <Description>
      This library contains Kiota-generated classes for communication
      with the Codebreaker games API service.
      See https://github.com/codebreakerapp for more information on
        the complete solution.
    </Description>
    <PackageReadmeFile>readme.md</PackageReadmeFile>
    <PackageIcon>codebreaker.jpeg</PackageIcon>
  </PropertyGroup>
```

Adding metadata to the package is specified with the `PackageId`, `PackageTags`, and `Description` elements.

It's also a good idea to define the version of the package. With the source code repository, the `VersionPrefix` element, which is defined within the `Directory.Build.props` file, specifies the first part of the version for all the projects found in subdirectories. Using a GitHub action, a `VersionSuffix` element is added dynamically that increments with every build. This versioning scheme is used with `alpha`, `beta`, and `prerelease` versions.

As soon as the library is released, the `Version` element is added to specify the complete version of the package. Adding the `Version` element to the project overrides `VersionPrefix` and `VersionSuffix`, and just this version is used. After the release, when the next beta versions are available, the `Version` element is removed again, and the `VersionPrefix` element is incremented to the next iteration.

Creating access tokens

To publish a package to GitHub Packages, a **personal access token (classic)** is required. At the time of writing, the new fine-grained personal access tokens cannot be used with GitHub Packages.

You can create a personal access token by clicking on the user icon in the top-right corner, selecting **Settings**, and then clicking on **Developer Settings** in the left pane. Select **Personal access tokens** and click **Tokens (classic)**. To create a new token, select **Generate new token (classic)**. Select an expiration date. The scope that is required to publish packages is **write:packages**. Selecting this scope also adds other scopes, such as reading packages and access to the repository. Click **Generate token**. You need to copy this generated token – it will not be visible again after you close the screen. Just make sure you store it in a safe place. You can create a new token in case you don't have one anymore, or when the token has expired.

For the GitHub action you wish to use, store this token alongside the secrets of the `PAT_PUBLISHPACKAGE` repository.

Now that we've stored this secret, let's use it with a GitHub action.

Creating a GitHub action to publish a GitHub package

The GitHub actions that are used to create a NuGet package have similarities with the GitHub actions we created previously. Check out the source code repository for details.

The `shared-create-nuget.yml` shared workflow builds the NuGet package and uploads it with GitHub artifacts. The following steps are completed in this workflow:

1. Check out the source code.
2. Set up .NET.
3. Calculate the build number (using a configured offset to the GitHub build number).
4. Build the library using `dotnet build`.
5. Test the library using `dotnet test`.
6. Create a NuGet package using `dotnet pack`.
7. Upload the package with GitHub artifacts.

The next shared workflow (`shared-githubpackages.yml`) uploads the package to GitHub Packages by following these steps:

1. Download the GitHub artifact.
2. Set up .NET.
3. Set the NuGet source with `dotnet nuget add source`.
4. Push the package to GitHub Packages with `dotnet nuget push`.

Pushing the package makes use of the configured access token.

The `kiota-lib.yml` workflow connects the two shared workflows and passes parameters. Upon running this workflow successfully, you can verify the packages with the organization of your GitHub repository, as shown in *Figure 8.14*:

Figure 8.14 – GitHub Packages

Using GitHub environments, you can enhance the creation of NuGet packages and define environments, such as those to publish to the publicly available NuGet server only after successfully using a private feed with GitHub Packages.

With modern deployments, there's more than just using development, staging, and production environments. We'll discuss this next.

Using modern deployment patterns

Using development, staging, and production environments is one of the "traditional" deployment patterns. Nowadays, other deployment patterns are used as well:

- When using **canary releases**, different versions of an application are available for the user to choose from. This is evident from the Edge browser, which offers a Beta channel that's updated monthly, a Dev channel that's updated weekly, and a Canary channel that's updated daily. The user can decide what version to test. See https://www.microsoft.com/en-us/edge/download/insider for more details.

- With **A/B testing**, users randomly receive one of two different user interfaces. When using this pattern, you can monitor which UI allows the user to be more productive.

- **Blue-green deployments** allow you to quickly roll back an installation by installing to a staging server, swapping staging with production. If something fails, an easy rollback can be done.

- **Dark launching** is a pattern that you can use to publish a new version of the application while ensuring that the new features are hidden until they are activated by turning on a switch. One example of this is when a feature should be available at a specific time. This switch can be turned on by a time event – there's no need to redeploy the application.

- **Feature toggles** allow you to turn each feature on/off. One option is to enable some of the features for a specific group of users, such as early adopters. Users themselves can also decide which of the new features they want to test. Such toggles are available with Microsoft Azure and Visual Studio.

In *Chapter 7*, you saw Azure App Configuration in action. This Azure service not only supports a central application configuration but also offers feature flags. This functionality of Azure App Configuration can be used with several of the modern deployment patterns by using different feature flag filters.

Configuring feature flags

Let's open the Azure App Configuration service that was created with Bicep scripts. In the left pane, within the **Operations** category, open **Feature manager**.

> **Note**
>
> You might not have access to this resource to add configuration data since the resource was created from Bicep scripts. Using **Access control (IAM)**, add your user to the **App Configuration Data Owner** or **Contributor** role. You might need to wait about 15 minutes before the role changes.

Create a new feature flag, as shown in *Figure 8.15*:

Create ···
Create a new feature flag

A feature flag is a variable with a binary state of on or off. It allows you to activate or deactivate features in your application without deploying new code. This can dynamically administer a feature's lifecycle. Learn more

Enable feature flag   ☑

Details

Feature flag name * ⓘ   Feature8x5Game ✓

Key ⓘ   .appconfig.featureflag/   Feature8x5Game ✓

Label   🔍 (No label)

Description   This feature enables the 8x5 game mode

Feature filters

A Feature filter consistently evaluates the state of a feature flag. Our feature management library supports two types of built-in filters: Targeting and TimeWindow. Custom filters can also be created based on different factors, such as device used, browser types, geographic location, etc. Learn more

☑ Use feature filter

+ Create

Name	Parameters
Microsoft.Targeting	Default percentage: 50%, included group(s): 0, included user(s): ... ✏️ ···

Figure 8.15 – Creating feature flags

Set the feature flag's name to `Feature8x5Game`, add a description, and check the **Enable feature flag** box. Before applying this configuration, click on **Create** under **Feature filters**. Select the targeting filter. Read the information about the evaluation flow, but leave the other values with their default settings. Add other feature flags for `FeatureGame6x4Mini`, `FeatureGame6x4`, and `FeatureGame5x5x4`. Don't add filters for the first two; just enable one of these. For the last one, add a time filter so that it can be enabled in the future, but don't set an expiration date.

The **targeting filter** allows you to open a feature for a specific user group (early adopters). It can also act as a percentage filter, so you can turn this feature on for a random percentage of users. The other built-in filter is the **time window filter**. Using this filter, you can specify start and end times when this feature should be enabled. You can also create a custom implementation for a filter.

Now that we've configured this feature flag, let's use this from the `game-apis` service.

DI and middleware configuration for feature flags

To use Feature Management, add the `Microsoft.FeatureManagement.AspNetCore` NuGet package to the `Codebreaker.GameAPIs` project. The DI container needs to be configured for feature flags, as shown here:

Codebreaker.GameAPIs/Program.cs

```
// code removed for brevity
builder.Services.AddFeatureManagement()
  .AddFeatureFilter<TargetingFilter>()
  .AddFeatureFilter<TimeWindowFilter>();
```

The `AddFeatureManagement` extension method registers types that are needed for feature flags. Every filter that is used is added using the `AddFeatureFilter` extension method.

> **Note**
>
> The Feature Management API can also be used without Azure App Configuration. Upon viewing the source code in this book's GitHub repository, you'll see that the Feature Management API can be configured without using Azure as well. In this case, an overload of the `AddFeatureManagement` API is invoked to pass an `IConfiguration` object. With this, feature flags can be configured with the .NET configuration options. Revisit *Chapter 7* for more information on configuration.

To connect Feature Management with Azure App Configuration, you must update the `AddAzureAppConfiguration` method:

Codebreaker.GameAPIs/Program.cs

```
builder.Configuration.AddAzureAppConfiguration(options =>
{
  options.Connect(new Uri(endpoint), credential)
    .Select("GamesAPI*")
    .ConfigureKeyVault(kv =>
    {
      kv.SetCredential(credential);
    })
    .UseFeatureFlags();
});
```

`UseFeatureFlags` is a method of the `AzureAppConfigurationOptions` class for connecting feature flags.

When using feature flags, the Azure App Configuration middleware also needs to be configured:

Codebreaker.GameAPIs/Program.cs

```
var app = builder.Build();
if (solutionEnvironment == "Azure")
{
  app.UseAzureAppConfiguration();
}
```

With the setup in place, we can check if feature flags have been set.

Using feature flags

Now, we can use the feature manager to check if features are available. We'll start by creating an extension method for the `IFeatureManager` interface:

Codebreaker.GameAPIs/Extensions/FeatureManagerExtensions.cs

```
public static class FeatureManagerExtensions
{
  private static List<string>? s_featureNames;
  public static async Task<bool> IsGameTypeAvailable(this
    IFeatureManager featureManager, GameType gameType)
  {
```

```
async Task<List<string>> GetFeatureNamesAsync()
{
  List<string> featureNames = [];
  await foreach (string featureName in featureManager.
    GetFeatureNamesAsync())
  {
    featureNames.Add(featureName);
  }
  return featureNames;
}

string featureName = $"Feature{gameType}";
if ((s_featureNames ?? await GetFeatureNamesAsync()).
Contains(featureName))
{
  return await featureManager.IsEnabledAsync(featureName);
}
else
{
  return true;
}
}
}
```

This method uses the `GetFeatureNamesAsync` and `IsEnabledAsync` methods defined by the `IFeatureManager` interface. On the first invocation of this method, the list of features registered with the feature manager is retrieved and added to the `_featureNames` collection. Not every game type is registered as a feature. For the game types that are not registered as features, the method returns `true` to inform us that this type is available. With all the game types registered as a feature, the `IsEnabledAsync` method is used to check if the feature is enabled.

Next, let's inject `IFeatureManager` with the minimal API:

Codebreaker.GameAPIs/Endpoints/GameEndpoints.cs

```
group.MapPost("/", async Task<Results<Created<CreateGameResponse>,
BadRequest<GameError>>> (
  CreateGameRequest request,
  IGamesService gameService,
  IFeatureManager featureManager,
  HttpContext context,
  CancellationToken cancellationToken) =>
  {
    Game game;
```

```
    try
    {
      bool featureAvailable = await featureManager.
        IsGameTypeAvailable(request.GameType);
      if (!featureAvailable)
      {
        GameError error = new(ErrorCodes.
          GameTypeCurrentlyNotAvailable, "Game type currently not
          available", context.Request.GetDisplayUrl());
        return TypedResults.BadRequest(error);
      }

      game = await gameService.StartGameAsync(request.GameType.
        ToString(), request.PlayerName, cancellationToken);}
  // code removed for brevity
```

On starting a game using the API, the IFeatureManagement interface is injected to check the requested game type for the feature to be enabled using the previously created extension method, IsGameTypeAvailable. Depending on the result, an error is returned, or a new game is created.

With this implementation, you can run the application and test these feature flags. The game-apis project contains an HTTP file that you can use to create all the different game types and see the results that were returned when using feature flags. You can test this locally on your developer system. Upon pushing an update to your GitHub repository, a workflow is ready to be triggered. Then, you just need to configure the link of your API service within the HTTP file to test the service that's running with Azure Container Apps.

Summary

After creating Azure services with Bicep scripts in *Chapter 6*, in this chapter, you learned how to use **continuous integration** (**CI**) and **continuous delivery** (**CD**) with GitHub Actions. Here, you changed the source code, created and merged a pull request, tested code, and deployed Azure Container Apps. Using GitHub Actions, you learned how to build NuGet packages and push them to GitHub Packages.

Using GitHub environments, you created multiple deployment environments where additional checks are required before deployment is extended to another stage.

After, you learned how to configure Azure App Configuration, as well as how to use feature flags, which are needed for modern deployment patterns such as A/B testing, blue-green deployments, and dark launching.

The next chapter covers another important topic: authentication and authorization. In *Chapter 7*, you learned how to run Azure services with managed identities. In *Chapter 9*, we'll restrict the applications that are allowed to invoke APIs, authenticate users to restrict functionality for anonymous users, and add APIs that are only allowed to be used by specific user groups.

Further reading

To learn more about the topics that were discussed in this chapter, please refer to the following links:

- *GitHub Actions documentation*: https://docs.github.com/en/actions

- *azd: configure a pipeline*: https://learn.microsoft.com/en-us/azure/developer/azure-developer-cli/configure-devops-pipeline

- *Using environments for deployments*: https://docs.github.com/en/actions/deployment/targeting-different-environments/using-environments-for-deployment

- *Creating custom protection rules*: https://docs.github.com/en/actions/deployment/protecting-deployments/creating-custom-deployment-protection-rules

- *Microsoft Feature Management documentation*: https://learn.microsoft.com/en-us/dotnet/api/microsoft.featuremanagement?view=azure-dotnet

Authentication and Authorization with Services and Clients

Not every user and application should be allowed to access all API services. Some APIs should only be accessible from specific applications, and others should be restricted to a group of users.

In this chapter, you'll learn how to use **business-to-consumer** (**B2C**) to allow users to register with our application and protect APIs. We'll use Azure **Active Directory** (**AD**) B2C for this. For an on-premises solution (which can also be used in the cloud), we'll be using ASP.NET Core Identity.

Instead of securing every API project, you'll learn about Microsoft **Yet Another Reverse Proxy** (**YARP**), a proxy that is put in front of the APIs that are available to restrict access to the services in the backend.

In this chapter, you'll learn how to do the following:

- Create an Azure AD B2C tenant
- Secure REST APIs
- Use Microsoft YARP
- Use ASP.NET Core Identity

Technical requirements

In this chapter, like the previous chapters, you'll need an Azure subscription, Docker Desktop, and .NET Aspire.

The code for this chapter can be found in this book's GitHub repository: `https://github.com/PacktPublishing/Pragmatic-Microservices-with-CSharp-and-Azure`.

The ch09 folder contains the following projects, along with their outputs:

- `Codebreaker.ApiGateway`: This is a new project that will act as an application gateway in front of the `game-apis` service and `bot-service` and secure the APIs with the help of YARP

- `WebAppAuth`: This is a new project for the client part that focuses on creating new users with Azure AD B2C, providing authentication from the client side, and invoking `bot-service` via the gateway

- `Codebreaker.ApiGateway.Identities`: This is a new project that can be used instead of `Codebreaker.ApiGateway` where instead of using Azure AD B2C, local users are created and managed

To help you go through the code with this chapter, start by using the code from the previous chapter.

Choosing an identity solution

Different options are available to authenticate users with .NET solutions. If you require a local database that can manage users, you can use **ASP.NET Core Identity**, which makes use of EF Core (see *Chapter 5*). It allows you to store local users and integrate user accounts, such as those from Microsoft, Facebook, and Google, with **OpenID Connect (OIDC)**. For the database, SQL Server and MySQL can be used, while the data schema is completely customizable.

To reduce the work required, and to enhance security, it's not necessary to implement this functionality with every service – here, Microsoft YARP can be used to forward the requests and send the required claims.

If external applications are accessing the identity management solution, an **OIDC** server should be used to manage identities. If storing user data in a cloud service is not an option, a third-party service such as Identity Server from Duende (`https://duendesoftware.com/products/communityedition`) can be used. This is free for small companies.

To store user data in a cloud service, many companies use **Microsoft Entra**. This can easily be integrated with .NET applications. This service offers **business-to-business (B2B)** functionality that allows you to add external users (**Entra External Identities**). Microsoft, Facebook, and Google accounts are on the list of supported external users. However, at the time of writing, **Microsoft Entra** does not allow users to register themselves. For this, **Azure AD B2C** is a great option. This service can also be used with services running on-premises and accessing authentication from the cloud.

> **Note**
> With its user data residency requirement, Azure AD B2C allows you to select a country when you're creating a directory and shows the location for the data. However, if, for example, the requirement is to keep the user data in Switzerland, it's stored in Europe, which might not be enough for the legal requirements of some businesses.

For the Codebreaker solution, we'll use Azure AD B2C and ASP.NET Core Identity.

Creating an Azure AD B2C tenant

The Codebreaker solution should allow users to register with the application and play different game types. Some limited game types are available to anonymous users. All the game types and more functionalities are available to registered users. Some parts of the solution should only be accessible to specific user groups – for example, bot-service should not be accessible from normal registered playing users. Specific user permissions (or claims) are required for differentiation.

To create a new AAD B2C tenant, open the Azure portal and click **Create a resource**. Select **Identity** from the left bar and choose **Azure Active Directory B2C**. Then, select **Create a new Azure AD B2C Tenant**. This will open the screen shown in *Figure 9.1*:

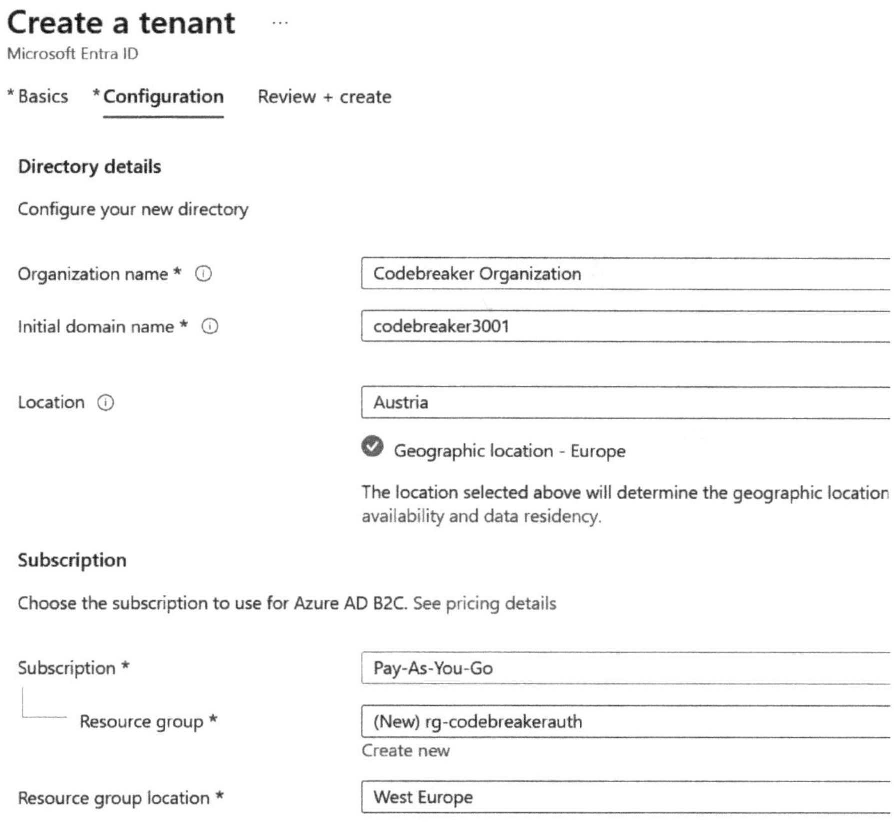

Create a tenant ...
Microsoft Entra ID

* Basics   * **Configuration**   Review + create

Directory details

Configure your new directory

Organization name * ⓘ   Codebreaker Organization

Initial domain name * ⓘ   codebreaker3001

Location ⓘ   Austria

✓ Geographic location - Europe

The location selected above will determine the geographic location availability and data residency.

Subscription

Choose the subscription to use for Azure AD B2C. See pricing details

Subscription *   Pay-As-You-Go

⎿ Resource group *   (New) rg-codebreakerauth
                                 Create new

Resource group location *   West Europe

Figure 9.1 – Creating an AAD B2C tenant

To create a new AAD B2C tenant, you need to enter the name of the organization, the name of the domain name (a domain name that does not exist yet), the location that will be used to define the region where the user data is stored, a subscription, and a resource group. Once you've done this, click **Review + Create**, then **Create**.

You'll need to wait a short time for the directory to be created. To list the directories available to you, and to switch directories, within the Azure portal, click the **Settings** button. Select the new directory and click **Switch** to change to it. Similarly, you can switch back to the directory where you run Azure resources.

In the next few sections, we'll do the following:

- Specify identity providers so that the user doesn't need to enter another password
- Configure user attributes to define what information the application needs from the user
- Define user flows to specify how the user information flows
- Create app registrations to define service applications that offer APIs and client applications for accessing APIs

Specifying identity providers

When you are in an Azure AD B2C, you can open the Azure AD B2C configuration. The B2C directory supports a large list of different identity providers. Users don't need to remember another password when they use identity providers. Within the Azure AD B2C configuration, in the **Manage** category in the left pane, select **Identity provider** (see *Figure 9.2*):

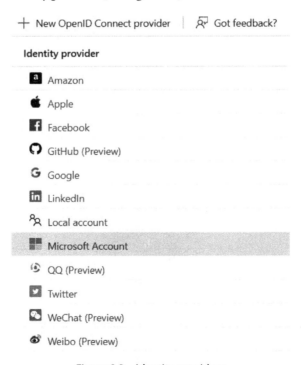

Figure 9.2 – Identity providers

By default, **Local account** is configured so that a password is stored locally with AAD B2C. You can configure Microsoft, Google, Facebook, and other accounts that support OIDC. The Codebreaker directory has GitHub configured as a provider because most developers already have a GitHub account.

For each provider, you're what needs to be done to configure it. You just need to click on the provider to get that information. With GitHub, for example, you need to create a GitHub OAuth application to get all the values you need to configure this provider. For authenticating services with AAD B2C, you can keep the default settings.

Configuring user attributes

No matter which provider you choose, you must have a way to identify users. To gather such information, you must ask your users for details. You can also create custom attributes that should be stored in the directory. Within the **Manage** category, select **User Attributes**, as shown in *Figure 9.3*:

+ Add   ⦻ Got feedback?

Name	Data Type	Description	Attribute type
City	String	The city in which the user is located.	Built-in
Country/Region	String	The country/region in which the user is located.	Built-in
Display Name	String	Display Name of the User.	Built-in
Email Addresses	StringCollection	Email addresses of the user.	Built-in
Gamer Name	String		Custom
Given Name	String	The user's given name (also known as first name).	Built-in
Identity Provider	String	The social identity provider used by the user to ac...	Built-in
Job Title	String	The user's job title.	Built-in
Legal Age Group Classification	String	The legal age group that a user falls into based on...	Built-in
Postal Code	String	The postal code of the user's address.	Built-in
State/Province	String	The state or province in user's address.	Built-in
Street Address	String	The street address where the user is located.	Built-in
Surname	String	The user's surname (also known as family name or...	Built-in
User is new	Boolean	True, if the user has just signed-up for your applic...	Built-in
User's Object ID	String	Object identifier (ID) of the user object in Azure AD.	Built-in

Figure 9.3 – User Attributes

Here, several built-in attributes, such as **Given Name**, **Surname**, **City**, **Country/Region**, and more, are available. You can also add custom attributes. Add a custom attribute called `Gamer Name` of the `String` type.

> **Note**
>
> Due to the **General Data Protection Regulation (GDPR)**, you need to ensure you only collect necessary data and keep it secured, allow the user to ask for the data that you've stored, and allow the user to delete that data if it doesn't need to be stored for legal reasons.

Defining user flows

With user flows, you define what information should be collected from the user when registering or editing the user profile, and what information should be sent to the application within **claims**.

Within the AAD B2C configuration, from the left pane, within the **Policies** category, click **User flows**. Create a new user flow for **Sign up and sign in**. The user flow's name is prefixed with B2C_1_. Add a name (for example, SUSI) and select the **Email** signup identity provider. You can also select social providers such as GitHub. Regarding the **User attributes and token claims** category, select the user attributes the user should enter when this dialogue is shown, as well as the claims that are passed to the application within a token, as shown in *Figure 9.4*:

Figure 9.4 – Creating a user flow

When defining what information to ask from the user, keep GDPR in mind.

Azure AD B2C allows you to customize user flow dialogues by specifying company branding, changing the page layout, returning custom pages, and adding API connectors for custom validators when a user registers.

> **Note**
> User attributes can be filled by creating user flows or **custom policies**. See the links in the *Further reading* section for more information. Also, check out the source code in the Codebreaker Backend repository (`https://github.com/codebreakerapp/Codebreaker.Backend`), which contains a custom policy for adding groups for privileged users.

As soon as an application has been registered (the next step), you can test the user flow, as shown in *Figure 9.5*:

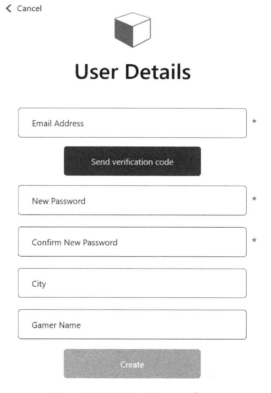

Figure 9.5 – Testing the user flow

Selecting the user attributes that should be collected defines the input elements of the dialogue. The icon, colors, and layout can be customized. It's even possible to create complete custom dialogues.

Creating app registrations

Next, we'll learn how to register apps. Here, we will register the application gateway that offers APIs and a client application. Other applications can be registered similarly.

In the **Manage** category in the left pane, click **App registrations**. This opens the **App registrations** page, as shown in *Figure 9.6*:

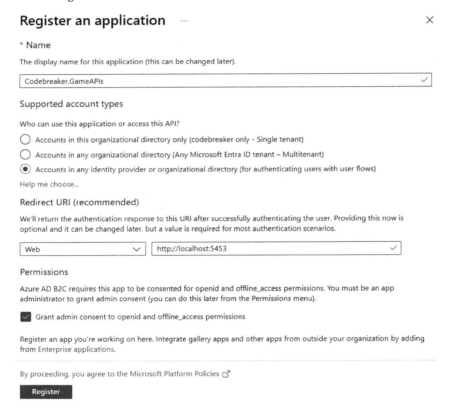

Figure 9.6 – Registering an app

Add multiple app registrations: the `Codebreaker.GameAPIs` application offers the game APIs, `Codebreaker.Bot` and `Codebreaker.Blazor` are web applications that need API permissions, and `Codebreaker.Client` is a client application that needs API permissions.

When you configure the app registration process, you specify what accounts are allowed to use this application. Here, we'll allow all accounts, externally registered users, and the redirect URI. To test the `game-apis` service from the local developer system, specify the port number that's used when running it locally, such as `http://localhost:5453`, and click the **Register** button. The link to the Azure container app needs to be added later.

Defining scopes

You can specify applications that offer APIs via the app registration process. In the **Manage** category, you'll see the **Expose an API** option. Click **Add a Scope**. The root scope can be the proposed GUID, but you can define a more readable name, such as games. Within this scope, add the Games.Play and Games.Query scopes, as shown in *Figure 9.7*:

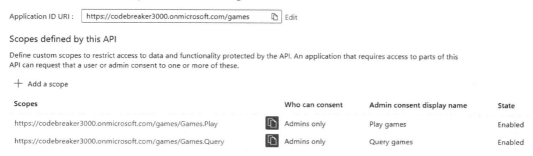

Figure 9.7 – Defining scopes

Creating a secret

To only allow applications that have been identified applications, you can add a certificate or a secret. Instead of using secrets, a better approach could be to run the application with a user that is allowed to access the service. Here, managed identities can be used. This is not possible in all scenarios.

Using the **Certificates and secrets** option from the left pane, create a client secret. Secrets cannot be read from the portal again, only after creation. Copy the secret before leaving the page.

Adding API permissions

For application registrations that invoke APIs, you need to configure **API permissions**, as shown in *Figure 9.8*:

Figure 9.8 – Adding API permissions

The `Codebreaker.Blazor` and `Codebreaker.Client` application registrations need the `Games.Play` and `Games.Query` application permissions. After adding these permissions, click **Grant admin consent**.

Evaluating the app registration process

With this configuration complete, open **Integration assistant** and check if the application registration supports the recommended configurations. Select the application type and check if the application invokes APIs (such as `bot-service` and the client applications). Then, click **Evaluate my app registration**, as shown in *Figure 9.9*:

Here's the integration assistant for Codebreaker.GamesAPI.Test

			⌀ Edit
Application type :	**Web API**		
Calls APIs :	**No**		
Supported account types:	**Accounts in any identity provider or organizational directory (for authenticating users with user flows)**		

Summary   Develop   Test   Release   Monitor

Recommended configurations

Item	Status	
Configure a unique Application ID URI.	✓ Complete	...
If expecting API requests on behalf of users, define scopes your API exposes.	✓ Complete	...
Grant admin consent for permissions.	✓ Complete	...
Add the 'offline_access' and 'openid' Microsoft Graph delegated permissions.	✓ Complete	...

Discouraged configurations

Item	Status	
Do not configure a credential (certificate/secret).	✓ Complete	...
If you are using the authorization code flow, disable the implicit grant settings.	✓ Complete	...

Figure 9.9 – Integration assistant results

The integration assistant offers great information for development, testing, releasing, and monitoring when you click the tabs above the recommended configurations. If you see some warnings or errors, click the ellipsis (…). From here, you can check the documentation and open a page where you can change your configuration.

With Azure AD B2C configured, let's implement some code so that we can make use of AAD B2C.

Securing an API

We can now secure every API project. However, there are different ways we can do this so that we can reduce the work we need to do. One option is to use Azure Container Apps to configure authentication. Instead of configuring this for every container app, let's create a new project that will be secured and routed to multiple services. For this, we'll use **YARP**.

Creating a new project with authentication

Create a new Web API project by using the .NET template with the -au authentication option:

```
dotnet new webapi -minimal -au IndividualB2C -o Codebreaker.ApiGateway
```

Using the .NET CLI, you can also pass all the values needed to configure the B2C service, such as --domain for the domain, --aad-b2c-instance to pass the domain link for logging in, --client-id for the application ID, --susi-policy-id for the signup user flow (before it was called *user flow*, it was called *policy*), and --default-scope to configure a scope. If you don't assign parameter values for these configurations, you just need to change them after they've been created in the appsettings.json file.

The NuGet packages related to authentication and authorization that have been added to this project are as follows:

- Microsoft.AspNetCore.Authentication.JwtBearer: This package supports authentication using **JSON Web Tokens (JWT)**

- Microsoft.AspNetCore.Authentication.OpenIdConnect: This package allows authentication with an OIDC against identity providers, such as Azure AD B2C

- Microsoft.Identity.Web: This package provides utilities and middleware for authentication flows and user authorization

- Microsoft.Identity.Web.DownstreamApi: This package helps call downstream APIs using the same authentication context

Next, we'll add YARP to this project.

Creating an application gateway with YARP

When creating a microservices solution, it's not necessary to implement authentication with every service. Instead, you can create a service that acts as a reverse proxy. Clients only call into the reverse proxy. This proxy forwards authenticated requests to other services. Here, we'll use Microsoft YARP. A reverse proxy sits in front of backend services and intercepts invocations from a client before it is sent to the service. The YARP proxy offers different features, such as load balancing, rate limiting, switching of protocols, selecting services based on different versions, and more. Based on Layer 7, the proxy can read HTTP requests to route based on links and HTTP headers, as well as change the protocol that's used. Here, we'll use a reverse proxy to deal with authentication and authorization before forwarding the requests to the backend service.

Figure 9.10 shows the new way to communicate with the services:

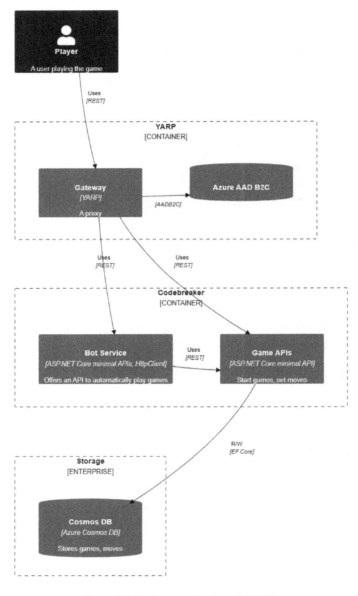

Figure 9.10 – Communication via YARP

The reverse proxy routes incoming requests to backend services. The backend services that have been routed are game-apis and bot-service. The client applications don't interact with these services; they just use the YARP gateway.

In addition to the NuGet packages we added earlier, we need to add the Yarp.ReverseProxy and Microsoft.Extensions.ServiceDiscovery.Yarp NuGet packages. The first one is the package for YARP, while the second one allows us to use.NET service discovery with YARP.

Mapping routes with YARP

How the proxy service communicates with the backend APIs can be configured both programmatically and using a configuration file. We'll use the second option with the appsettings.json file. First, let's configure the addresses of the game-apis service and bot-service:

Codebreaker.ApiGatewayIntro/appsettings.json

```
{
  "ReverseProxy": {
    "Clusters": {
      "gamesapicluster": {
        "Destinations": {
          "gamescluster/destination1": {
            "Address": "http://gameapis"
          }
        }
      }
      "botcluster": {
        "Destinations": {
          "botcluster/destination1": {
            "Address": "http://bot"
          }
        }
      },
    },
    // configuration removed for brevity
  }
}
```

The complete reverse proxy configuration is added to the ReverseProxy section, The configuration section, called Clusters, defines the list of systems that are available for the game-apis service and bot-service. With every service, multiple addresses can be added. Using the service discovery YARP package, we can use the .NET Aspire named endpoints.

The following code configures the routes that use the cluster configuration:

Codebreaker.ApiGatewayIntro/appsettings.json

```json
{
  "ReverseProxy": {
    // configuration removed for brevity
    "Routes": {
      "gamesRoute": {
        "ClusterId": "gamesapicluster",
        "Match": {
          "Path": "/games/{*any}"
        }
      },
      "botRoute": {
        "ClusterId": "botcluster",
        "Match": {
          "Path": "/bot/{*any}"
        }
      }
    }
  }
}
```

The `Routes` configuration contains a list of routes. `gamesRoute` references the previously specified `gamesapicluster`, while `botRoute` references the hosts defined by `botcluster`. The `Match` configuration specifies the `Path` that's used to map the request to the corresponding cluster.

We just need to make a small update to the startup code to activate this reverse proxy library:

Codebreaker.ApiGateway/Program.cs

```csharp
var builder = WebApplication.CreateBuilder(args);
builder.Services.AddReverseProxy()
  .LoadFromConfig(
    builder.Configuration.GetSection("ReverseProxy"));
var app = builder.Build();
app.MapReverseProxy();
app.Run();
```

The `AddReverseProxy` method registers the services that are needed by the reverse proxy to the DI container. The `LoadFromConfig` method retrieves the configuration values from the previously specified configuration. The `MapReverseProxy` method configures the middleware to forward the requests, as defined by the configuration.

With the `AppHost` project, after adding a reference to the gateway project, the gateway can be added to the app model:

Codebreaker.AppHost/Program.cs

```
var gameAPIs = builder.AddProject<Projects.Codebreaker_
GameAPIs>("gameapis")
  .WithReference(cosmos)
  .WithEnvironment("DataStore", dataStore);
var bot = builder.AddProject<Projects.CodeBreaker_Bot>("bot")
  .WithReference(gameAPIs);
builder.AddProject<Projects.Codebreaker_ApiGateway>("gateway")
  .WithReference(gameAPIs)
  .WithReference(bot)
  .WithExternalHttpEndpoints();
// code removed for brevity
```

The API gateway needs to reference the game-apis service and bot-service. Here, external HTTP endpoints are no longer needed. Only the gateway needs references from outside when it's deployed to the Azure Container Apps environment, thus only the gateway configuration uses the `WithExternalHttpEndpoints` method.

With this in place, you can start the application and invoke the two services via the gateway. The requests are forwarded to the specific service.

Next, we'll add authentication to the gateway.

Adding authentication to the gateway

Using the .NET template for the Web API with Identity already added some code for authentication and authorization. We will enhance this code now.

We can configure the DI container to authenticate users by invoking the `AddAuthentication` method:

Codebreaker.ApiGateway/Program.cs

```
var builder = WebApplication.CreateBuilder(args);
builder.Services.AddAuthentication(JwtBearerDefaults.
AuthenticationScheme)
  .AddMicrosoftIdentityWebApi(
    builder.Configuration.GetSection("AzureAdB2C"));
// code removed for brevity
```

The AddAuthentication method registers services that are needed for authentication. The JwtBearerDefaults.AuthenticationScheme argument returns **Bearer** as the authentication scheme. Bearer tokens are used with most REST APIs because they can be used easily and don't require encryption but need to perform HTTPS encryption to secure it.

AddMicrosoftIdentityWebApi is an extension method that extends AuthenticationBuilder and protects the API using the Microsoft Identity platform. AzureAdB2C is a configuration section that specifies the values from AADB2C from appsettings.json:

Codebreaker.ApiGateway/appsettings.json

```
{
  // configuration removed for brevity
  «AzureAdB2C»: {
    «Instance»: «https://<domain>.b2clogin.com»,
    «Domain»: «<domain>.onmicrosoft.com»,
    "ClientId": "<app-id>",
    "SignedOutCallbackPath": "/signout/B2C_1_SUSI",
    "SignUpSignInPolicyId": "B2C_1_SUSI"
  }
}
```

With the appsettings.json configuration file, you need to configure your Azure AD B2C domain name, the application ID, and the previously configured user flow.

The AddAuthentication method specifies the authentication configuration:

Codebreaker.ApiGateway/Program.cs

```
var builder = WebApplication.CreateBuilder(args);
// code removed for brevity
builder.Services.AddAuthentication(JwtBearerDefaults.
AuthenticationScheme)
  .AddMicrosoftIdentityWebApi(
    builder.Configuration.GetSection("AzureAdB2C"));
builder.Services.AddAuthorization(options =>
{
  options.AddPolicy("playPolicy", config =>
  {
    config.RequireScope("Games.Play");
  });
  options.AddPolicy("queryPolicy", config =>
  {
```

```
    config.RequireScope("Games.Query");
    config.RequireAuthentication();
  }
});
```

The `AddAuthorization` method allows configuration with an `AuthorizationOptions` delegate. The options allow you to specify a default policy and named policies. The preceding code snippet defines the `playPolicy` and `queryPolicy` policies. `playPolicy` requires the `Games.Play` scope to be set, whereas `queryPolicy` requires the `Games.Query` scope to be set. `queryPolicy` also requires the user to be authenticated. You can define a claim to be passed with the token by using the `RequireClaim` method.

With the policies in place, routes can be restricted to the required policies:

Codebreaker.ApiGateway/appsettings.json

```
// configuration removed for brevity
  "ReverseProxy": {
    "Routes": {
      "botRoute": {
        "ClusterId": "botcluster",
        "AuthorizationPolicy": "botPolicy",
        "Match": {
          "Path": "/bot/{*any}"
        }
      },
```

Using the `AuthorizationPolicy` configuration, alongside `botRoute`, `botPolicy` is referenced to require authenticated users and the application to send the correct scope.

Now that we've configured the DI container, the middleware needs to be configured:

Codebreaker.GameAPis/Program.cs

```
var app = builder.Build();
app.UseAuthentication();
app.UseAuthorization();
// code removed for brevity
```

The `UseAuthentication` method adds authentication middleware, while the `UseAuthorization` method adds authorization middleware.

> **Note**
>
> If the minimal API needs to be restricted directly, the `RequireAuthorization` extension method can be used. A policy can be passed as an argument to check for the policy's requirements. Upon injecting `ClaimsPrincipal` as an argument to a minimal API method, information about the user and claim information can be retrieved programmatically. This allows us to check for restrictions based on values that are retrieved with the API.

To test this out, we'll update our client application.

Authentication using Microsoft Identity with ASP.NET Core web applications

To authenticate using Azure AD B2C, we'll use the Microsoft Identity platform. In this section, we'll focus on creating accounts with Azure AD B2C, logging in, and invoking secured REST APIs with ASP.NET Core web applications.

Like with the minimal API we created earlier, a .NET template can be used. Invoke this command to create a new project:

```
dotnet new webapp -au IndividualB2C -o WebAppAuth
```

In creating this project, several NuGet packages are added for identities and authentication. These were discussed when we secured the API. An additional package that hasn't been used before is `Microsoft.Identity.Web.UI`. This package integrates with `Microsoft.Identity.Web` and offers pre-built UI elements for login, logout, and profile management.

With the DI container configuration, authentication is added. So, we need to customize it for calling APIs:

WebAppAuth/Program.cs

```
IConfigurationSection scopeSections = builder.Configuration
  .GetSection("AzureAdB2C").GetSection("Scopes");
String[] scopes = scopeSection.Get<string[]>() == [];

builder.Services.AddAuthentication(OpenIdConnectDefaults.
AuthenticationScheme)
  .AddMicrosoftIdentityWebApp(
    builder.Configuration.GetSection("AzureAdB2C"))
  .EnableTokenAcquisitionToCallDownstreamApi(scopes)
  .AddInMemoryTokenCaches();
```

To use Azure AAD B2C, `AddAuthentication` is invoked using the configuration from the `AzureAdB2C` section within `appsettings.json`. `AddMicrosoftIdentityWeb` is an extension method from the `Microsoft.Identity.Web` NuGet package. This configures supporting cookies and `OpenIdConnect`. The `EnableTokenAcquisitionToCallDownstreamApi` method allows us to pass tokens that have been received from the application so that we can forward them to the APIs that have been invoked by the application via `HttpClient`. When using this method, the `ITokenAcquisition` interface is registered in the DI container. This can be used to retrieve the tokens and pass them to the HTTP headers of `HttpClient`.

For the Microsoft Identity user interface, the `AddMicrosoftIdentityUI` method needs to be configured with the DI container:

WebAppAuth/Program.cs

```
builder.Services.AddRazorPages()
  .AddMicrosoftIdentityUI();
```

This method configures `AccountController` (based on ASP.NET Core MVC) with `SignIn` and `SignOut` methods in the `MicrosoftIdentity` area.

> **Note**
>
> When creating Blazor client applications, AD B2C support is built-in with .NET 7 templates, but not with .NET 8. Support has been planned for .NET 9. You can add AD B2C integration manually.
>
> Some differences in authentication can be implemented with different client technologies. Check out the links in the *Further reading* section for more information. Also, check out the Codebreaker GitHub (`https://github.com/codebreakerapp`) for implementations for Blazor, WinUI, .NET MAUI, WPF, and Uno Platform.

Specifying authentication with Azure Container Apps

Instead of needing to manage authentication with the service itself, we can do this directly with Azure Container Apps. After selecting the deployed games API, within the Azure portal, choose **Authentication** from the **Settings** category in the left pane. Here, you can add an **identity provider**. By selecting **Microsoft**, you can configure **Workforce** or **Customer** tenant types. **Workforce** is for B2B scenarios. Here, you can directly create an app registration within Microsoft Entra. For B2C, select **Customer**.

Using ASP.NET Core Identity to store user information in a local database

If Azure AD B2C is not an option for you, you can use **ASP.NET Core Identity**, which .NET offers for storing users in a local database. We'll use this as an alternative way to run the solution without the need to configure Azure AD B2C.

With **ASP .NET Core Blazor**, a template is available to create the core code needed to create an application that allows users to register, store user information in a database, and manage users. Use this template with the `-au Individual` option:

```
dotnet new blazor -au Individual -int Auto -o Codebreaker.ApiGateway.
Identities
```

This creates two projects: `Codebreaker.ApiGateway.Identities` and `Codebreaker.ApiGateway.Identities.Client`. The second project is a library that contains **Razor components** that can be run on the client with **interactive WebAssembly rendering**, as well as **interactive server rendering**. This library was referenced in the first project, which hosts the Blazor application and contains Razor components that support interactive server rendering. This project contains a huge list of Razor components for registering users to help users with forgotten passwords, as well as components for managing user information.

Let's cover some important parts of this application, beginning with the database.

Customizing the EF Core configuration

With this project, user information is stored in a relational database via EF Core. By default, MySQL is used. This can easily be changed to SQL Server, but using MySQL for this scenario is great as well.

What information about users is stored is defined with the `ApplicationDbContext` class:

Codebreaker.ApiGateway.Identities/Data/ApplicationDbContext.cs

```
public class
ApplicationDbContext(DbContextOptions<ApplicationDbContext> options) :
  IdentityDbContext<ApplicationUser>(options)
{
}
```

`ApplicationDbContext` is an EF Core context with a hierarchy of base classes. The body of this class is empty as it was created from the template. Adding custom `DbSet` properties allows you to add additional tables to the database. The base class, `IdentityDbContext`, uses the `ApplicationUser` class as a generic parameter to define what information to store about the user:

Codebreaker.ApiGateway.Identities/Data/ApplicationUser.cs

```
public class ApplicationUser : IdentityUser
{
}
```

Adding properties to this class allows you to customize the `users` table with additional columns. To see the defined properties, you need to follow the base classes, starting with `IdentityUser`.

`IdentityUser` derives from `IdentityUser<string>`. The generic string parameter specifies the use of GUID values for the key. The generic `IdentiyUser` type defines the `UserName`, `Email`, `PasswordHash`, and `PhoneNumber` properties, among others, to map to columns.

`IdentityDbContext<TUser>` has some more base classes, such as `IdentityUserContext<TUser, TRole, TKey, TUserClaim, TuserLogin, and TUserToken>`, to define several tables that are used.

The EF Core context needs to be configured with the DI container:

Codebreaker.ApiGateway.Identities/Program.cs

```
// code removed for brevity
builder.AddMySqlDbContext<ApplicationDbContext>("usersdb");
```

Here, the EF Core configuration has been changed to use the `Aspire.Pomelo.EntityFrameworkCore.MySql` NuGet package with the MySQL Entity Framework .NET Aspire component.

With that, the EF Core context has been configured with ASP.NET Core Identity.

Configuring ASP.NET Core Identity

When configuring ASP.NET Core Identity, EF Core must be mapped:

Codebreaker.ApiGateway.Identities/Program.cs

```
// code removed for brevity
builder.Services.AddIdentityCore<ApplicationUser>(options =>
  options.SignIn.RequireConfirmedAccount = true)
  .AddEntityFrameworkStores<ApplicationDbContext>()
  .AddSignInManager()
  .AddDefaultTokenProviders();
```

The `AddIdentityCore` method configures the `ApplicationUser` class (the same class that was used with the EF Core model) for ASP.NET Core Identity. When the user registers, before using the account, it needs to be confirmed by setting the `RequireConfirmedAccount` property (discussed next). With the invocation of `AddEntityFrameworkStores`, the EF Core context, `ApplicationDbContext`, is mapped to ASP.NET Core Identity. The `AddSignInManager` method registers the `SignInManager` class with the DI container. `SignInManager` can be used to log the user in and out, retrieve claims, and work with two-factor authentication options. The `AddDefaultTokenProviders` method registers token providers by implementing the `IUserTwoFactorTokenProvider` interface to return and validate tokens for two-factor authentication, such as email, phone, and so on.

To confirm an account, the IEmailSender interface needs to be registered with the DI container:

Codebreaker.ApiGateway.Identities/Data/ApplicationUser.cs

```
builder.Services.AddSingleton<IEmailSender<ApplicationUser>,
IdentityNoOpEmailSender>();
```

With the default configuration, a no-op IdentityNoOpEmailSender class is implemented. This is practical for testing purposes but needs to be changed to verify a user's email address.

Now, let's configure the project with the .NET Aspire AppHost project:

CodebreakerAppHost/Program.cs

```
string startupMode = Environment.GetEnvironmentVariable("STARTUP_
MODE") ?? "Azure";
bool useAzureADB2C = startupMode == "Azure";
// code removed for brevity
if (startupMode == "OnPremises")
{
  var usersDbName = "usersdb";
  var mySqlPassword = builder.AddParameter("mysql-password", secret:
true);

  var usersDb = builder.AddMySql("mysql", password: mySqlPassword)
    .WithEnvironment("MYSQL_DATABASE", usersDbName)
    .WithDataVolume()
    .WithPhpMyAdmin()
    .AddDatabase(usersDbName);

  var gateway = builder.AddProject<Projects.Codebreaker_ApiGateway_
    Identities>("gateway-identities")
    .WithReference(gameAPIs)
    .WithReference(bot)
    .WithReference(usersDb)
    .WithExternalHttpEndpoints();
```

Here, the AppHost project uses multiple launch profiles to either start the solution with Azure AD B2C (the Azure launch profile), or with the local database (the OnPremises launch profile). When it comes to the different launch profile settings, the STARTUP_MODE environment variable is configured, which is then used to differentiate the projects to be started and how they are configured. When launching OnPremises mode, the newly created project is configured to reference the MySQL database running in a container via the Aspire.Hosting.MySql NuGet package. The WithDataVolume method creates a named Docker volume (see *Chapter 5*) to have persistence, while the WithPhpMyAdmin method adds an admin UI.

If we run the solution now, we can register a new user, as shown in *Figure 9.11*:

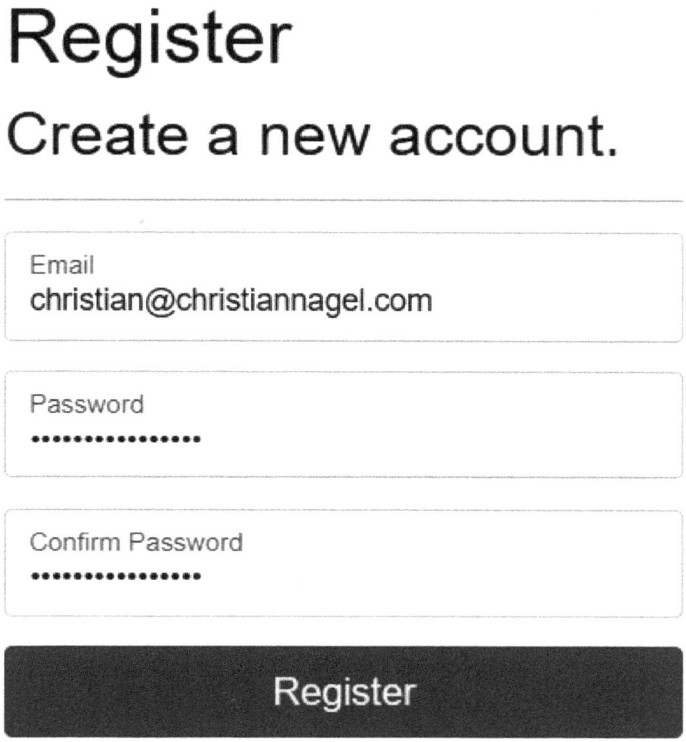

Register

Create a new account.

Email
christian@christiannagel.com

Password
•••••••••••••••

Confirm Password
•••••••••••••••

Register

Figure 9.11 – Registering a local user

When registering the user, you might need to apply EF Core migrations to create the database. On receiving the registration confirmation, choose **Click here to confirm your account** to approve the email. Then, click the **Login** button on the left pane. After logging in, the email will be shown on the **Auth Required** page.

> **Note**
>
> Instead of having the user remember another password, with ASP.NET Core Identity, it's also possible to add external providers, such as Microsoft, Facebook, and Google accounts, as shown at `https://learn.microsoft.com/en-us/aspnet/core/security/authentication/social`.

With phpMyAdmin enabled, you can open the management UI and see the tables that have been created, as shown in *Figure 9.12*:

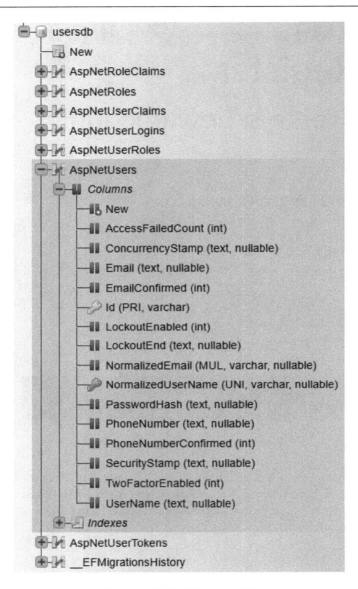

Figure 9.12 – MySQL admin UI

Using this admin UI, you can perform SQL queries and easily change and delete records.

With ASP.NET Core Identity in place, users can now register with this application and manage their accounts. This option is great if user data isn't stored within a managed cloud service and can be implemented easily. What about using desktop client applications? They can use an API to access this data. We'll learn how to add this API in the next section.

Creating identity API endpoints

.NET 8 offers identity API endpoints that use the ASP.NET Core Identity infrastructure.

With the EF Core configuration for ASP.NET Core Identity in place, all we need to do is configure the identity endpoints with the DI container and the middleware. The DI container must be configured first:

Codebreaker.Gateway.Identity/Program.cs

```
// code removed for brevity
builder.Services
  .AddIdentityApiEndpoints<ApplicationUser>()
  .AddEntityFrameworkStores<ApplicationDbContext>();
```

The `AddIdentityApiEndpoints` method adds authentication with a Bearer token and identity cookies, as well as options and validators to validate allowed passwords and usernames, register `UserManager`, and provide a factory for user claims. `IEmailSender`, which is used to validate correct emails, is configured to use `NoOpEmailSender`. When you have a real implementation of `IEmailSender` in place (using your email provider), you need to make sure you register this class after the invocation of `AddIdentityApiEndpoints` to overwrite `NoOpEmailSender` with your configuration. The `AddEntityFrameworkStores` method is an extension method for the returned `IdentityBuilder` object and adds the EF Core store for user and role data.

The middleware can be configured using the `MapIdentityApi` method:

Codebreaker.Gateway.Identity/Program.cs

```
// code removed for brevity
app.MapGroup("/identity")
  .MapIdentityApi<ApplicationUser>();
```

The `MapGroup` method is used to add a common prefix for the identity API. `MapIdentityApi` itself defines several URIs, such as `/register` to register a new user by using `RegisterRequest` with the body of a POST request, and `/login` to log a user in while passing `LoginRequest`, which can include username, password, and two-factor codes, links to reset a forgotten password, confirmation of the email, and more.

Some of these APIs allow anonymous access (for example, when registering or logging in), while with others, authentication is required. The API group with the `/manage` link is configured to require authentication.

When Swagger is enabled, you'll see all these APIs, as shown in *Figure 9.13*. This means you can test them before using them from a client application:

Figure 9.13 – Identity API endpoints

Try invoking the /register API and create a new user passing two values. With a success, HTTP status code 200 is returned with an empty HTTP body if you didn't create a custom implementation of the IEmailSender interface. If this is the case, you can use the MySQL admin UI to approve the user (or change the ASP.NET Core Identity configuration so that it doesn't require confirmed accounts) before logging in; otherwise, login will be denied.

With a successful login, Bearer tokens are returned. You receive access and refresh tokens and expiration information that is set to 3,600 seconds by default. The refresh token can be used with the /refresh API to get new access and refresh tokens.

> **Note**
>
> To learn how to use SendGrid to implement IemailSender, take a look at the following article: https://learn.microsoft.com/en-us/aspnet/core/security/authentication/accconfirm.

Summary

In this chapter, you learned how to authenticate users with Azure AD B2C using Microsoft Identities and ASP.NET Core Identity. With Azure AD B2C, you added custom user attributes, specified user flows, and registered applications.

Instead of implementing protection with every API, you created a reverse proxy using Microsoft YARP and protected the APIs with a gateway service. Using YARP, we defined routes to map different backend services and configured policies with routes to require authenticated clients.

You also learned to use ASP.NET Core Identity as an alternative option for authentication and authorization with built-in ASP.NET Core functionality but a simpler feature set.

The next chapter covers how to test microservices solutions, from unit tests to integration tests, including testing services with Microsoft Playwright.

Further reading

To learn more about the topics that were discussed in this chapter, please refer to the following links:

- *Azure AD B2C Claims Schema*: https://learn.microsoft.com/en-us/azure/active-directory-b2c/claimsschema

- *Enrich tokens with claims from external sources using API connectors*: https://learn.microsoft.com/en-us/azure/active-directory-b2c/add-api-connector-token-enrichment

- *ASP.NET Core Middleware*: https://learn.microsoft.com/en-us/aspnet/core/fundamentals/middleware

- *How to use Identity to secure a Web API backend for SPAs*: https://learn.microsoft.com/en-us/aspnet/core/security/authentication/identity-api-authorization

- *GitHub repository for YARP*: https://github.com/microsoft/reverse-proxy

- *Securing a Blazor WASM app with Azure AD B2C*: https://learn.microsoft.com/en-us/aspnet/core/blazor/security/webassembly/hosted-with-azure-active-directory-b2c

- *Securing a WPF desktop app with Aure AD B2C*: https://learn.microsoft.com/en-us/azure/active-directory-b2c/configure-authentication-sample-wpf-desktop-app

- *Choosing an identity management solution*: https://learn.microsoft.com/en-us/aspnet/core/security/how-to-choose-identity-solution

- *Azure API Management*: https://learn.microsoft.com/en-us/azure/api-management/

Part 3: Troubleshooting and Scaling

In this part, the focus shifts towards ensuring the smooth operation of the application and promptly addressing any emerging issues. Emphasis is placed on early issue detection through unit testing. You will delve into creating integration tests using .NET Aspire libraries and implementing end-to-end testing with Microsoft Playwright. The importance of logs, metrics, and distributed tracing, facilitated by Open Telemetry and supported by .NET Aspire, will be explored. Monitoring service interactions, performance metrics, memory consumption, and more during development will be facilitated by the .NET Aspire dashboard. Within the Azure environment, Azure Log Analytics and Application Insights will be utilized, alongside alternative options like **Prometheus** and **Grafana** that can be deployed in both on-premises and cloud environments. When scaling services, insights gained from previous chapters will be leveraged, with caution advised when using Azure Load Testing to prevent exceeding budget limits. Before scaling up and out, potential performance enhancements will be identified and implemented.

This part has the following chapters:

- *Chapter 10, All about Testing the Solution*
- *Chapter 11, Logging and Monitoring*
- *Chapter 12, Scaling Services*

All About Testing the Solution

When creating microservices and using **continuous integration and continuous delivery (CI/CD)**, finding errors early is an important part. Having errors in production is costly, and it's best to find them as early as possible. Testing helps reduce costs by finding errors early.

This chapter covers different kinds of tests needed with microservices solutions. We start creating unit tests, which should be the major tests used because issues are found fast, followed by integration tests, where multiple components of the solution are tested in collaboration. Integration tests can be done in-process where HTTP requests are simulated and in an environment where services are running on the systems, which allows you to test the environment under load.

In this chapter, you'll learn how to do the following:

- Create unit tests
- Create .NET Aspire integration tests
- Create end-to-end .NET Playwright tests

Technical requirements

With this chapter, as with the previous chapters, you need an Azure subscription and a Docker Desktop.

The code for this chapter can be found in this GitHub repository: `https://github.com/PacktPublishing/Pragmatic-Microservices-with-CSharp-and-Azure/`.

In the `ch10/final` folder, you'll see these projects with the final result of this chapter.

These projects are unchanged from previous chapters:

- `Codebreaker.AppHost` – The .NET Aspire host project
- `Codebreaker.ServiceDefaults` – Common service configuration
- `Codebreaker.Bot` – The bot service to run games

These projects are unchanged from previous chapters, but of special interest for the tests:

- `Codebreaker.Analyzers` – This is the project that contains analyzers to verify game moves and return results
- `Codebreaker.GameApis` – The games API service project

These projects are new:

- `Codebreaker.Analyzers.Tests` – Unit tests for the analyzer library
- `Codebreaker.Bot.Tests` – Unit tests for the bot service library
- `Codebreaker.GameAPIs.Tests` – Unit tests for the games services project
- `Codebreaker.GameAPIs.IntegrationTests` – In-memory integration tests
- `Codebreaker.GameAPIs.Playwright` – Tests with Microsoft Playwright

Working through the code with this chapter, you can start using the `start` folder, which contains the same projects without the test projects.

To easily deploy the solution to Microsoft Azure, check out the README file in the source code repo of this chapter.

Creating unit tests

Unit tests are tests that test a small piece of testable software. Does this functionality behave as expected? These tests should be fast, used directly on the developer system (and run with CI as well). With the **Visual Studio Live Unit Testing** feature (part of Visual Studio Enterprise), unit tests run while the code is updated, even before saving the source code.

The cost of bugs grows during the **software development life cycle** (**SDLC**). When bugs are found late (for example, in production), the cost grows exponentially. For fixing bugs early (for example, while typing the code), Visual Studio can give hints and show errors; as we are already working on the code, it's not necessary to take the time to dive into the functionality as we are already working on it. For finding bugs with other test types (for example, integration or load tests), the fixes are more expensive – but of course, a lot less expensive than finding a bug in production.

A goal should be to reduce cost, and thus if some functionality can be verified with unit tests and other test types, prefer unit tests.

Before we start creating a unit test, what is the heart of the games service in need of unit tests? It's the analyzer library. There's some complexity with the game rules, and it's easy to make some logical errors writing the code. It's also a place where some refactoring could be done to increase performance and reduce memory needs. After refactoring, the application should function in the same way.

> **Note**
>
> When I was initially developing the games analyzer library, I created unit tests beforehand and enhanced the unit tests while developing the algorithms. With **test-driven development (TDD)**, unit tests are created before the functionality.
>
> Before fixing bugs, I also created new unit tests. Why did the bug occur? Why was it not covered by a test? With many different projects, I see bugs that have been fixed come back with a later version. If there's a unit test to verify the functionality, the same issue cannot resurface with a new version.

Next, let's dive into the Codebreaker code, which needs many unit tests.

Exploring the games analyzer library

Let's explore the GameGuessAnalyzer class in the Codebreaker analyzer library:

Codebreaker.GameAPIs.Analyzers.Tests/Analyzers/GameGuessAnalyzer.cs

```
public abstract class GameGuessAnalyzer<TField, TResult> :
  IGameGuessAnalyzer<TResult>
  where TResult : struct
{
  protected readonly IGame _game;
  private readonly int _moveNumber;
  protected TField[] Guesses { get; private set; }
  protected GameGuessAnalyzer(IGame game, TField[] guesses, int
moveNumber)
  {
    _game = game;
    Guesses = guesses;
    _moveNumber = moveNumber;
  }

  protected abstract TResult GetCoreResult();

  private void ValidateGuess()
  {
    // code removed for brevity
  }

  protected abstract void SetGameEndInformation(TResult result);

  public TResult GetResult()
```

```
    {
      ValidateGuess();
      TResult result = GetCoreResult();
      SetGameEndInformation(result);
      return result;
    }
  }
```

The GetResult method is the heart of this class. With the constructor of the GameGuessAnalyzer abstract base class, the game and the guesses are passed with parameters. The GetResult method uses the codes of the game and uses the guesses to return the result – the number of colors that are in the correct position and the number of colors that are correct but in the wrong position. The implementation of the GetResult method is just an invocation of four methods. The ValidateGuess method analyses the correctness of guesses and throws an exception if the guesses are not correct. The GetCoreResult method is abstract and needs to be implemented by a derived class.

One of the classes deriving from the GameGuessAnalyzer class is the ColorGameGuessAnalyzer class. This is used by the Game6x4 and Game8x5 game types (six colors with four codes and eight colors with five codes):

Codebreaker.Analyzers.Tests/Analyzers/ColorGameGuessAnalyzer.cs

```
public class ColorGameGuessAnalyzer(
  IGame game, ColorField[] guesses, int moveNumber) :
  GameGuessAnalyzer<ColorField, ColorResult>(game, guesses,
moveNumber)
{
  protected override ValidateGuessValues()
  {
    // code removed for brevity
  }

  protected override ColorResult GetCoreResult()
  {
    // code removed for brevity
  }
}
```

This class overrides the ValidateGuessValues and the GetCoreResult methods. ValidateGuessValues validates the input data and throws exceptions if the data is not valid. The GetCoreResult method implements the algorithm for the Codebreaker game, finds if the guesses are correctly placed and if the guesses are correct but incorrectly placed, and returns the result accordingly.

Let's create a unit test project for this library.

Creating a unit test project

Using the .NET CLI, we can create a new xUnit test project:

```
dotnet new xunit -o Codebreaker.Analyzers.Tests
cd Codebreaker.Analyzers.Tests
dotnet add reference ..\Codebreaker.Analyzers
```

This command creates a `Codebreaker.Analyzers.Tests` project with references to xUnit NuGet packages and a project reference to the analyzer project.

Note

I'm mainly using xUnit for unit tests. It's a matter of choice whether to use **MSTest**, **NUnit**, or **xUnit**; you can use any of these frameworks for unit tests, and all of these are greatly integrated within the .NET tools. I myself switched from MSTest to xUnit with early betas of .NET Core 1.0 when xUnit was available, but MSTest was not ready for the new .NET – and most unit tests from the .NET team themselves are done using xUnit.

Before creating the first tests, some preparations need to be done.

Mocking the IGame interface

With the constructor of the `ColorGameGuessAnalyzer` class, an object implementing the `IGame` interface is required with the constructor. A unit test should only test a small functionality without testing dependencies that are covered by their own unit tests. When testing the `ColorGameGuessAnalyzer` class, we don't want to add a dependency to the `Game` class while testing the analyzer. What's needed by the `ColorGameGuessAnalyzer` class is the `IGame` interface. To allow the test to run, the `IGame` interface is implemented by a mocking class:

Codebreaker.GameAPIs.Analyzers.Tests/MockColorGame.cs

```
public class MockColorGame : IGame
{
  public Guid Id { get; init; }
  public int NumberCodes { get; init; }
  public int MaxMoves { get; init; }
  public DateTime? EndTime { get; set; }
  public bool IsVictory { get; set; }
  // code removed for brevity
}
```

The `MockColorGame` class is just a simple data holder to implement the `IGame` interface, thus we don't need to use any mocking library. With another unit test implementation done later, we'll use a mocking library to mock functionality that should not be tested by the unit test.

Creating test helpers

To define common functionality needed by multiple unit tests, helper methods are created within the `ColorGame6x4AnalyzerTests` test class:

Codebreaker.GameAPIs.Analyzers.Tests/Analyzers/ColorGame6x4Analyz-erTests.cs

```
private static MockColorGame CreateGame(string[] codes) => new()
{
  GameType = GameTypes.Game6x4,
  NumberCodes = 4,
  MaxMoves = 12,
  IsVictory = false,
  FieldValues = new Dictionary<string, IEnumerable<string>>()
  {
    [FieldCategories.Colors] = [.. TestData6x4.Colors6]
  },
  Codes = codes
};

private static ColorResult AnalyzeGame(
  string[] codes,
  string[] guesses,
  int moveNumber = 1)
{
  MockColorGame game = CreateGame(codes);
  ColorGameGuessAnalyzer analyzer = new(game, [.. guesses.
    ToPegs<ColorField>()], moveNumber);
  return analyzer.GetResult();
}
```

The `AnalyzeGame` method receives a string array representing the valid code, a string array representing the guesses, and the move number. This information is used to create a mocked game instance and to invoke the `GetResult` method of the analyzer class. The result of the analysis is returned with a `ColorResult` type. This helper method can now be used to easily create unit tests.

Creating a simple unit test

The first unit test is implemented with the GetResult_Should_ReturnThreeWhite method:

Codebreaker.GameAPIs.Analyzers.Tests/ColorGame6x4AnalyzerTests.cs

```
[Fact]
public void GetResult_Should_ReturnThreeWhite()
{
    ColorResult expectedKeyPegs = new(0, 3);
    ColorResult? resultKeyPegs = AnalyzeGame(
        [Green, Yellow, Green, Black],
        [Yellow, Green, Black, Blue]
    );

    Assert.Equal(expectedKeyPegs, resultKeyPegs);
}
```

Using xUnit, the Fact attribute declares a method to be a unit test. A unit test consists of three parts: **arrange, act, and assert** (**AAA**). With *arrange*, the expected result is defined using the expected-KeyPegs variable. Invoking the AnalyzeGame method is the act. The Green – Yellow – Green – Black code is passed as valid code, and Yellow – Green – Black – Blue as a guess. With this guess, no color is in the correct position, but three colors are correct in the wrong positions, thus three whites should be returned. If this result is correct, this is verified using the Assert.Equal method.

Passing test data to unit tests

With this scenario, it's useful to just define one method that's used with different test data to verify the different outcomes:

Codebreaker.GameAPIs.Analyzers.Tests/ColorGame6x4AnalyzerTests.cs

```
[InlineData(1, 2, Red, Yellow, Red, Blue)]
[InlineData(2, 0, White, White, Blue, Red)]
[Theory]
public void GetResult_ShouldReturn_InlineDataResults(
    int expectedBlack, int expectedWhite,
    params string[] guessValues)
{
    string[] code = [Red, Green, Blue, Red];
    ColorResult expectedKeyPegs = new (expectedBlack, expectedWhite);
```

```
    ColorResult resultKeyPegs = AnalyzeGame(code, guessValues);
    Assert.Equal(expectedKeyPegs, resultKeyPegs);
}
```

With xUnit, using the `Theory` attribute instead of the `Fact` attribute allows the test method to be invoked multiple times, passing different test data. The `GetResult_ShouldReturn_InlineResults` method uses arguments that are specified with the `InlineData` attribute. With every `InlineData` attribute, the parameter values for the arguments defined with the method are passed. Here, two tests are covered with one implementation. This feature allows for quickly extending test cases by just adding new `InlineData` attributes.

Instead of using the `InlineDataAttribute` class, a class can be created implementing `IEnumerable<object[]>` to supply test data:

Codebreaker.GameAPIs.Analyzers.Tests/ColorGame6x4AnalyzerTests.cs

```
public class TestData6x4 : IEnumerable<object[]>
{
  public static readonly string[] Colors6 = [Red, Green, Blue, Yellow,
    Black, White];

  public IEnumerator<object[]> GetEnumerator()
  {
    yield return new object[]
    {
      new string[] { Green, Blue,  Green, Yellow },
      new string[] { Green, Green, Black, White },
      new ColorResult(1, 1) // expected
    };
    yield return new object[]
    {
      new string[] { Red,   Blue,  Black, White },
      new string[] { Black, Black, Red,   Yellow },
      new ColorResult(0, 2)
    };
    // code removed for brevity - more test cases here
  }

  IEnumerator IEnumerable.GetEnumerator() => GetEnumerator();
}
```

`object []` defines all the values for one method invocation. The first argument passed defines the valid code for the game, the second argument the guess data, and the third argument the expected result. With every iteration of `IEnumerable`, a new test run is done. The next code snippet shows the test method implementation using the data class:

Codebreaker.GameAPIs.Analyzers.Tests/ColorGame6x4AnalyzerTests.cs

```
[Theory]
[ClassData(typeof(TestData6x4))]
public void GetResult_ShouldReturn_UsingClassdata(
  string[] code,
  string[] guess,
  ColorResult expectedKeyPegs)
{
  ColorResult actualKeyPegs = AnalyzeGame(code, guess);
  Assert.Equal(expectedKeyPegs, actualKeyPegs);
}
```

Instead of using the `InlineData` attribute, here `ClassData` is used. Using an object returning the test data is more flexible. The `InlineData` attribute requires constant values that are stored by the compiler. With the `ClassData` attribute, data can also be created dynamically.

Expecting exceptions with a unit test

Another test case where we expect an exception to be thrown is shown in the next code snippet:

Codebreaker.GameAPIs.Analyzers.Tests/ColorGame6x4AnalyzerTests.cs

```
[Fact]
public void GetResult_Should_ThrowOnInvalidGuessValues()
{
  Assert.Throws<ArgumentException>(() =>
    AnalyzeGame(
      ["Black", "Black", "Black", "Black"],
      ["Black", "Der", "Blue", "Yellow"] // "Der" is wrong
  ));
}
```

`Assert.Throws` defines the exception type that should be thrown by the implementation when the test data is passed. If an exception is not thrown, the test fails.

Using a mocking library

With some classes that should be tested, it's great to have a mocking library. The GamesService class injecting the IGamesRepository interface is shown in the following code snippet:

Codebreaker.GameAPIs/Services/GamesService.cs

```
public class GamesService(IGamesRepository dataRepository) :
IGamesService
{
  public async Task<(Game Game, Move Move)> SetMoveAsync(
    Guid id, string gameType, string[] guesses,
    int moveNumber,
    CancellationToken cancellationToken = default)
  {
    Game? game = await dataRepository.GetGameAsync(id,
    cancellationToken);
    CodebreakerException.ThrowIfNull(game);
    CodebreakerException.ThrowIfEnded(game);
    CodebreakerException.ThrowIfUnexpectedGameType(game, gameType);

    Move move = game.ApplyMove(guesses, moveNumber);

    await dataRepository.AddMoveAsync(game, move, cancellationToken);

    return (game, move);
  }
  // code removed for brevity
}
```

With the GamesService class, the IGamesRepository interface is injected using constructor injection. When testing the SetMoveAsync method, the implementation of the IGamesRepository interface should not be part of this test. There's another test for the games repository. Instead, a mocking implementation of this class is used for the unit test. The SetMoveAsync method invokes the GetGameAsync method of the IGamesRepository interface. The real implementation of this method should not be part of the test, but we need some different results that can be used with the methods used afterward. When this method returns null because it didn't find the game, CodebreakerException.ThrowIfNull should throw an exception. If the method returns a game that already ended, the next method should throw an exception because a new move cannot be set to a game that already ended. The ThrowIfUnexpectedGameType method should throw an exception if the game type passed is different from the game type of the game retrieved. This can easily be solved by using a mocking library.

Let's create another xUnit test project named `Codebreaker.GameAPIs.Tests` to test the `GamesService` class. To mock the `IGamesRepository` interface, add the moq NuGet package.

The following code snippet shows fields for games and game IDs that are used by the unit test:

Codebreaker.GameAPIs.Tests/GamesServiceTests.cs

```
public class GamesServiceTests
{
  private readonly Mock<IGamesRepository> _gamesRepositoryMock =
new();
  private readonly Guid _endedGameId = Guid.Parse("4786C27B-3F9A-4C47-
9947-F983CF7053E6");
  private readonly Game _endedGame;
  private readonly Guid _running6x4GameId = Guid.Parse("4786C27B-3F9A-
4C47-9947-F983CF7053E7");
  private readonly Game _running6x4Game;
  private readonly Guid _notFoundGameId = Guid.Parse("4786C27B-3F9A-
4C47-9947-F983CF7053E8");
  private readonly Guid _running6x4MoveId1 = Guid.Parse("4786C27B-
3F9A-4C47-9947-F983CF7053E9");
  private readonly string[] _guessesMove1 = ["Red", "Green", "Blue",
"Yellow"];
```

The `IGamesRepository` interface is mocked creating a new instance by using the generic `Mock` type. After this, games are predefined for a game not found in the repository (`_notFoundGameId`), a game that already ended (`_endedGame`), and a running game that is active (`_running6x4Game`).

The constructor of the `GamesServiceTests` class initializes the game objects:

Codebreaker.GameAPIs.Tests/GamesServiceTests.cs

```
public GamesServiceTests()
{
  _endedGame = new(_endedGameId, "Game6x4", "Test", DateTime.Now, 4,
12)
  {
    Codes = ["Red", "Green", "Blue", "Yellow"],
    FieldValues = new Dictionary<string, IEnumerable<string>>()
    {
      { FieldCategories.Colors, ["Red", "Green", "Blue", "Yellow",
        "Purple", "Orange"] }
    },
    EndTime = DateTime.Now.AddMinutes(3)
  };
  // code removed for brevity
```

```
_gamesRepositoryMock.Setup(repo => repo.GetGameAsync(_endedGameId,
    CancellationToken.None)).ReturnsAsync(_endedGame);
_gamesRepositoryMock.Setup(repo => repo.GetGameAsync
(_running6x4GameId, CancellationToken.None)).ReturnsAsync
(_running6x4Game);
_gamesRepositoryMock.Setup(repo => repo.AddMoveAsync
(_running6x4Game, It.IsAny<Move>(), CancellationToken.None));
}
```

With the constructor, instances of the different game types are created. The game already ended has the EndTime property set. To specify the behavior of the mocking implementation, the Setup method is invoked. With this, if the GetGameAsync method receives the ended game ID with the parameter, it returns the configured game instance that already ended. Passing the game ID of the running game, the corresponding instance is returned. With the third invocation of the Setup method, it's defined that the AddMoveAsync method contains an implementation when passing the running game. It.IsAny<Move> allows us to invoke this method with any Move instance.

Now, we can implement unit tests. The first unit test is to verify that the SetMoveAsync method throws an exception if the game already ended:

Codebreaker.GameAPIs.Tests/GamesServiceTests.cs

```
[Fact]
public async Task SetMoveAsync_Should_ThrowWithEndedGame()
{
    GamesService gamesService = new(_gamesRepositoryMock.Object);
    await Assert.ThrowsAsync<CodebreakerException>(async () =>
    {
        await gamesService.SetMoveAsync(_endedGameId, "Game6x4", ["Red",
        "Green", "Blue", "Yellow"], 1, CancellationToken.None);
    });

    _gamesRepositoryMock.Verify(repo => repo.GetGameAsync(_endedGameId,
        CancellationToken.None), Times.Once);
}
```

In the *arrange* step, the GamesService class is instantiated with the mocking object of the IGamesRepository implementation. With the unit test act – as already used before – Assert. ThrowAsync is used to check if an exception was thrown when invoking the SetMoveAsync method with the specified game that already ended. Another check that is done here is using the Verify method on the Mock class to check if the method is exactly called once.

The SetMoveAsync_Should_ThrowWithUnexpcectedGameType and SetMoveAsync_Should_ThorwWithNotFoundGameType unit test methods are very similar, thus are not listed here. Check the source code repo for details.

The test method to test the normal flow is shown here:

```
[Fact]
public async Task SetMoveAsync_Should_UpdateGameAndAddMove()
{
  GamesService gamesService = new(_gamesRepositoryMock.Object);
  var result = await gamesService.SetMoveAsync(_running6x4GameId,
    "Game6x4", ["Red", "Green", "Blue", "Yellow"], 1,
    CancellationToken.None);

  Assert.Equal(_running6x4Game, result.Game);
  Assert.Single(result.Game.Moves);

  _gamesRepositoryMock.Verify(repo => repo.GetGameAsync
  (_running6x4GameId, CancellationToken.None), Times.Once);
  _gamesRepositoryMock.Verify(repo => repo.AddMoveAsync
  (_running6x4Game, It.IsAny<Move>(), CancellationToken.None), Times.
  Once);
}
```

The SetMoveAsync_Should_UpdateGameAndAddMove method verifies that the GetGameAsync and AddMoveAsync methods are called once, and with the first move in the game, the Moves property contains exactly one value.

Running unit tests

To start unit tests, you can use the dotnet test .NET CLI command to run all the tests. Using Visual Studio, the **Test** menu is available to run all the tests. Using **Test Explorer**, as shown in *Figure 10.1*, you can start testing by test, a group of tests, or all tests, see the outcome of every test, debug tests, run tests until they fail, define a playlist of tests, and more:

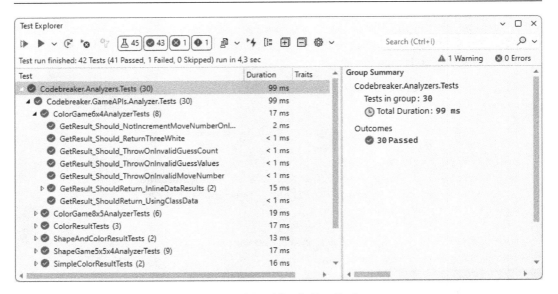

Figure 10.1 – Visual Studio Test Explorer

When using Visual Studio 2022 Enterprise Edition, you can start Live Unit Testing. With Live Unit Testing, a unit test runs while you change the source code. Here, you can also monitor which code lines are covered by a unit test and which lines are missed. *Figure 10.2* shows the Visual Studio Code editor with Live Unit Testing turned on, and code line *53* missed from all unit tests:

```
22
23              // check black
24    > ✓       for (int i = 0; i < guessPegsToCheck.Count; i++)[...]
34
35              // check white
36    > ✓       foreach (ColorField value in guessPegsToCheck)[...]
49
50      ✓       ColorResult resultPegs = new ColorResult(Correct: black, WrongF
51    ✓ ✓       if (resultPegs.Correct + resultPegs.WrongPosition > _game
52              {
53      —           throw new InvalidOperationException(message: "More key
54              }
55
56      ✓       return resultPegs;
57            }
```

Figure 10.2 – Visual Studio Live Unit Testing

After working through some unit tests, let's move over to other test types.

Creating .NET Aspire integration tests

While unit tests should be the primary tests, integration tests not only test a small functionality but include testing of multiple components within one test, such as including infrastructure – for example, a database.

.NET Aspire contains a library and test templates using xUnit, which easily allows creating integration tests to directly access the app model.

Let's create a .NET Aspire test project using .NET Aspire and xUnit, named `Codebreaker.IntegrationTests`:

```
dotnet new aspire-xunit -o Codebreaker.IntegrationTests
```

This project includes references to the `Aspire.Hosting.Testing` NuGet package, as well as `Microsoft.NET.Test.Sdk`, xUnit, and `xunit.runner.visualstudio`. To allow accessing the app model, add a project reference to `Codebreaker.AppHost`. With the integration test we implement, we need types from the game APIs project, thus we also add a reference to `Codebreaker.GameAPIs`.

Creating asynchronous initialization

With all the integration tests of the game APIs, we need an `HttpClient` instance. xUnit allows for asynchronous initialization by implementing the `IAsyncLifetime` interface:

Codebreaker.IntegrationTests/GameAPIsTests.cs

```csharp
public class GameAPIsTests : IAsyncLifetime
{
    private DistributedApplication? _app;
    private HttpClient? _client;

    public async Task InitializeAsync()
    {
        // code removed for brevity
    }

    public async Task DisposeAsync()
    {
        if (_app is null) throw new InvalidOperationException();
        await _app.DisposeAsync();
    }
    // code removed for brevity
```

Renaming the class created from the `IAsyncLifetime` interface defines the `InitializeAsync` and `DisposeAsync` methods. Field members that will be initialized within the InitalizeAsync method are the `DistributedApplication` and `HttpClient` classes. You already know the `DistributedApplication` class from the app model in the `AppHost` project. You'll see how this will be used in the `InitalizeAsync` method.

While we didn't dispose of the `DistributedApplication` instance in the `AppHost` project (because there's only one instance running for the lifetime of the application, and the resources are freed on the application end), it's important to dispose of it with unit tests, as it initialized providers and file watchers. With many tests, the user limit of `INotify` instances and the process limit of open file descriptors can be reached – thus, don't forget to dispose of this resource in test projects.

Let's see how to create `DistributedApplication` and `HttpClient` classes:

Codebreaker.IntegrationTests/GameAPIsTests.cs

```
public async Task InitializeAsync()
{
    var appHost = await DistributedApplicationTestingBuilder.
      CreateAsync<Projects.Codebreaker_AppHost>();
    _app = await appHost.BuildAsync();
    await _app.StartAsync();
    _client = _app.CreateHttpClient("gameapis");
}
```

Using `DistributedApplicationTestingBuilder` (defined in the `Aspire.Hosting.Testing` namespace), invoking the `CreateAsync` method, a new instance of `DistributedApplicationTestingBuilder` is returned. The generic parameter references the `Codebreaker AppHost` project. Similar to what you've seen when using the generic parameter with the projects referenced in the `AppHost` project, the same mechanism is used here, referencing the `AppHost` project itself. Invoking the `BuildAsync` method returns a `DistributedApplication` instance that we may forget to dispose of. Using this instance, we can access the app model definition. In the app model specified by the `Codebreaker.AppHost` project, we have gameapis defined, which is the name of the `Codebreaker.GameAPIs` project. `CreateHttpClient` returns an `HttpClient` object to reference this service. Both the `HttpClient` and the `DistributedApplication` objects returned are assigned to field members. Now, we are ready to create tests.

Creating a test to verify an HTTP bad request status

With the first test, let's verify if the correct status code is returned when an invalid move number is sent. First, we need to start a new game:

Codebreaker.IntegrationTests/GameAPIsTests.cs

```
[Fact]
public async Task SetMove_Should_ReturnBadRequest_
WithInvalidMoveNumber()
{
  if (_client is null) throw new InvalidOperationException();

  CreateGameRequest request = new(GameType.Game6x4, "test");
  var response = await _client.PostAsJsonAsync("/games", request);
  var gameResponse = await response.Content.
    ReadFromJsonAsync<CreateGameResponse>();
  Assert.NotNull(gameResponse);
  // code removed for brevity
```

Starting the game, we already use the `HttpClient` instance and invoke an HTTP `POST` request, passing the `CreateGameRequest` object. `CreateGameRequest` is available in the test project because we added a project reference to the `Codebreaker.GameAPIs` project on creation of the test project.

Continue the implementation of this method by setting a game move:

Codebreaker.IntegrationTests/GameAPIsTests.cs

```
[Fact]
public async Task SetMove_Should_ReturnBadRequest_
WithInvalidMoveNumber()
{
  // code removed for brevity
  int moveNumber = 0;
  UpdateGameRequest updateGameRequest = new(gameResponse.Id,
  gameResponse.GameType, gameResponse.PlayerName, moveNumber)
  {
    GuessPegs = ["Red", "Red", "Red", "Red"]
  };

  string uri = $"/games/{updateGameRequest.Id}";
  var updateGameResponse = await _client.PatchAsJsonAsync(uri,
    updateGameRequest);
```

```
   Assert.Equal(HttpStatusCode.BadRequest, updateGameResponse.
     StatusCode);
}
```

We use `HttpClient` once more – this time to send a PATCH request. Passing `moveNumber` with a 0 value specifies an incorrect move. The first correct move starts with 1. This way, we expect to receive a `BadRequest` result, which is verified using `Assert.Equal`.

Let's create another test to play a complete game.

Creating a test to play a complete game

The following code snippet shows an integration test setting multiple moves:

Codebreaker.IntegrationTests/GameAPIsTests.cs

```
[Fact]
public async Task SetMoves_Should_WinAGame()
{
  // code removed for brevity
  int moveNumber = 1;
  UpdateGameRequest updateGameRequest = new(gameResponse.Id,
    gameResponse.GameType, gameResponse.PlayerName, moveNumber)
  {
    GuessPegs = ["Red", "Red", "Red", "Red"]
  };

  string uri = $"/games/{updateGameRequest.Id}";
  response = await _client.PatchAsJsonAsync(uri, updateGameRequest);
  var updateGameResponse = await response.Content.
    ReadFromJsonAsync<UpdateGameResponse>();
  Assert.NotNull(updateGameResponse);
  // code removed for brevity
```

Starting the game is the same as before, thus the code is not shown here. Sending the first move is just a little bit different in that we send the correct move number. From there, we continue sending a GET request:

Codebreaker.IntegrationTests/GameAPIsTests.cs

```
  // code remove for brevity
  if (!updateGameResponse.IsVictory)
  {
```

```
Game? game = await _client.GetFromJsonAsync<Game?>(uri);
Assert.NotNull(game);

moveNumber = 2;
updateGameRequest = new UpdateGameRequest(gameResponse.Id,
  gameResponse.GameType, gameResponse.PlayerName, moveNumber)
{
  GuessPegs = game.Codes
};
response = await _client.PatchAsJsonAsync(uri, updateGameRequest);
```

Before sending a GET request, we check if the game was won with the first move. This should happen about once in 1,296 invocations; thus, it will happen when running the test often. We don't want to fail the test if the game was won with the first move. If the game has not been won yet, a GET request is done to find out the correct values, then the correct values are used to make the second move.

Sending the correct move, we should get a successful result:

Codebreaker.IntegrationTests/GameAPIsTests.cs

```
    // code removed for brevity
    Assert.True(response.IsSuccessStatusCode);
    updateGameResponse = await response.Content.
      ReadFromJsonAsync<UpdateGameResponse>();

    Assert.NotNull(updateGameResponse);
    Assert.True(updateGameResponse.Ended);
    Assert.True(updateGameResponse.IsVictory);
  }
  // delete the game
  response = await _client.DeleteAsync(uri);
  Assert.True(response.IsSuccessStatusCode);
}
```

After sending the second move, the result is verified. Finally, the game is deleted. In between all these invocations, results are verified.

Run all the integration tests either using the dotnet test .NET CLI command or with Visual Studio Test Explorer, just as before with the unit tests. Just remember not to use integration tests with Live Unit Testing.

With CI, as covered in *Chapter 8*, all these tests should run as well. This can simply be done using dotnet test.

Using .NET Aspire testing for integration tests has the advantage that the server doesn't need to be started. However, creating load tests, testing the solution before a switch to the production environment, and directly sending HTTP requests should be done from a test environment as well. We'll do this in the next section.

Creating end-to-end .NET Playwright tests

Microsoft Playwright (`https://playwright.dev`) offers tools and libraries from Microsoft for web tests, which include tests on web APIs.

Playwright offers several tools (including generating tests by recording actions with web pages, inspecting web pages, generating selectors, and viewing traces), tests across different platforms, and test libraries for TypeScript, JavaScript, Python, .NET, and Java. With UI automation, Playwright can replace manual testers! Here, we'll use Playwright to test APIs – using .NET!

Creating a test project with Playwright

Let's start creating a test project with Playwright. Because xUnit has a focus on unit tests and there's an issue with limiting concurrent test runs, Playwright supports NUnit and MSTest. Here, we use NUnit:

```
dotnet new nunit -o Codebreaker.GameAPIs.Playwright
cd Codebreaker.GameAPIs.Playwright
dotnet add package Microsoft.Playwright.NUnit
dotnet build
```

Using `dotnet new`, we create a new .NET project, this time using NUnit for the testing framework. `Microsoft.Playwright.NUnit` is the Playwright package for NUnit. After `dotnet build`, a `playwright.ps1` PowerShell script file is created in the `bin/debug/net8.0` folder that installs required browsers:

```
pwsh bin/debug/net8.0/playwright.ps1 install
```

Creating a context

Playwright has its own API for creating HTTP requests. This needs to be initialized, together with some housekeeping:

Codebreaker.GameAPIs.Playwright/GamesAPITests.cs

```
[assembly: Category("SkipWhenLiveUnitTesting")]
namespace Codebreaker.APIs.PlaywrightTests;

[Parallelizable(ParallelScope.Self)]
```

```
public class GamesApiTests : PlaywrightTest
{
  private IAPIRequestContext? _requestContext;
```

Because this test class shouldn't participate in live unit testing, the `Category` assembly attribute is used to mark the complete assembly with `SkipWhenLiveUnitTesting`. Contrary to xUnit where the `AssemblyTrait` attribute was used, NUnit uses the `Category` attribute.

With Playwright, the test class needs to derive from the `PlaywrightTest` base class. The field of type `IAPIRequestContext` is Playwright's API to create HTTP requests. This field is initialized with the next source code snippet:

Codebreaker.GameAPIs.Playwright/GamesAPITests.cs

```
[SetUp]
public async Task SetupApiTesting()
{
  ConfigurationBuilder configurationBuilder = new();
  configurationBuilder.SetBasePath(
    Directory.GetCurrentDirectory());
  configurationBuilder.AddJsonFile("appsettings.json");
  var config = configurationBuilder.Build();
  if (!int.TryParse(config["ThinkTimeMS"], out _thinkTimeMS))
  {
    _thinkTimeMS = 1000;
  }
  Dictionary<string, string> headers = new()
  {
    { "Accept", "application/json" }
  };
  _requestContext = await Playwright.APIRequest.NewContextAsync(new()
  {
    BaseURL = config["BaseUrl"] ?? "http://localhost",
    ExtraHTTPHeaders = headers
  });
}
```

The `SetupAPITesting` method is invoked before every test. With NUnit, such an initialization method needs to be annotated with the `Setup` attribute. To initialize `IAPIRequestContext`, the `Playwright.APIRequest.NewContextAsync` method is invoked. Here, the HTTP headers and the base address for the service are specified. To allow this to be configured with the `appsettings.json` file, the `ConfigurationBuilder` class is used. To simulate a think time, `_thinkTimeMS` is retrieved from the configuration, which is then used before setting every game move.

As the API context is created, it also needs to be disposed of:

Codebreaker.GameAPIs.Playwright/GamesAPITests.cs

```
[TearDown]
public async Task TearDownAPITesting()
{
  if (_requestContext != null)
  {
    await _requestContext.DisposeAsync();
  }
}
```

A method that's invoked after the test has run is annotated with the TearDown attribute. The context needs to be disposed of after use.

After the preparation, let's create our test.

Playing a game with Playwright

Tests created with NUnit are annotated with the Test attribute:

Codebreaker.GameAPIs.Playwright/GamesAPITests.cs

```
[Test]
[Repeat(20)]
public async Task PlayTheGameToWinAsync()
{
  // code removed for brevity
  string playerName = "test";
  (Guid id, string[] colors) = await CreateGameAsync(playerName);

  int moveNumber = 1;
  bool gameEnded = false;

  while (moveNumber < 10 && !gameEnded)
  {
    await Task.Delay(_thinkTimeMS);
    string[] guesses = [.. Random.Shared.GetItems<string>(colors, 4)];
    gameEnded = await SetMoveAsync(id, playerName, moveNumber++,
    guesses);
  }

  if (!gameEnded)
```

```
    {
      await Task.Delay(_thinkTimeMS);
      string[] correctCodes = await GetGameAsync(id, moveNumber - 1);
      gameEnded = await SetMoveAsync(id, playerName, moveNumber++,
        correctCodes);
    }

    Assert.That(gameEnded, Is.True);
}
```

NUnit uses the `Test` attribute to specify a test. The `Repeat` attribute can be used to specify the number of runs the test should be repeated running one test. This attribute is useful in generating a longer load on the server. The `PlayTheGameToWin` method defines the flow with the API. First, a new game is created invoking the `CreateGameAsync` method. After this, for up to 10 moves, moves are placed with the `SetMoveAsync` method. If – with the randomly chosen guesses – the game is already finished, we are done. Otherwise, information about the game is retrieved using `GetGameAsync`, and one more time, `SetMoveAsync` is invoked – this time with the correct move.

One of these invocations is shown in the next code snippet. For the other ones, check the source code repository:

Codebreaker.GameAPIs.Playwright/GamesAPITests.cs

```
private async Task<bool> SetMoveAsync(Guid id, string playerName, int
moveNumber, string[] guesses)
{
  Dictionary<string, object> request = new()
  {
    ["id"] = id.ToString(),
    ["gameType"] = "Game6x4",
    ["playerName"] = playerName,
    ["moveNumber"] = moveNumber,
    ["guessPegs"] = guesses
  };

  var response = await _requestContext.PatchAsync($"/games/{id}", new()
  {
    DataObject = request
  });

  Assert.That(response.Ok, Is.True);

  var json = await response.JsonAsync();
  JsonElement results = json.Value.GetProperty("results");
```

```
Assert.Multiple(() =>
{
  Assert.That(results.EnumerateArray().Count(),
    Is.LessThanOrEqualTo(4));
  Assert.That(results.EnumerateArray().All(x => x.ToString() is
    "Black" or "White"));
});

bool hasEnded = bool.Parse(json.Value.GetProperty("ended").
  ToString());
return hasEnded;
}
```

The `SetMoveAsync` method sets a move by using the `IAPIRequestContext` interface's `PatchAsync` method. Depending on the HTTP verb used, `GetAsync`, `PostAsync`... methods are available. The HTTP body that is sent to the service is specified with the `DataObject` property. The `PatchAsync` method returns an `IAPIResponse` response. Using this response, the JSON data can be retrieved using the `JsonAsync` method. The `Ok` property that is used with an `Assert` verification returns `true` with a status code in the range of 200 to 299.

With this test in place, we can run the test using `dotnet test` or with Test Explorer within Visual Studio. Just this time, the service needs to be running!

Creating test loads

The Playwright tests can now be used to simulate a user load, to run multiple users concurrently. For this, just compute resources are needed to run the needed load. By reducing the delay time, a few "virtual users" can be used to simulate the load of a bigger number of real users. How long real users are thinking between moves needs to be analyzed monitoring the solution in production.

> **Note**
>
> Reducing the delay between moves, you can use fewer compute resources to simulate a large number of real users with just a few virtual users. There's also a good reason to increase the delay time for the time used by real users. In *Chapter 12*, we'll enhance the solution with caching. What if the cached game is not available after a user has a long delay between moves? Does the application still behave correctly? You should also run such integration tests.

Using the **Microsoft Playwright Testing** cloud service, compute resources are available to test web applications. This service is – at the time of this writing – not available to test REST APIs. Another service to run load tests is **Azure Load Testing**. With this tool, you can write **JMeter** scripts to run the tests or specify web requests from the web portal. In *Chapter 12*, we'll use this service to create load tests to increase the replica count of the games API. This tool not only runs the load but also gives a great report to show information about all resources interacting with the requests.

> **Note**
>
> All the tests covered here can also be used with GitHub actions. After building the .NET libraries and applications, `dotnet test` should be triggered to start all the unit tests. After deploying the services to the test environment, integration tests should run before the solution is deployed to the next environment – for example, the staging environment. Automated load tests should make sure the solution is working under load.
>
> Continuously – for workflows triggered on a timely basis – you should check if new security issues are found in dependencies, and these dependencies should be updated. For this, with GitHub, just *Dependabot* needs to be configured.

Summary

In this chapter, you learned to create *unit tests* to test simple functionality. These tests can be used with Live Unit Testing where test errors immediately show up during development. With unit tests, you learned to use a *mocking library* to replace functionality that is not in the scope of the unit test and is covered by a different unit test.

You learned how .NET Aspire makes *integration tests* simple using `Aspire.Hosting.Testing`. There's no need to start the service, as the handler of `HttpClient` is replaced to send requests to the service in-process.

Using Microsoft Playwright, you created an integration test that makes HTTP requests to the API and can be used to test the solution under load.

While you monitored metrics data in this chapter, the next chapter expands on this so that you can create your own metric counts and add logging and distributed tracing to the microservices solution.

Further reading

To learn more about the topics discussed in this chapter, you can refer to the following links:

- Martin Fowler on testing microservices: `https://martinfowler.com/articles/microservice-testing`
- Live Unit Testing: `https://learn.microsoft.com/en-us/visualstudio/test/live-unit-testing`
- Integration tests: `https://learn.microsoft.com/en-us/aspnet/core/test/integration-tests`

- *Testing .NET Aspire apps*: `https://learn.microsoft.com/en-us/dotnet/aspire/fundamentals/testing`

- Microsoft Playwright: `https://playwright.dev/`

- *Guide to Secure .NET Development with OWASP Top 10*: `https://learn.microsoft.com/en-us/training/modules/owasp-top-10-for-dotnet-developers/`

- *Working with Dependabot*: `https://docs.github.com/en/code-security/dependabot/working-with-dependabot`

11
Logging and Monitoring

With a microservices solution, many services can interact with each other. When one service fails, the complete solution should not break. In the previous chapter, we covered different kinds of tests to find issues early. Here, we'll look at finding issues in production as early as possible – probably before a user sees a problem.

To find issues when the application is running and see how the application runs successfully, the solution needs to be enhanced to offer telemetry data. With **logging**, we see what's going on; based on different log levels, we can differentiate between informational logs and errors. With **metrics** data, we can monitor counters, such as memory and CPU consumption, and the number of HTTP requests. We will also write custom counters to see the number of games played and the number of game moves needed for a win. **Distributed tracing** gives information on how services interact. Who is making calls to this service? Where does this error originate from?

OpenTelemetry is an industry standard – a collection of APIs that allows different languages and tools to instrument, generate, collect, and export telemetry data. The .NET APIs for logging, metrics, and distributed tracing support OpenTelemetry, and this is what this chapter is about. We'll use **Prometheus** and **Grafana**, which have great graphical views for an on-premises solution, as well as **Azure Application Insights** for the solution to run with the Microsoft Azure cloud.

In this chapter, you'll learn how to do the following:

- Add log messages

- Use and create metrics data

- Use distributed tracing

- Monitor with Azure Application Insights

- Monitor with Prometheus and Grafana

Technical requirements

With this chapter, as with the previous chapters, you need an Azure subscription and Docker Desktop. To create all the Azure resources for the solution, you can use the Azure Developer CLI – `azd up` creates all the resources. Check the README file of this chapter in the repository for details.

The code for this chapter can be found in this GitHub repository: `https://github.com/PacktPublishing/Pragmatic-Microservices-with-CSharp-and-Azure/`.

In the `ch11` folder, you'll see these projects with the result of this chapter. This chapter adds the Prometheus launch profile with the `launchsettings.json` file of the `AppHost` project. This also sets the `ASPNETCORE_ENVIRONMENT` and `DOTNETCORE_ENVIRONMENT` environment variables to `Prometheus`. The default launch profile uses services running with Microsoft Azure. The Prometheus launch profile is used to run Prometheus and Grafana, which can be used easily in an on-premises environment.

These are the important projects for this chapter:

- **Common projects**

  - `Codebreaker.AppHost` – The .NET Aspire host project.

  - `Codebreaker.ServiceDefaults` – Common service configuration. This project is enhanced with service configurations for monitoring.

- **Services**

  - `Codebreaker.GamesAPI` – The service project is enhanced with logging, metrics, and distributed tracing.

  - `Codebreaker.Bot` – This project has monitoring information included and will be used to play games that can be monitored.

- **Configuration folders**

  - The `grafana` folder contains configuration files that are used within the Grafana Docker container.

  - The `prometheus` folder contains a configuration file that is used by the Prometheus Docker container.

You can start with the source code from the previous chapter to integrate the features from this chapter.

Adding log messages

To see what's going on successfully or not when running the solution, we add log messages. The important parts of understanding the concept of logging are the following:

- **The source**: Who writes log information – what is the category name?
- **The log provider**: Where is log information written to?
- **The log level**: What is the level of the log message? Is it just information or an error?
- **Filtering**: What information is logged?

The source is defined using the `ILogger<T>` generic interface. With this generic interface, the category name is taken from the class name of the generic parameter type. In case you use the `ILoggerFactory` interface instead of `ILogger<T>`, the category name is passed by invoking the `CreateLogger` method. Examples of category names used by .NET are `Microsoft.EntityFrameworkCore.Database.Command`, `System.Net.Http.HttpClient`, and `Microsoft.Hosting.Lifetime`. Having hierarchical names helps with common configuration settings.

To define where log messages are written, log providers are configured with the startup of the application. The `CreateBuilder` method of the `WebApplication` class configures multiple log providers:

- `ConsoleLogProvider` to write log messages to the console
- `DebugLoggerProvider`, which only writes messages to the debug output window when a debugger is attached
- `EventSourceLoggerProvider`, which writes log messages using **Event Tracing for Windows** (**ETW**) on Windows and the **Linux Trace Toolkit: next generation** (**LTTng**) on Linux

With the AOT ASP.NET Core application that was created in *Chapter 5*, the `CreateSlimBuilder` method was used. `CreateSlimBuilder` only configures the console provider; other providers need to be added manually.

The `ILogger` interface defines a Log method with a `LogLevel` enum value containing `Trace` (0) – `Debug` – `Information` – `Warning` – `Error` – `Critical` (5) – `None` (6) values. With this, we can configure to only write `Warning`-level messages and higher or write every message specifying the `Trace` level and higher). This configuration can be different based on the provider and the source.

The following snippet shows a customized configuration with a JSON configuration file:

```json
{
    "Logging": {
        "LogLevel": {
            "Default": "Information",
            "Microsoft.AspNetCore": "Warning",
            "Microsoft.EntityFrameworkCore": "Warning",
            "Codebreaker": "Trace"
        },
        "EventSource": {
            "LogLevel": {
                "Default": "Warning"
            }
        }
    }
}
```

The `Logging` section with the configuration is accessed within the implementation of the `CreateBuilder` method at startup. Here, we can customize the logging configuration. With this configuration file, the default log level is specified with the `LogLevel:Default` key. Here, logging is set to `Information`, thus `Debug` and `Trace` log messages are not written. This default configuration is changed with log categories that start with `Microsoft.AspNetCore`. With this category, only warnings, errors, and critical messages are written. With the `LogLevel` key as a subkey to `Logging`, all log providers are configured, unless the configuration for the provider is overwritten. Here, this is done for the `EventSource` log provider. The `Default` log level is set to `Warning`.

Next, let's add logging to the games API.

Creating strongly typed log messages

The games API service makes use of **strongly typed logging**. Strongly typed logging gives us methods with arguments that should be written to the log output. Instead of using .NET-defined `ILogger` extension methods such as `LogError` and `LogInformation`, when writing log messages, we use custom log methods as shown next. Let's add a Log class to define all log messages with the games API project:

Codebreaker.GameAPIs/Infrastructure/Log.cs

```csharp
public static partial class Log
{
```

```
  [LoggerMessage(
    EventId = 3001,
    Level = LogLevel.Warning,
    Message = "Game {GameId} not found")]
  public static partial void GameNotFound(this ILogger logger,
    Guid gameId);
  // code removed for brevity
  [LoggerMessage(
    EventId = 4001,
    Level = LogLevel.Information,
    Message = "The move {Move} was set for {GameId} with result
{Result}")]
  public static partial void SendMove(this ILogger logger, string
    move, Guid gameId, string result);
  // code removed for brevity
}
```

The LoggerMessage attribute is used by a source generator. For methods that are annotated with this attribute, the logger source generator creates an implementation. The method needs to be void with an ILogger parameter. The method can also be defined as an extension method, as is the case here. Parameter names need to match the expressions used within the Message property, such as gameId, move, and result.

Before the log message is written, the generated logging code checks if the log level is enabled. Sometimes, it can be useful to create custom methods that make use of generated methods, as shown in the next code snippet:

Codebreaker.GameAPIs/Infrastructure/Log.cs

```
[LoggerMessage(
  EventId = 4003,
  Level = LogLevel.Information,
  Message = "Game lost after {Seconds} seconds with game {Gameid}")]
private static partial void GameLost(this ILogger logger, int seconds,
Guid gameid);

public static void GameEnded(this ILogger logger, Game game)
{
  if (logger.IsEnabled(LogLevel.Information))
  {
    if (game.IsVictory)
    {
      logger.GameWon(game.Moves.Count, game.Duration?.Seconds ?? 0,
        game.Id);
```

```
    }
    else
    {
      logger.GameLost(game.Duration?.Seconds ?? 0, game.Id);
    }
  }
}
```

The `GameEnded` method checks the `Game` object to see if it's a victory or not, and depending on this, either the `GameWon` or the `GameLost` logging method is invoked. Before using any CPU and memory for this process (logging could also need to enumerate collections to produce useful log messages), it's good practice to verify if this should be done at all – if the log level is enabled. This is checked using the `logger.IsEnabled` method and passing the log level.

Note

Writing logs, don't use interpolated strings such as `logger.LogInformation($"log message {expression}");`. Instead, use `logger.LogInformation("log message {expression}", expression);`. The second form supports structured logging. The message string passed is a template. With this template, the content within the curly braces can be used to create an index, and (depending on the log collector) you can query for all log entries containing this term. Also, using the formatted string allocates a new string that needs to be garbage-collected. With the second version, there's just one string for all log entries written.

Check the GitHub repo for more methods defined with the `Log` class. Next, let's use this class to write log messages.

Writing log messages

Log messages are mainly written from the `GamesService` class, thus we need to change the constructor:

Codebreaker.GameAPIs/Services/GamesService.cs

```
public class GamesService(
  IGamesRepository dataRepository,
  ILogger<GamesService> logger) : IGamesService
{
  // code removed for brevity
}
```

With the updated constructor, the generic version of the `ILogger` interface is injected. The type parameter specifies the category name for logging.

> **Note**
>
> In this chapter, the GamesService class is enhanced with logging, distributed tracing, and metrics functionality. That's why you see all these changes in the final code in the source code repository.

The StartGameAsync method is enhanced with logging:

Codebreaker.GameAPIs/Services/GamesService.cs

```csharp
public async Task<Game> StartGameAsync(
    string gameType,
    string playerName,
    CancellationToken cancellationToken = default)
{
    Game game;
    try
    {
        game = GamesFactory.CreateGame(gameType, playerName);

        await dataRepository.AddGameAsync(game, cancellationToken);
        logger.GameStarted(game.Id);
    }
    catch (CodebreakerException ex) when (ex.Code is
        CodebreakerExceptionCodes.InvalidGameType)
    {
        logger.InvalidGameType(gameType);
        throw;
    }
    catch (Exception ex)
    {
        logger.Error(ex, ex.Message);
        throw;
    }
    return game;
}
```

The GamesFactory class can throw an exception of type CodebreakerException. This is caught to write a log message and to re-throw the exception. The exception will be dealt with by the endpoint implementation to finally return a specific HTTP result. Here, we just want to log this information and re-throw the exception.

For generic exceptions, a strongly typed `Error` method is defined by the `Log` class and used to write this message. The `InvalidGameType` method writes a log message with the `Warning` level. Here, the client probably sent an invalid (or currently not accepted) game type. While this shouldn't happen, it's usually an issue with the client, and we don't have to deal with it on the service side. It's just good to know about such clients. The `Error` method writes a log message with the `Error` level. It could be useful to check for more specific error types and create additional messages.

On a successful invocation, a log message is written by invoking the `GameStarted` method of the `Log` class, which has the `Informational` level set.

Let's check the log messages with .NET Aspire next.

Viewing logs with the .NET Aspire dashboard

The .NET Aspire-generated `Codebreaker.ServiceDefaults` library contains logging configuration:

Codebreaker.ServiceDefaults/Extensions.cs

```
public static IHostApplicationBuilder ConfigureOpenTelemetry(this
IHostApplicationBuilder builder)
{
  builder.Logging.AddOpenTelemetry(logging =>
  {
    logging.IncludeFormattedMessage = true;
    logging.IncludeScopes = true;
  });
  // code removed for brevity
}
```

The `AddOpenTelemetry` method adds the `OpenTelemetry` logger to the logger factory. This provider is configured to include formatted messages and to include logging scopes. Setting `IncludeFormattedMessage` to `true` specifies that if log templates are used (which we did), formatted messages are also included when creating log records for OpenTelemetry. By default, this would not be the case. Setting `IncludeScopes` to `true` specifies to include logging scope IDs with logs, which allows us to see a hierarchy of log messages when using the `BeginScope` method of the `ILogger` interface to define scopes. The `ConfigureOpenTelemetry` method is invoked from within the `AddServiceDefaults` method, which in turn is invoked both from the games API and the bot service.

With this logging configuration in place, it's time to start the services locally, running the .NET Aspire dashboard. Start the application and solution and let the bot service play some games. Then, open the .NET Aspire dashboard and select **Console Logs** within the **Monitoring** category. Here, you'll see log outputs for games that have been started, as shown in *Figure 11.1*. You can also see log outputs from **Entity Framework Core (EF Core)**, including queries done and ASP.NET Core logs – unless the level is set to not show informational messages:

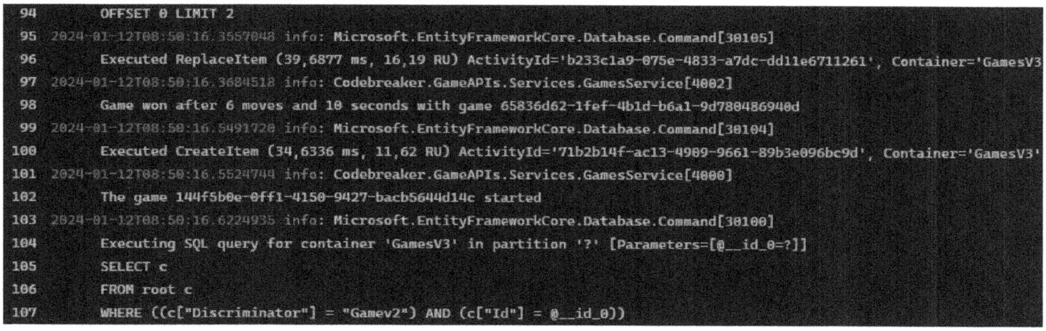

```
 94       OFFSET 0 LIMIT 2
 95  2024-01-12T08:50:16.3557048 info: Microsoft.EntityFrameworkCore.Database.Command[30105]
 96       Executed ReplaceItem (39,6877 ms, 16,19 RU) ActivityId='b233c1a9-075e-4833-a7dc-dd11e6711261', Container='GamesV3
 97  2024-01-12T08:50:16.3684518 info: Codebreaker.GameAPIs.Services.GamesService[4002]
 98       Game won after 6 moves and 10 seconds with game 65836d62-1fef-4b1d-b6a1-9d7804869400d
 99  2024-01-12T08:50:16.5491720 info: Microsoft.EntityFrameworkCore.Database.Command[30104]
100       Executed CreateItem (34,6336 ms, 11,62 RU) ActivityId='71b2b14f-ac13-4989-9661-89b3e096bc9d', Container='GamesV3'
101  2024-01-12T08:50:16.5524744 info: Codebreaker.GameAPIs.Services.GamesService[4000]
102       The game 144f5b0e-0ff1-4150-9427-bacb5644d14c started
103  2024-01-12T08:50:16.6224935 info: Microsoft.EntityFrameworkCore.Database.Command[30100]
104       Executing SQL query for container 'GamesV3' in partition '?' [Parameters=[@__id_0=?]]
105       SELECT c
106       FROM root c
107       WHERE ((c["Discriminator"] = "Gamev2") AND (c["Id"] = @__id_0))
```

Figure 11.1 – Logs with the .NET Aspire dashboard

Also, open the logs from the bot service. The bot service writes log output with every move set after the result is received to show how successful the move was and how many remaining options are available, as shown in *Figure 11.2*:

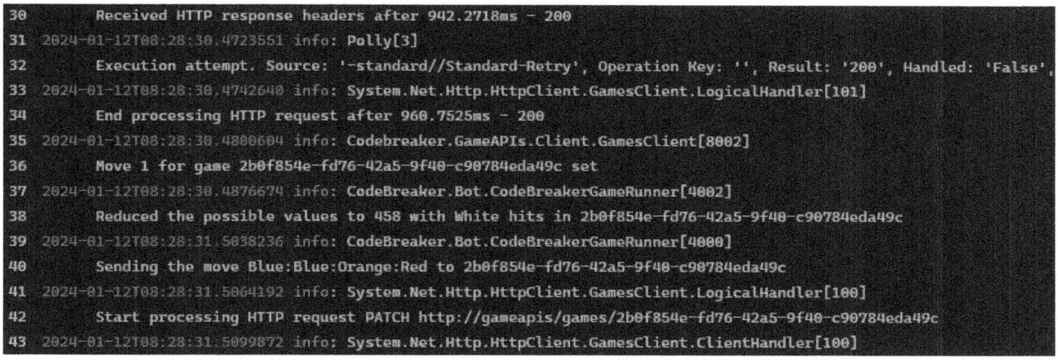

```
30       Received HTTP response headers after 942.2718ms - 200
31  2024-01-12T08:28:30.4723551 info: Polly[3]
32       Execution attempt. Source: '-standard//Standard-Retry', Operation Key: '', Result: '200', Handled: 'False',
33  2024-01-12T08:28:30.4742640 info: System.Net.Http.HttpClient.GamesClient.LogicalHandler[101]
34       End processing HTTP request after 960.7525ms - 200
35  2024-01-12T08:28:30.4800604 info: Codebreaker.GameAPIs.Client.GamesClient[8002]
36       Move 1 for game 2b0f854e-fd76-42a5-9f40-c90784eda49c set
37  2024-01-12T08:28:30.4876674 info: CodeBreaker.Bot.CodeBreakerGameRunner[4002]
38       Reduced the possible values to 458 with White hits in 2b0f854e-fd76-42a5-9f40-c90784eda49c
39  2024-01-12T08:28:31.5038236 info: CodeBreaker.Bot.CodeBreakerGameRunner[4000]
40       Sending the move Blue:Blue:Orange:Red to 2b0f854e-fd76-42a5-9f40-c90784eda49c
41  2024-01-12T08:28:31.5064192 info: System.Net.Http.HttpClient.GamesClient.LogicalHandler[100]
42       Start processing HTTP request PATCH http://gameapis/games/2b0f854e-fd76-42a5-9f40-c90784eda49c
43  2024-01-12T08:28:31.5099872 info: System.Net.Http.HttpClient.GamesClient.ClientHandler[100]
```

Figure 11.2 – Logs from the bot service

When you open **Structured Logs** in the dashboard, you can see the logs of the bot service and the games API on one screen – or just select the service from which you want to look at the logs. With a log entry, clicking on **Details** shows every detail information, such as the `GameId` placeholder we've written with the `GameStarted` event, as shown in *Figure 11.3*. Other data, such as the `RequestPath` placeholder, is coming from .NET activities, which we'll look at later in the *Using distributed tracing* section:

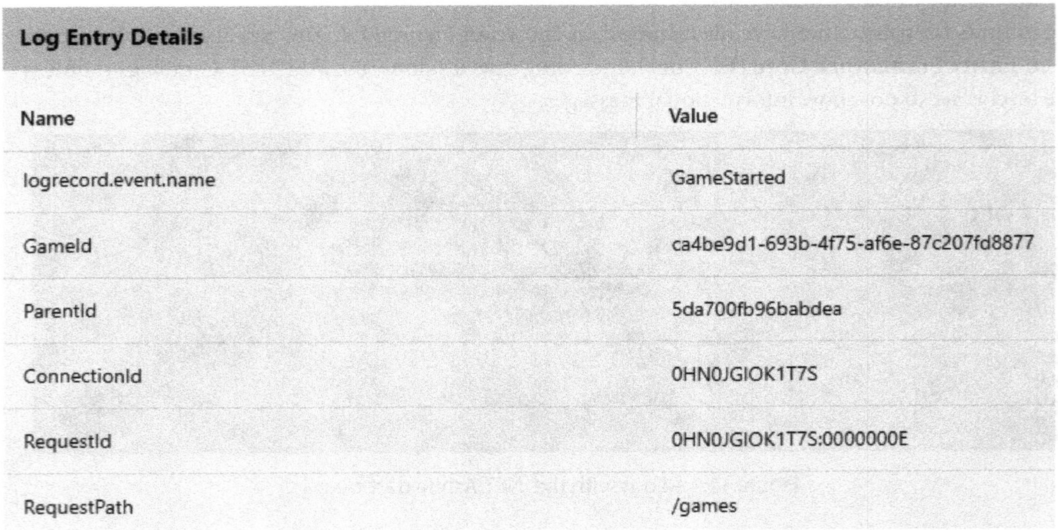

Log Entry Details	
Name	**Value**
logrecord.event.name	GameStarted
GameId	ca4be9d1-693b-4f75-af6e-87c207fd8877
ParentId	5da700fb96babdea
ConnectionId	0HN0JGIOK1T7S
RequestId	0HN0JGIOK1T7S:0000000E
RequestPath	/games

Figure 11.3 – Structured logging

With the **Structured Logs** view, you can add a filter, select any term that has been used with placeholders such as `GameId`, set the game identifier as the value, and read all the logs related to this game. Here, you easily can follow a single gameplay with the moves set, as shown in *Figure 11.4*:

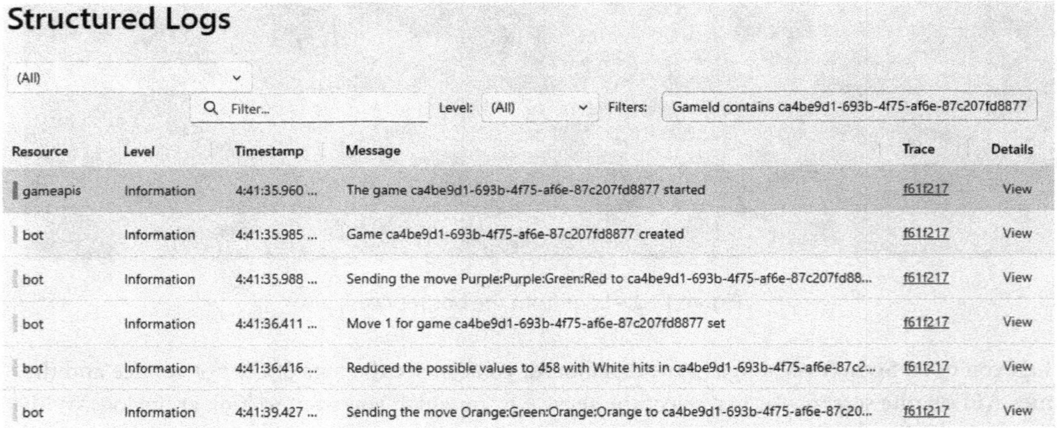

Structured Logs

(All) ⌄						
	Q Filter...		Level: (All) ⌄	Filters:	GameId contains ca4be9d1-693b-4f75-af6e-87c207fd8877	
Resource	**Level**	**Timestamp**	**Message**		**Trace**	**Details**
gameapis	Information	4:41:35.960 ...	The game ca4be9d1-693b-4f75-af6e-87c207fd8877 started		f61f217	View
bot	Information	4:41:35.985 ...	Game ca4be9d1-693b-4f75-af6e-87c207fd8877 created		f61f217	View
bot	Information	4:41:35.988 ...	Sending the move Purple:Purple:Green:Red to ca4be9d1-693b-4f75-af6e-87c207fd88...		f61f217	View
bot	Information	4:41:36.411 ...	Move 1 for game ca4be9d1-693b-4f75-af6e-87c207fd8877 set		f61f217	View
bot	Information	4:41:36.416 ...	Reduced the possible values to 458 with White hits in ca4be9d1-693b-4f75-af6e-87c2...		f61f217	View
bot	Information	4:41:39.427 ...	Sending the move Orange:Green:Orange:Orange to ca4be9d1-693b-4f75-af6e-87c20...		f61f217	View

Figure 11.4 – Structured logging with a GameId filter

After writing logs, let's get started with metrics data.

Using metrics data

Metrics data is used to monitor counts such as CPU and memory consumption or the length of an HTTP queue. This information can be used to analyze resources needed by services and can scale the services accordingly.

With metrics data, we get some counts. Such counts can be used for scaling services, based on memory or CPU consumption, or the length of an HTTP queue.

Let's check the built-in metrics data before we add custom metrics.

Monitoring built-in .NET metrics

As mentioned, .NET offers much built-in metrics data that can be monitored using the `dotnet counters` .NET tool (install it via `dotnet tool install dotnet-counters -g` as a global tool), and many counts are already available from the .NET Aspire dashboard by opening the **Metrics** view. *Figure 11.5* shows the .NET-managed heap size of the games API service at a time the bot played several games in parallel:

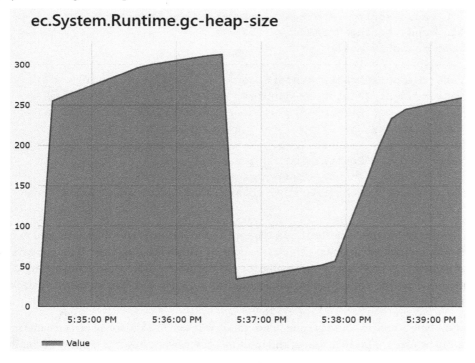

Figure 11.5 – Metrics

With many applications, you don't need to create custom metrics data – but some custom counts might be interesting to see, and it's not hard to add these, as shown in the next section.

Creating custom metrics data

With the `Codebreaker` solution, we are interested in knowing about the number of active games that are just played, the time it takes from one game move to another, the time it takes to complete a game, and how many games are won versus lost.

> **Note**
>
> With all the data we collect, we need to pay attention to the **General Data Protection Regulation (GDPR)**. Not storing any user-related data with logging and metrics information, we are on the safe side.

Let's create a new `GamesMetrics` class that contains all the counters needed:

Codebreaker.GameAPIs/Infrastructure/GamesMetrics.cs

```
public sealed class GamesMetrics : IDisposable
{
  public const string MeterName = "Codebreaker.Games";
  public const string Version = "1.0";
  private readonly Meter _meter;

  private readonly UpDownCounter<long> _activeGamesCounter;
  private readonly Histogram<double> _gameDuration;
  private readonly Histogram<double> _moveThinkTime;
  private readonly Histogram<int> _movesPerGameWin;
  private readonly Counter<long> _invalidMoveCounter;
  private readonly Counter<long> _gamesWonCounter;
  private readonly Counter<long> _gamesLostCounter;
  private readonly ConcurrentDictionary<Guid, DateTime> _moveTimes =
new();
```

The fields defined within the `GameMetrics` class are for the `Meter` class, which is needed to create all the different metric instruments. This class is defined within the `System.Diagnostics.Metrics` namespace. This class is responsible for creating all the different instruments needed to monitor metrics data. The `Meter` type needs a name that is used to specify what metrics data we are interested in. The value for the version is optional.

The `Counter` type is used to count the number of games won and lost and for all invalid game moves. `Counter` can be used for positive values, and most metrics viewers show the number of counts per second but can also show cumulative values. The `UpDownCounter` type is used for positive and negative values. We use this for the number of games active. Every time a game ends, a decrement will be made. The `Histogram` type is of special interest. This metrics instrument can be used to show arbitrary values. Here, this instrument is used to show the time it takes to complete a game, the time a user needs between game moves, and the number of moves needed to win a game.

With the constructor of the GamesMetrics class, the Meter class and the instruments are created and initialized:

Codebreaker.GameAPIs/Infrastructure/GamesMetrics.cs

```
public GamesMetrics(IMeterFactory meterFactory)
{
  _meter = meterFactory.Create(MeterName, Version);

  _activeGamesCounter = _meter.CreateUpDownCounter<long>(
    "codebreaker.active_games",
    unit: "{games}",
    description: "Number of games that are currently active on the
      server.");

  _gameDuration = _meter.CreateHistogram<double>(
    "codebreaker.game_duration",
    unit: "s",
    description: "Duration of a game in seconds.");
  // code removed for brevity
}
```

IMeterFactory is a new interface since .NET 8. This allows the creation of metrics types via **dependency injection** (DI). IMeterFactory is injected with the GamesMetrics constructor to create a Meter instance and instruments. CreateCounter, CreateUpDownCounter, and CreateHistogram are the methods to create the different metric instruments. The name of the instrument, the unit, and the description are assigned upon creating the instruments.

Before using these counters, let's add tags.

Creating tags

Writing metrics data, **tags** can be added with every record written to an instrument. Tags allow us to filter metrics data – somehow similar to the filtering we've used with structured logging. A tag consists of a key-value pair. To make tags easier to add, these methods are added to the GameMetrics class:

Codebreaker.GameAPIs/Infrastructure/GamesMetrics.cs

```
private static KeyValuePair<string, object?> CreateGameTypeTag(string
gameType) =>
  KeyValuePair.Create<string, object?>("GameType", gameType);
```

```
private static KeyValuePair<string, object?> CreateGameIdTag(Guid id)
=>
   KeyValuePair.Create<string, object?>("GameId", id.ToString());
```

CreateGameTypeTag is a helper method to create a tag with the name GameType and set the value passed with the method parameter. Similarly, CreateGameIdTag is a method to create a tag for GameId.

Now, we are ready to create methods using the instruments.

Creating strongly typed methods for metrics data

The GameStarted method is for writing metrics data on creating a new game:

Codebreaker.GameAPIs/Infrastructure/GamesMetrics.cs

```
public void GameStarted(Game game)
{
   if (_moveThinkTime.Enabled)
   {
      _moveTimes.TryAdd(game.Id, game.StartTime);
   }

   if (_activeGamesCounter.Enabled)
   {
      _activeGamesCounter.Add(1, CreateGameTypeTag(game.GameType));
   }
}
```

When nobody listens to metrics data, there's no need to take any action. Before writing values to an instrument, it should be verified that the instrument is enabled. If nobody listens to the meter, the counters are disabled.

To write the delta time between moves, we need to remember the time of the previous move. For this, the GameMetrics class holds the _moveTimes dictionary. This dictionary uses the game ID for the key and the last time for the latest move (or game start) value. Calculating this information is only necessary when the _moveThinkTime instrument is used.

The counter that's incremented at the start of the game is _activeGamesCounter. With UpDownCounter, the Add method is used to change the counter value. The second – optional – argument of the Add method allows passing tags. Here, a tag for the game type is added. This allows us to check the metrics data filtered based on the game type. It's interesting to compare the active game counts based on the different game types.

To write a histogram value, we implement the `MoveSet` method:

Codebreaker.GameAPIs/Infrastructure/GamesMetrics.cs

```
public void MoveSet(Guid id, DateTime moveTime, string gameType)
{
  if (_moveThinkTime.Enabled)
  {
    _moveTimes.AddOrUpdate(id, moveTime, (id1, prevTime) =>
    {
      _moveThinkTime.Record((moveTime - prevTime).TotalSeconds,
        [CreateGameIdTag(id1), CreateGameTypeTag(gameType)]);
      return moveTime;
    });
  }
}
```

With the implementation of `MoveSet`, for the received game ID, we get the previous recorded time from the dictionary, calculate the delta with the new time, use the `Record` method of the `Histogram` instrument to write the data, and write the new received time to the dictionary.

On ending the game, the `GameEnded` method is implemented. Here, multiple instruments are used, but this method just needs a simple implementation to check for every instrument to be enabled and write the counts accordingly. Check the source code repository for the complete code.

Next, we can change the implementation of the `GamesService` class to use a `GamesMetrics` instance.

Injecting and using metrics

Let's update the `GamesService` class for metrics data:

Codebreaker.GameAPIs/Services/GamesService.cs

```
public class GamesService(
  IGamesRepository dataRepository,
  ILogger<GamesService> logger,
  GamesMetrics metrics) : IGamesService
{
  public async Task<Game> StartGameAsync(
    string gameType,
    string playerName,
    CancellationToken cancellationToken = default)
  {
    Game game;
```

```
    try
    {
      game = GamesFactory.CreateGame(gameType, playerName);
      await dataRepository.AddGameAsync(game, cancellationToken);
          metrics.GameStarted(game);
          logger.GameStarted(game.Id);
        }
      // code removed for brevity
    return game;
}
```

All that needs to be done is to inject the GamesMetrics class and invoke the GameStarted method.

Of course, the GamesMetrics class needs configuration within the **DI container (DIC)**:

Codebreaker.GameAPIs/ApplicationServices.cs

```
builder.Services.AddMetrics();
builder.Services.AddSingleton<GamesMetrics>();
builder.Services.AddOpenTelemetry()
  .WithMetrics(m => m.AddMeter(GamesMetrics.MeterName));
```

The AddMetrics extension method registers an implementation for the IMeterFactory interface. The GamesMetrics class is registered as a singleton – to create the instruments once. We also configure the GamesMetrics class with OpenTelemetry – this way, we have a listener, and these metrics will be shown with the .NET Aspire dashboard.

With this, we could run the application. However, the unit tests for the GamesService class no longer compile because of this additional parameter. Let's update this before we continue.

Updating unit tests to inject metrics types

The GamesService class uses a concrete type – it injects the GamesMetrics type. This cannot be mocked directly, but we can mock the IMeterFactory interface to create a GamesMetrics instance.

The following code snippet shows an implementation of the IMeterFactory interface to be used for unit tests:

Codebreaker.GameAPIs.Tests/TestMeterFactory.cs

```
internal sealed class TestMeterFactory : IMeterFactory
{
  public List<Meter> Meters { get; } = [];

  public Meter Create(MeterOptions options)
```

```
  {
    Meter meter = new(options.Name, options.Version, Array.
      Empty<KeyValuePair<string, object?>>(), scope: this);
    Meters.Add(meter);
    return meter;
  }

  public void Dispose()
  {
    foreach (var meter in Meters)
    {
      meter.Dispose();
    }
    Meters.Clear();
  }
}
```

To implement the `IMeterFactory` interface, `Create` and `Dispose` methods need to be implemented. With the `Create` method, a new `Meter` instance is created using name, version, and tag information.

This `TestMeterFactory` class can now be used to create an instance of the `GamesService` class for the unit test:

Codebreaker.GameAPIs.Tests/GamesServiceTests.cs

```
private GamesService GetGamesService()
{
  IMeterFactory meterFactory = new TestMeterFactory();
  GamesMetrics metrics = new(meterFactory);
  return new GamesService(
    _gamesRepositoryMock.Object,
    NullLogger<GamesService>.Instance,
    metrics);
}
```

Creating a new `GamesMetrics` instance, the `TestMeterFactory` class is created. The unit test for the `GamesService` class now builds successfully again.

> **Note**
>
> The GitHub contains an additional parameter, `ActivitySource`, when invoking the `GamesService` constructor. The `ActivitiySource` is added in the section *Using distributed tracing*, and requires an adaption of the unit tests.

There's also a unit test for the `GamesMetrics` class needed, which we'll do next.

Creating unit tests to verify metrics

Metric data can be important business information easily shown on monitors in the office. What's going on with the application? How active are users? Is an error rate going up? While metrics information is not important for orders coming in and being processed, if metrics data is not written, it can easily be missed that something is not working – thus creating unit tests for metrics data should be part of creating custom metric types.

First, let's create a skeleton to return an `IMeterFactory` instance and a `GamesMetrics` instance.

Meter factory skeleton

The following code snippet defines the skeleton used by the `GamesMetrics` unit tests:

Codebreaker.GameAPIs.Tests/GamesMetricsTests.cs

```
private static IServiceProvider CreateServiceProvider()
{
  ServiceCollection services = new();
  service.AddMetrics();
  services.AddSingleton<GamesMetrics>();
  return serviceCollection.BuildServiceProvider();
}


private static (IMeterFactory MeterFactory, GamesMetrics Metrics)
CreateMeterFactorySkeleton()
{
  var container = CreateServiceProvider();
  GamesMetrics metrics = container.GetRequiredService<GamesMetrics>();
  IMeterFactory meterFactory = container.
GetRequiredService<IMeterFactory>();
  return (meterFactory, metrics);
}
```

Here, we need the real implementation of the `IMeterFactory` Interface. This is configured with the DIC for the unit test – along with the `GamesMetrics` singleton. The `CreateMeterFactorySkelton` method now gets the `IMeterFactory` and `GameMetrics` instances from the DIC.

Unit tests

Using this skeleton, we can create unit tests for all GameMetrics methods:

Codebreaker.GameAPIs.Tests/GamesMetricsTests.cs

```csharp
public class GamesMetricsTests
{
  private Guid _gameId = Guid.Parse("DBDF4DD9-3A02-4B2A-87F6-
FFE4BA1DCE52");
  private DateTime _gameStartTime = new DateTime(2024, 1, 1, 12, 10,
5);
  private DateTime _gameMove1Time = new DateTime(2024, 1, 1, 12, 10,
15);

  [Fact]
  public void MoveSet_Should_Record_ThinkTime()
  {
    // arrange
    (IMeterFactory meterFactory, GamesMetrics metrics) =
      CreateMeterFactorySkeleton();
    MetricCollector<double> collector = new(meterFactory,
      GamesMetrics.MeterName, "codebreaker.move_think_time");
    var game = GetGame();
    metrics.GameStarted(game);

    // act
    metrics.MoveSet(game.Id, _gameMove1Time, "Game6x4");

    // assert
    var measurements = collector.GetMeasurementSnapshot();
    Assert.Single(measurements);
    Assert.Equal(10, measurements[0].Value);
  }
  // code removed for brevity
```

For easy unit testing of metrics classes, the MetricCollector class defined in the Microsoft.Extensions.Diagnostics.Testing NuGet package in the Microsoft.Extensions.Diagnostics.Metrics.Testing namespace can register as a listener for the metrics data and collect this information. It's also great for debugging purposes to have the metrics instruments enabled.

After the IMeterFActory and GamesMetrics objects are returned from the skeleton, a collector is created. You need to create a collector for every instrument that needs to be tested. The generic type parameter and the name of the instrument need to match. The MoveSet method of the GamesMetrics class records the time between the previous move (or the game start) and the current move. Using `Assert.Single,` it's verified that exactly one measurement is written to the collector. With `Assert.Equal`, it's checked that this one recording contains the value 10. If you calculate the values from `_gameStartTime` and `_gameMove1Time`, this matches the time difference passed with the test data.

As the `GameMetrics` class tests successfully, let's go to the .NET Aspire dashboard to see the custom metrics data.

Viewing metrics data with the .NET Aspire dashboard

We injected metrics and configured our custom `GamesMetrics` class with OpenTelemetry in the game APIs project. Now, we can use the .NET Aspire dashboard to see the games played!

Running the services and starting the bot to run multiple games in parallel, we can see interesting outcomes. Sometimes, the bot doesn't find an answer within 12 moves because it lost the game, as shown in *Figure 11.6*:

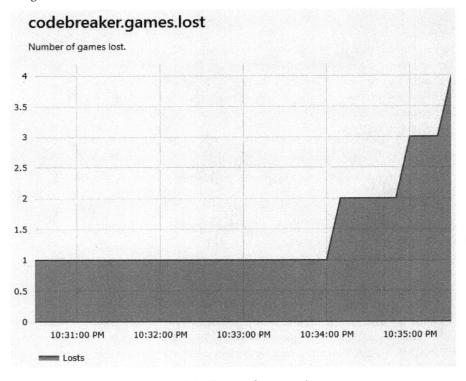

Figure 11.6 – Counter for games lost

Figure 11.7 shows the number of games won by the bot is much higher. This figure also shows the filter for the game type that can be selected because of the tag specified:

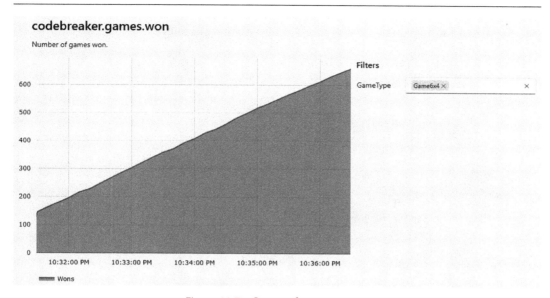

Figure 11.7 – Counter for games won

While games won and lost used simple counters, *Figure 11.8* shows the up-down counter with the number of active games:

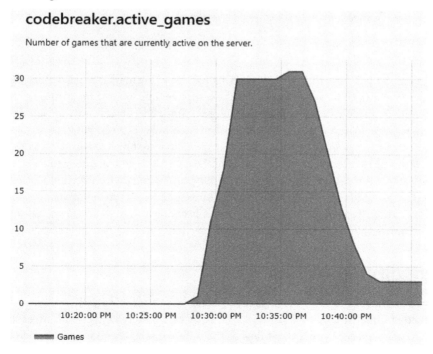

Figure 11.8 – Up-down counter for active games

Figure 11.9 shows a histogram that allows checking the duration of games:

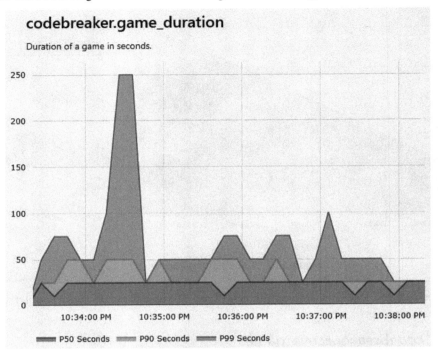

Figure 11.9 – Histogram showing the game duration

A histogram graph shows **P50**, **P90**, and **P99** values. These names are marks for percentiles. 50% of the finished games are within the lowest values: over the complete time, 50% of the games were finished within 25 seconds. The next higher line marks 90% of the game runs. Over time, with some peaks, we can see that sometimes 90% of the games were finished within 25 seconds, but it also took up to 50 seconds. To get to a higher number of games, 99% of the games were finished within 50 to 75 seconds. At peak time, it took 250 seconds. If a user is playing the game, this can be expected. We need more time than the bot to solve this. But here, it was just the bot playing. With different games, the bot was configured to take different times between game moves. However, the bot was never configured to take that long. Thus, this needs to be another issue, probably too high a load on the service. To find the reason more easily for such behavior, the next section covering distributed tracing will help.

Using distributed tracing

If errors happen on the service, where is this request coming from, and from where does it originate? With distributed tracing, we can see the interaction of services and resources and can easily follow information on how requests from a client flow to the different services and see when errors occur, going from the error up to the stack.

Using .NET, we use `ActivitySource` and `Activity` classes to specify information for distributed tracing.

Creating an ActivitySource class with the DIC

When writing trace information, you'll usually have one `ActivitySource` class in a project that's used by all classes that write trace information. With the games client library, an `ActivitySource` class is used as a static member. Using an `ActivitySource` class from an executable project such as the games API, we can register this in the DIC:

Codebreaker.GameAPIs/ApplicationServices.cs

```
public static void AddApplicationTelemetry(this
IhostApplicationBuilder builder)
{
  // code removed for brevity
  const string ActivitySourceName = "Codebreaker.GameAPIs";
  const string ActivitySourceVersion = "1.0.0";
  builder.Services.AddKeyedSingleton(ActivitySourceName, (services, _)
=>
    new ActivitySource(ActivitySourceName,
      ActivitySourceVersion));
```

The ASP.NET Core initialization already registers an `ActivitySource` class with the name `Microsoft.AspNetCore` as a singleton for the `ActivitySource` type. We don't want to overwrite this setting with our name. The ASP.NET Core features injecting this `ActivitySource` instance should still get this instance, but for our own trace messages, the `Codebreaker.GameAPIs` activity source should be used. With .NET 8 enhancements on the DIC, we can register a named service within the DIC by invoking the `AddKeyedSingleton` method and specifying the name and version strings. One instance is created using the factory defined with the lambda expression.

Next, we can inject this singleton instance with the `GamesService` class.

Writing trace messages

With the constructor of the `GamesService` class, we can now inject the configured `ActivitySource` class:

Codebreaker.GameAPIs/Services/GamesService.cs

```
public class GamesService(
  IGamesRepository dataRepository,
  ILogger<GamesService> logger,
  GamesMetrics metrics,
```

```
    [FromKeyedServices("Codebreaker.GameAPIs")] ActivitySource
      activitySource) :
      IGamesService
{
    // code removed for brevity
```

Using the `FromKeyedServices` attribute, we get the named instance from the DIC.

Next, let's update the creation of a new game by creating an `Activity` object:

Codebreaker.GameAPIs/Services/GamesService.cs

```
public async Task<Game> StartGameAsync(string gameType, string
playerName, CancellationToken cancellationToken = default)
{
    Game game;
    using var activity = activitySource.CreateActivity("StartGame",
      ActivityKind.Server);
    try
    {
      game = GamesFactory.CreateGame(gameType, playerName);
      activity?.AddTag(GameTypeTagName, game.GameType)
        .AddTag(GameIdTagName, game.Id.ToString())
        .Start();

      await dataRepository.AddGameAsync(game, cancellationToken);
      metrics.GameStarted(game);
      logger.GameStarted(game.Id);
      activity?.SetStatus(ActivityStatusCode.Ok);
    }
    catch (CodebreakerException ex) when (ex.Code is
CodebreakerExceptionCodes.InvalidGameType)
    {
      logger.InvalidGameType(gameType);
      activity?.SetStatus(ActivityStatusCode.Error, ex.Message);
      throw;
    }
    catch (Exception ex)
    {
      logger.Error(ex, ex.Message);
      activity?.SetStatus(ActivityStatusCode.Error, ex.Message);
      throw;
    }
    return game;
}
```

The `Activity` object is created by invoking the `CreateActivity` method. The parameters used here are the name of the activity and the activity kind. The service specifies `ActivityKind.Server`, whereas the client library uses `ActivityKind.Client`. Other types available are `Producer` and `Consumer`.

The methods creating an `Activity` object might return `null`. If no one adds a listener to the `ActivitySource` class, the `CreateActivity` method returns `null`. This reduces the overhead but also means we always need to check for `null` values before using an `Activity` object. Using the `null` conditional operator, this is easy to do.

Using the `StartActivity` method instead of the `CreateActivity` method would immediately start the activity. Here, we want to add some data to the activity that is shown with the log output. The `AddTag` method is used to add the game type and the game ID. This method adds key-value pairs to the log entries, which allows filtering and searching. The `SetBaggage` method allows adding information not only to this activity output – this information is passed to the child activities. Baggage information is used across processes and thus needs to be serializable. Invoking the `Start` method starts the activity – this writes the first log record along with tag and baggage information.

An activity ends when the `Stop` method is invoked. Here, the `using` declaration is used to dispose of the activity when the `activity` variable goes out of scope. This stops the activity implicitly.

Before the activity ends, the `SetStatus` method is invoked. This method specifies the outcome of the activity and is written when the activity ends. With a successful start of the game, `ActivityStatusCode.Ok` is the status of the activity. In case of errors, the status code is `ActivityStatusCode.Error` and an exception message is written.

Check the other source code repo for the other activities created with the `GamesService` class.

With this implementation, we just need to configure the service defaults library to monitor the custom `ActivitySource` class.

Viewing distributed traces with the .NET Aspire dashboard

First, let's add a custom `ActivitySource` class to the configuration:

Codebreaker.ServiceDefaults/Extensions.cs

```
public static IHostApplicationBuilder ConfigureOpenTelemetry(this
IHostApplicationBuilder builder)
{
  // code removed for brevity
  builder.Services.AddOpenTelemetry()
    .WithTracing(tracing =>
    {
      if (builder.Environment.IsDevelopment())
```

```
{
    tracing.SetSampler(new AlaysOnSampler());
}
tracing.AddSource(
    "Codebreaker.GameAPIs.Client",
    "Codebreaker.GameAPIs")
    .AddAspNetCoreInstrumentation()
    .AddGrpcClientInstrumentation()
    .AddHttpClientInstrumentation();
```

The `WithTracing` method configures the distributed trace settings. The `AddSource` method of the `TracerProviderBuilder` class sets the sources that should be subscribed to. `Codebreaker.GameAPIs` is the activity source name that has been configured with the games API service. `Codebreaker.GameAPIs.Client` is the activity source name used from the client library, which is referenced by the bot service. The next methods invoked configure the built-in sources with ASP.NET Core, gRPC, and `HttpClient`.

Now, running a few games using the bot, you can see *traces* in the .NET Aspire dashboard, as shown in *Figure 11.10*:

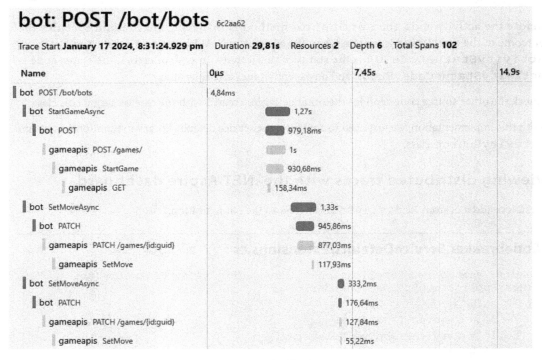

Figure 11.10 – Tracing

Sending a POST request to the bot to play multiple games, the request immediately returns (the one request in the figure lists 4,84 ms). The figure also shows activities started from this request from within a background task. This task is to play games. How this relates to multiple activities (or **spans**, using the OpenTelemetry term, as shown in the figure) is greatly visible. The first activity, bot POST bot/bots, is an activity created from ASP.NET Core. The next activity, bot StartGameAsync, is a custom activity that was created from the games client library. bot POST is the next activity from HttpClient. From there, we switch over to the gameapis service. The custom activities created with the games API are gameapis StartGame and gameapis SetMove.

With every one of these activities, you can dig deeper and get the data written, including the tags for the game ID and the game type, as shown in *Figure 11.11*:

gameapis: StartGame fc7365e	
Service **gameapis** Duration **930,68ms** Start Time **3,04s**	
Name	**Value**
SpanId	fc7365e7d55af475
Name	StartGame
Kind	Server
Status	Ok
codebreaker.gameId	a85d6d72-a673-4f3d-b3d0-9bfddd458fc0
codebreaker.gameType	Game6x4

Application

Name	**Value**
service.name	gameapis
service.instance.id	106f7a2e-7d4f-4162-9c8d-b1cc8a2ea7aa

Figure 11.11 – Trace data

While monitoring this data, also switch to **Structured Logs**. With these logs, you can see a **Trace** identifier. Clicking on this, you switch from the log to the trace information. It's also working the other way around. While opening a trace, you can switch to the log information associated with the trace.

Using the .NET Aspire dashboard is now great in the development environment. For production environments, we have different needs. Let's switch over to using .NET Azure services for monitoring.

Monitoring with Azure Application Insights

Creating an Azure Container Apps environment (starting with *Chapter 6*) also creates an **Azure Log Analytics** resource. In this chapter, we add an **Azure Application Insights** resource, and add a Log Analytics respource explicitly to the app model. Log Analytics and Application Insights are both part of the **Azure Monitor** service.

Log Analytics is used to monitor the amount of log data created and the cost associated and gives reasons when there is a higher-than-expected usage. Application Insights has a focus on application telemetry data and user data.

Configuring the .NET Aspire host for Application Insights

To add Application Insights, add the NuGet package `Aspire.Hosting.Azure.Application-Insights`, and update the `Program.cs` file of the .NET Aspire AppHost project:

Codebreaker.AppHost/Program.cs

```
var builder = DistributedApplication.CreateBuilder(args);
// code removed for brevity
var logs = builder.AddAzureLogAnalyticsWorkspace("logs");
var appInsights = builder.AddAzureApplicationInsights("insights",
logs);

var cosmos = builder.AddAzureCosmosDB("codebreakercosmos")
  .AddDatabase("codebreaker");

var gameAPIs = builder.AddProject<Projects.Codebreaker_
GameAPIs>("gameapis")
    .WithReference(cosmos)
    .WithReference(appInsights);

var bot = builder.AddProject<Projects.CodeBreaker_Bot>("bot")
    .WithReference(gameAPIs)
    .WithReference(appInsights);

builder.Build().Run();
// code removed for brevity
```

The Application Insights resource is added invoking the AddAzureApplicationInsights method. This method requires a name and a log analytics workspace resource which is created invoking the AddAzureLogAnalytics method. Both the games API and the bot services will use this resource, thus it is forwarded using Aspire orchestration by using the WithReference method.

With this, we can configure the services to use Application Insights.

Configuring the services to use Application Insights

To use Application Insights, the common configuration project can be updated:

Codebreaker.ServiceDefaults/Extensions.cs

```
private static IHostApplicationBuilder AddOpenTelemetryExporters(
  this IHostApplicationBuilder builder)
{
  builder.Services.AddOpenTelemetry()
    .UseAzureMonitor(options =>
    {
      options.ConnectionString = builder.Configuration[
        "APPLICATIONINSIGHTS_CONNECTION_STRING"];
    });
  // code removed for brevity
  return builder;
}
```

The AddOpenTelemetryExporters method is invoked from the AddServiceDefaults method in the same class. AddServiceDefaults is called with the WebApplication configuration of every service project. The AddOpenTelemetry method is part of the OpenTelementry SDK to configure services for logging, metrics, and distributed tracing. The UseAzureMonitor extension method configures providers to write this information to Azure Monitor.

This is all that we need to do. Let's check the information we get from the application.

Monitoring the solution with Application Insights

As before, upon running the application when monitoring using the Aspire dashboard, use the bot to run multiple games. As Application Insights is configured, it's not necessary to have the solution deployed to Azure; monitoring information will be available within Azure when running the application locally as well. After starting a few game runs using the bot, open the Azure Application Insights resource within the Azure portal.

After playing a few games, check the Azure portal. With Application Insights, in the **Investigate** category on the left pane, select **Application map**. Here, you see how the different services communicate, as shown in *Figure 11.12*:

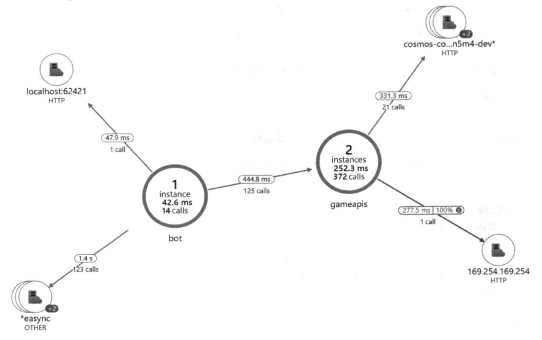

Figure 11.12 – Application map

The previous figure shows the bot service invokes the game API service, and this communicates with the Azure Cosmos database. The count of messages, the time used, and errors can easily be seen from this information. Clicking on a connection, it's possible to investigate performance by seeing the slowest calls and directly accessing log information.

With the **Investigate** category in the Azure portal, you can dig into other interesting information such as **Performance**, where you get insights into slow operations, and **Failures**, where you can dive into errors and exceptions. Smart detection can give you notifications (alerts) when services are not behaving with usual, expected values.

Within the **Monitoring** category in the left pane, you see the **Metrics** entry. Selecting this, you can select metric data, including custom metrics such as codebreaker.active_games and codebreaker.game_moves-per-win, where you can select the metric name and the aggregation type to calculate values and see graphical results, as shown in *Figure 11.13*:

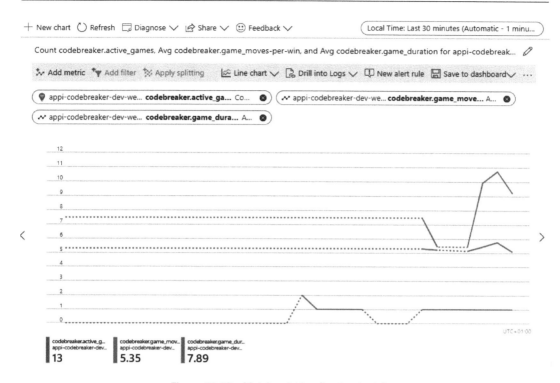

Figure 11.13 – Metrics via Application Insights

With this figure, the first line shows the average duration of a game (seconds), the second line shows the average number of moves needed, and the third line shows the number of games currently active.

Clicking **Logs**, you can see Application Insights tables stored. Here, you can use **Kusto Query Language (KQL)** to query all logged data and not only see text but also a graphic output. Clicking on the customMetrics table, you can query all the metrics data.

Use this KQL query to get the average EF Core queries on a time chart:

```
customMetrics
 | where name == "ec.Microsoft.EntityFrameworkCore.queries-per-second"
 | summarize avg(value) by bin(timestamp, 5min)
 | render timechart
```

This query accesses the customMetrics table filtered by the name column to get the EF Core queries by second. This result (you can see every record fitting this query by clicking on **Results**) is then used to show the average values and a timestamp rendered on a time chart. Your result might look like the one shown in *Figure 11.14*:

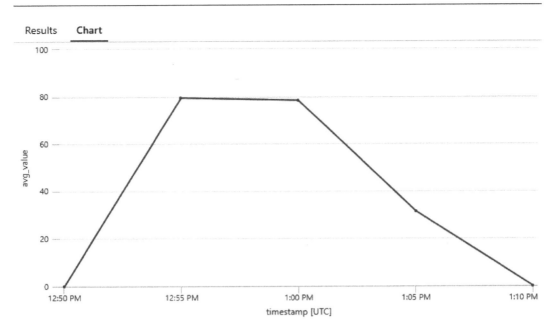

Figure 11.14 – EF Core average queries using KQL

The Application Insights resource can be used no matter where your services are running if they have access to this resource on Microsoft Azure. In case the complete solution needs to run on-premises, you can use Prometheus and Grafana, as we'll do next.

Monitoring with Prometheus and Grafana

For monitoring microservices solutions, Prometheus and Grafana are often used. **Prometheus** uses a pull model to collect data from services. Using this data, the **PromQL** query language analyzes this information. **Grafana** can access the data collected from Prometheus to show a graphical view.

We'll use Docker containers running Prometheus and Grafana. Microsoft Azure also offers managed services for Prometheus and Grafana that can be used as well.

Let's configure Docker containers for Prometheus and Grafana next.

Adding Docker containers for Prometheus and Grafana

To use the solution with Prometheus and Grafana, you need to select the launch profile – using Visual Studio, select **OnPremises** in the toolbar after the project selection.

Using the command line, use the `--launch-profile` parameter:

```
dotnet run --project Codebreaker.AppHost.csproj --launch-profile
OnPremises
```

To also use SQL Server within a Docker container, set the `DataStore` configuration to `SQLServer` (see *Chapter 3* for details); you could also use in-memory or Cosmos instead with the same build configuration.

For using a Prometheus Docker container, we need to change the .NET Aspire `AppHost` implementation:

Codebreaker.AppHost/Program.cs

```
var builder = DistributedApplication.CreateBuilder(args);

var sqlServer = builder.AddSqlServer("sql")
  .WithDataVolume()
  .PublishAsContainer()
  .AddDatabase("CodebreakerSql");

var prometheus = builder.AddContainer("prometheus", "prom/prometheus")
  .WithBindMount("../prometheus", "/etc/prometheus", isReadOnly: true)
  .WithHttpEndpoint(9090, hostPort: 9090);
// code removed for brevity
```

The first Docker container we used was the container for SQL Server. For using the Docker image for SQL Server, the `AddSqlServerContainer` extension method was used. To use the Prometheus Docker image, we need to use the `AddContainer` generic method, which allows adding any Docker image. The Docker image for Prometheus can be pulled from Docker Hub using `prom/prometheus`. For the Prometheus configuration, we use the `prometheus` folder, which is stored outside of the container. Using `WithBindMount`, the `prometheus` host directory is mapped to the `/etc/prometheus` folder within the container. Prometheus used port `9090` to access its services. This port is mapped to the host port `9090` with the `WithHttpEndpoint` method to make Prometheus available.

Next, add a Docker container for Grafana with the same file:

Codebreaker.AppHost/Program.cs

```
var grafana = builder.AddContainer("grafana", "grafana/grafana")
  .WithBindMount("../grafana/config", "/etc/grafana",
    isReadOnly: true)
  .WithBindMount("../grafana/dashboards",
    "/var/lib/grafana/dashboards", isReadOnly: true)
  .WithHttpEndpoint(containerPort: 3000, hostPort: 3000,
    name: "grafana-http");
```

```
var gameAPIs = builder.AddProject<Projects.Codebreaker_
GameAPIs>("gameapis")
    .WithReference(sqlServer)
    .WithEnvironment("DataStore", dataStore)
    .WithEnvironment("GRAFANA_URL",
      grafana.GetEndpoint("grafana-http"));
    .WithEnvironment("StartupMode", startupMode);

    builder.AddProject<Projects.CodeBreaker_Bot>("bot")
      .WithReference(gameAPIs);
      .WithEnvironment("StartupMode", startupMode);
// code removed for brevity
```

The Docker image for Grafana is pulled from Docker Hub with the name grafana/grafana. This container needs mounts within the /etc/grafana and /var/lib/grafana/dashboards Docker container directories for the Grafana configuration and dashboard configurations used. For both mounts, we'll have a local grafana directory. The Grafana service will be available on port 3000.

After configuring the Docker containers, let's add the configuration for Prometheus.

Configuring Prometheus

Prometheus is configured with this YML file in the prometheus folder that's referenced with the Docker container configuration specified previously:

prometheus/prometheus.yml

```
global:
  scrape_interval: 1s

scrape_configs:
  - job_name: 'codebreakergames'
    static_configs:
      - targets: ['host.docker.internal:9400']
  - job_name: 'codebreakerbot'
    static_configs:
      - targets: ['host.docker.internal:5141']
```

Prometheus pulls telemetry data from services. How often this is done from Prometheus is defined by the scrape_interval parameter. To run tests and get fast information, here, 1 second is configured. On a production system, you might increase this value to, for example, 30 seconds to decrease the load on the services. The services that are accessed from Prometheus are the games API and the bot service. Make sure to configure the ports to the port numbers of your service projects. You can see these values with the `Properties/launchsettings.json` files.

To add an API endpoint that's accessed by Prometheus for scraping telemetry data, the `MapDefaultEndpoints` method of the `Extensions` class needs an update:

Codebreaker.ServiceDefaults/Extensions.cs

```
public static WebApplication MapDefaultEndpoints(this WebApplication
app)
{
  if (Environment.GetEnvironmentVariable("StartupMode") ==
"OnPremises")
  {
    app.MapPrometheusScrapingEndpoint();
  }
// code removed for brevity
  return app;
}
```

The `MapPrometheusScrapingEndpoint` method configures an endpoint and maps the `PrometheusExporterMiddleware` middleware.

The `AddOpenTelemetryExporters` method needs an update as well:

Codebreaker.ServiceDefaults/Extensions.cs

```
private static IHostApplicationBuilder AddOpenTelemetryExporters(this
IHostApplicationBuilder builder)
{
  // code removed for brevity
  builder.Services.AddOpenTelemetry()
    .WithMetrics(metrics => metrics.AddPrometheusExporter());
  return builder;
}
```

With Azure Application Insights, we used the `UseAzureMonitor` method. For Prometheus, we use the `WithMetrics` method, and `AddPrometheusExporter` adds the exporter for Prometheus.

After the configuration of Prometheus, let's configure Grafana.

Configuring Grafana

With this configuration of the Grafana Docker container, we defined using the `grafana` host folder. Here, we need to create a `grafana.ini` configuration file:

grafana/grafana.ini

```
[auth.anonymous]
enabled = true
org_name = Main Org.
org_role = Admin
hide_version = false


[dashboards]
default_home_dashboard_path = /var/lib/grafana/dashboards/aspnetcore.
json
min_refresh_interval = 1s
```

For simple tests locally, we allow anonymous authentication and specify that non-authenticated users have admin access to change settings and customize dashboards. The home dashboard that's used is the ASP.NET Core dashboard.

To access Prometheus from Grafana, the data source needs to be specified with a YML file:

grafana/config/provisioning/datasources/default.yaml

```
apiVersion: 1

datasources:
  - name: Prometheus
    type: prometheus
    access: proxy
    url: http://host.docker.internal:9090
    uid: PBFA97CFB590B2093
```

The `datasources` folder contains the configuration file for the Prometheus data source. This uses port `9090` used by Prometheus.

The default dashboard is configured with a YML file as well:

grafana/config/provisioning/dashbaords/default.yml

```
apiVersion: 1

providers:
```

```
- name: Default
  folder: .NET
  type: file
  options:
    path:
      /var/lib/grafana/dashboards
```

The dashboards themselves are stored within the `grafana/dashboards` folder. You can get pre-built dashboards at `https://grafana.com/grafana/dashboards`. The ASP.NET Core team provides the dashboards for .NET 8 named *ASP.NET Core Endpoint* (ID: 19925) and *ASP.NET Core* (ID: 19924) to monitor request durations, error rates, current connections, total requests... for ASP.NET Core metrics. Both dashboards are copied to the final solution of this chapter.

When this configuration is in place, we are ready to start the solution again – running all services on the local system.

Monitoring the solution with Prometheus and Grafana

When you run the application now, three Docker containers are running: SQL Server, Prometheus, and Grafana, and the bot and game APIs projects, as shown in *Figure 11.15*. Grafana lists an endpoint that's accessible from the host:

Resources

Type	Name	State	Start Time	Source	Endpoints
Container	grafana 525e3c72	Running	21/01/2024 17:27:20	grafana/grafana:latest 3000	http://localhost:3000
Container	prometheus 0283e...	Running	21/01/2024 17:27:20	prom/prometheus:latest 9090	None
Container	sql 06af01e3	Running	21/01/2024 17:27:20	mcr.microsoft.com/mssql/server:2022-latest	None
Project	bot 34580	Running	21/01/2024 17:27:20	D:\codebreaker\Pragmatic...ot\CodeBreak...	http://localhost:5141/swa...
Project	gameapis 34108	Running	21/01/2024 17:27:20	D:\codebreaker\Pragmatic...debreaker.Ga...	http://localhost:9400/swa...

Figure 11.15 – Aspire dashboard with Prometheus and Grafana

By accessing the bot service again to let it play some games, we can access the configured dashboards from the Grafana Docker container, as shown in *Figure 11.16*:

Dashboards

New ∨

Create and manage dashboards to visualize your data

Q Search for dashboards and folders

◇ Filter by tag   ∨   ☐ Starred   ☐   ≔   ↑≡ Sort   ∨

	Name	Tags
☐	∨ ▷ .NET	
☐	⊞ ASP.NET Core	`aspnetcore` `dotnet` `prometheus`
☐	⊞ ASP.NET Core Endpoint	`aspnetcore` `dotnet` `prometheus`

Figure 11.16 – Grafana Dashboards

This page is opened by selecting **Dashboards** in the left pane. There, you can open both configured ASP.NET Core dashboards to see these metrics.

You can also see the custom metric counts written, as shown in *Figure 11.17*:

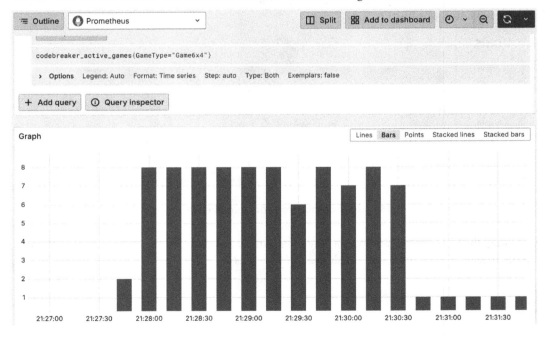

Figure 11.17 – Active games with Grafana

The preceding figure shows the active game count (all started from the bot). To see this screen, open the **Explore** menu in the left pane. Then, select the value of the metric from a combobox and click the **Query** button. With the graph, you can select from different display types.

In the previous section, you've seen Azure services to monitor the solution. With Prometheus and Grafana, you've seen services that can easily be used in an on-premises environment. In case you prefer Prometheus and Grafana while running within Microsoft Azure, one way of use is to run these services in Azure Container Apps. Azure-managed services are available as well: Azure offers Managed Grafana and Azure Monitor Managed Service for Prometheus. Using these services, the same service is available but with reduced management needs.

Summary

In this chapter, you learned about offering telemetry data from a microservice solution covering logging, metrics, and distributed tracing. With logging, you used high-performance, strongly typed logging to write information-level logs as well as errors. For metrics, you created custom metric data using the `Meter` class with instruments created. For distributed tracing, you used the `ActivitySource` and `Activity` classes.

To monitor all this telemetry data, you used the .NET Aspire dashboard, Azure Application Insights, and Prometheus with Grafana.

In the next chapter, we'll look into how to use metrics data to scale the services running with Azure Container Apps. We'll find out about the memory and CPU usage of services using load tests we created in *Chapter 10*, combine this with metrics information from this chapter, learn about scaling services to be ready no matter how demand grows, and implement health checks to recover services when not healthy.

Further reading

To learn more about the topics discussed in this chapter, you can refer to the following links:

- OpenTelemetry: `https://opentelemetry.io/`
- High-performance logging in .NET: `https://learn.microsoft.com/en-us/dotnet/core/extensions/high-performance-logging`
- Application Insights: `https://learn.microsoft.com/en-us/azure/azure-monitor/app/app-insights-overview`
- KQL repo: `https://github.com/microsoft/Kusto-Query-Language`
- Prometheus: `https://prometheus.io/`
- Grafana repo: `https://github.com/grafana/grafana`
- Grafana Dashboards: `https://grafana.com/grafana/dashboards/`

12
Scaling Services

How fast is the service responding? Is the service limited to CPU cores or memory? Based on user load, when is it useful to start more server instances? If you run too many compute resources, or if they're too big, you pay more than is necessary. If the resources you use are too small, the response time increases or the applications might not be available at all. With this, you lose customers, and your income is reduced. You should know how to find bottlenecks and know what good knobs to turn to scale the resources as needed.

In *Chapter 10*, we created load tests to see how the service behaves under load, while in *Chapter 11*, we extended the service by adding telemetry data. Now, we'll use both load tests and telemetry data to find out what scaling option is best.

In this chapter, we'll start reducing the response time with the help of telemetry data before analyzing the load, which can be run with one instance. Finally, we'll define rules so that we can scale out to multiple instances. To automatically restart instances when the service is not responding, we'll add health checks.

In this chapter, you'll learn how to do the following:

- Increase performance using caching
- Simulate users with Azure Load Testing
- Scale up and scale out services
- Use scale rules
- Implement health checks

Technical requirements

In this chapter, like the previous chapters, you'll need an Azure subscription, the Azure Developer CLI (`winget install Microsoft.Azd`), and Docker Desktop.

The code for this chapter can be found in this book's GitHub repository: `https://github.com/PacktPublishing/Pragmatic-Microservices-with-CSharp-and-Azure`.

The `ch12` folder contains the projects we'll need for this chapter, as well as their output. To add the functionality from this chapter, you can start with the source code from the previous chapter.

Here are the projects we'll be implementing in this chapter:

- `Codebreaker.AppHost`: The .NET Aspire host project. This project has been enhanced by adding a Redis resource for caching.

- `Codebreaker.ServiceDefaults`: Here, we use a common health check configuration for all the services.

- `Codebreaker.GameAPIs`: With this project, we implement caching games to reduce database access and add a custom health check.

To learn how to publish the resources to Microsoft Azure, check out the README file for this chapter.

> **Note**
> While working on this chapter, we created load tests with many users and changed the scale of the Azure Cosmos database. The duration of these tests and the number of virtual users you can use with them depends on the amount of money you want to spend. If you increase the RU/s with the database, make sure you delete the resources after running the tests, or at least reduce the number of RU/s again after running the tests. You might also skip running the tests with larger user numbers and just read the results.

Increasing performance with caches

Before we analyze the application's CPU and memory needs, let's look at where easy wins are possible to return faster responses to the client. By checking telemetry information (as we did in the previous chapter), we can see that when using distributed tracing to send a game move, several requests are made to the database. *Figure 12.1* shows the bot sending the SetMoveAsync request:

bot SetMoveAsync	164,98ms
bot PATCH	83,8ms
gameapis PATCH /games/{id:guid}	63,74ms
gameapis ➜ 127.0.0.1,54323 DATA ...	13,33ms
gameapis SetMove	15,65ms
gameapis ➜ 127.0.0.1,54323 DAT...	6,9ms

Figure 12.1 – Tracing a move set

As shown in the preceding figure, when receiving a PATCH request, the game ID is used to retrieve the game from the database to verify the correctness of the data that's received. After the move is calculated, the resulting game is written to the database. Trace information from EF Core is shown with the DATA keyword, along with the time needed for access.

Performance might be good enough, but this also depends on the database load. When using the SQL Server database, having many writes can reduce the read performance because of locks with write operations. With higher database loads, increasing the number of Request Units (RU) or using bigger machines (which increases the price) can be a solution for higher loads. A better option is to cache data. Many of the database reads can be replaced by reading objects from a memory cache.

An initial idea might be to store the game in the memory of the process. If it is not there, retrieve it from the database. However, if multiple instances of the service are running, the client could invoke one move with server A and another move with server B. Because the game contains the last move number, reading it from the local cache could result in an older version of the game, and thus the request fails. One option around this would be to use sticky sessions. With this, one client always gets the same service instance to fulfill a request. This requirement can easily be avoided by using a distributed memory cache.

> **Note**
>
> With a sticky session, a client always connects to the same service instance. The biggest disadvantage of sticky sessions is when the service goes down. Without sticky sessions, the client can immediately switch to another service instance, and no downtime is detected. With sticky sessions, all the session data is lost for the client. This is not the only disadvantage. What if another instance is started because of low performance? The new service instance only receives the traffic from new clients. Existing ones stick with the servers they already communicate with. There's a delayed server utilization (only from new clients). With sticky sessions, the load can be unevenly distributed between service instances. The best thing to do is try to avoid them.

When using a distributed memory cache, multiple options are available. With Microsoft Azure, Azure Cache for Redis can be used. This service offers Standard, Premium, Enterprise, and Enterprise Flash offerings based on your availability and memory size needs. Using Azure Cosmos DB, an integrated in-memory cache built into the Azure Cosmos DB gateway, can be used. One feature of this service is an item cache for point reads, which fulfills the purpose of reading the item several times while the game is running. This reduces the cost with Azure Cosmos DB because the RU/s needed to read from the cache are 0.

Here, we'll use a Docker container for Redis that can be used in the local Docker environment, as well as to run the solution with Azure Container Apps.

Reading and writing from the cache

The API of the IDistributedCache interface supports writing byte arrays and strings – the data needs to be sent across the network to a Redis cluster. For this, we'll create methods to convert the Game class to and from bytes:

Codebreaker.GameAPIs/Models/GameExtensions.cs

```
public static class GameExtensions
{
  public static byte[] ToBytes(this Game game) =>
    JsonSerializer.SerializeToUtf8Bytes(game);

  public static Game? ToGame(this byte[] bytes) =>
    JsonSerializer.Deserialize<Game>(bytes);
}
```

The System.Text.Json serializer supports serializing the data not only to JSON but also to a byte array. The Game class already supports serialization with this serializer, so no other changes need to be made to the Game and Move model types.

We can access the cache from the GamesService class:

Codebreaker.GameAPIs/Services/GamesService.cs

```
public class GamesService(
  IGamesRepository dataRepository,
  IDistributedCache distributedCache,
  ILogger<GamesService> logger,
  GamesMetrics metrics,
  [FromKeyedServices("Codebreaker.GameAPIs")]
  ActivitySource activitySource) : IGamesService
{
```

No matter what technology is used for the distributed memory cache, we can inject the `IDistrib-utedCache` interface.

To update the `Game` class with the cache, we can implement the following method:

Codebreaker.GameAPIs/Services/GamesService.cs

```
private async Task UpdateGameInCacheAsync(Game game, CancellationToken
cancellationToken = default)
{
    await distributedCache.SetAsync(game.Id.ToString(), game.ToBytes(),
    cancellationToken);
}
```

The game ID is used as a key to retrieve the game object from the cache. Invoking the `SetAsync` method adds the object to the cache. If the object has already been cached, it is updated with the new value. With an additional parameter of the `DistributedEntryCacheOptions` type, the object can be configured to specify the time the object should stay in the cache. Here, we need to use a typical time the user needs from one move to another. With every retrieval and update, the **sliding expiration** starts anew. Instead of specifying this here, we can configure default values.

The `UpdateGameInCacheAsync` method needs to be invoked from the `GamesService` class when the game (`StartGameAsync`) is created, as well as after setting the game move (`SetMoveAsync`).

The implementation within the `StartGameAsync` method is shown here:

Codebreaker.GameAPIs/Services/GamesService.cs

```
game = GamesFactory.CreateGame(gameType, playerName);
activity?.AddTag(GameTypeTagName, game.GameType)
    .AddTag(GameIdTagName, game.Id.ToString())
    .Start();
await Task.WhenAll(
    dataRepository.AddGameAsync(game, cancellationToken),
    UpdateGameInCacheAsync(game, cancellationToken));
metrics.GameStarted(game);
```

Writing to the database and the cache can be done in parallel. We don't need to wait until the database write is completed to add the game object to the cache to return a faster answer. If the database fails, it doesn't matter if the game is cached or not.

To read the data from the cache, we need to implement GetGameFromCacheOrDataStoreAsync:

Codebreaker.GameAPIs/Services/GamesService.cs

```
// code removed for brevity
private async Task<Game?> GetGameFromCacheOrDataStoreAsync(
  Guid id, CancellationToken cancellationToken = default)
{
  byte[]? bytesGame = await distributedCache.GetAsync(id.ToString(),
  cancellationToken);
  if (bytesGame is null)
  {
    return await dataRepository.GetGameAsync(id, cancellationToken);
  }
  else
  {
    return bytesGame.ToGame();
  }
}
```

The GetAsync method of the cache returns a byte array of the cached data, which is then converted using the ToGame method. If the data is not available within the cache (the item might have been removed from the cache because too much memory was already allocated, or if the user was thinking about their next move for too long), we get the game from the database. The code in the source code repository includes a flag where you can switch off reading from the cache to easily try out not using the cache with different loads that are used to check the results.

GetGameFromCacheOrDataStoreAsync needs to be invoked from the SetMoveAsync and GetGameAsync methods.

Configuring the Aspire Redis component

Regarding the game-apis project, we need to add the **.NET Aspire StackExchange Redis component** to configure the DI container:

Codebreaker.GameAPIs/ApplicationServices.cs

```
public static void AddApplicationServices(this IHostApplicationBuilder
builder)
{
  // code removed for brevity
```

```
    builder.Services.AddScoped<IGamesService, GamesService>();

    builder.AddRedisDistributedCache("redis");
}
```

The `AddRedisDistributedCache` method uses the cache name that needs to be configured with the Aspire App Host project to get the connection string and configuration values. With this method, it's also possible to specify the configuration values programmatically.

Finally, a Docker container for the Redis resource is configured with `app-model` in the AppHost project:

Codebreaker.AppHost/Program.cs

```
var redis = builder.AddRedis("redis")
    .WithRedisCommander()
    .PublishAsContainer();

var cosmos = builder.AddAzureCosmosDB("cosmos")
    .AddDatabase("codebreaker");

var gameAPIs = builder.AddProject<Projects.Codebreaker_
GameAPIs>("gameapis")
    .WithReference(cosmos)
    .WithReference(redis)
    .WithReference(appInsights)
    .WithEnvironment("DataStore", dataStore);
```

The `AddRedis` method configures using the `redis` Docker image for this service. This needs to be configured both with **Prometheus** and the default Azure environment. With the Azure environment, this creates an Azure Container App running Redis. As an alternative, you can add the `PublishAsAzureRedis` API instead of `PublishAsContainer`. This method configures the PaaS offering for **Azure Cache for Redis**. `WithRedisCommander` adds a management UI for Redis to `app-model`.

With this configuration in place, running games via the bot provides the results shown in *Figure 12.2*. Even when using a low load on the local system, writing to SQL Server took 5.96 ms, and writing to the cache took 1.83 ms. Both were running in a Docker container:

Figure 12.2 – Set move with a distributed cache

Next, let's add some load to the game-apis project to see the resource consumption.

Simulating users with Azure Load Testing

In *Chapter 10*, we created Playwright tests that were used to create load tests. These Playwright tests allowed us to use .NET code to easily create a complete flow so that we could play a game from a test. Using Microsoft Azure, we can use another service to create tests and get integrated analysis with Azure services: Azure Load Testing.

> **Note**
>
> At the time of writing, the **Microsoft Playwright Testing** cloud service is great for testing the load of web applications. However, it doesn't support load testing APIs, so we'll use Azure Load Testing here. You can still use Azure compute (for example, Azure Container Instances) to run Playwright tests, but Azure Load Testing has a better report configuration and report functionality.

Before creating the load test, make sure you deploy the solution to Microsoft Azure using azd up. Check the README file for this chapter for more details about the different azd versions.

After creating the Azure resources, open the `game-apis` Azure Container App in the Azure portal and select **Application | Containers** from the left bar. The container's resource allocation will be shown as *0.5 CPU cores* and *1 Gi memory*.

Now, let's make sure the first tests use just one replica.

Scaling to one replica

Scales and replicas can scale up to 300 instances. The default configuration is to scale from 1 to 10. Creating a load with many users would automatically scale out and start multiple instances. To see what the limits of one instance are, change the scale to just one instance for both Min replicas and Max replicas, as shown in *Figure 12.3*. Clicking Create creates a new revision of the app and deprovisions the existing revision afterward:

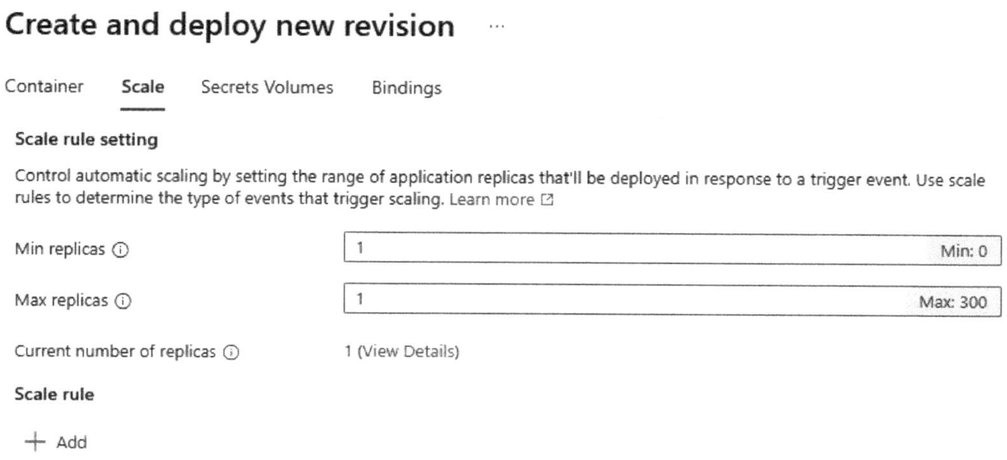

Create and deploy new revision ⋯

Container   **Scale**   Secrets Volumes   Bindings

Scale rule setting

Control automatic scaling by setting the range of application replicas that'll be deployed in response to a trigger event. Use scale rules to determine the type of events that trigger scaling. Learn more ☐

Min replicas ⓘ

| 1 | Min: 0 |

Max replicas ⓘ

| 1 | Max: 300 |

Current number of replicas ⓘ   1 (View Details)

Scale rule

+ Add

Figure 12.3 – Changing replicas with Azure Container Apps

To specify scaling at deployment time, create YAML templates that specify the configuration for Azure Container Apps. Start a terminal with the current directory set to the solution and run the following command from the Azure Developer CLI (after you've initialized the solution with azd init):

```
azd infra synth
```

This tool uses the app-model manifest to create Bicep files to deploy the Azure resources of app-model (in the root infra folder). The infra folder of the AppHost project contains YAML templates that describe every Azure Container App that's been created (from projects and Docker images). See *Chapter 6* for more details on Bicep files.

In the AppHost project, you'll see that a `<app>.tmpl.yaml` file has been generated to specify the settings for Azure Container Apps.

By default, the minimum number of replicas is set to `1`. With bot-service, you can change the configuration:

```
  template:
    containers:
# code removed for brevity
    scale:
      minReplicas: 0
      maxReplicas: 1
```

With bot-service, to reduce cost, you can define the scale from 0 to 1. When the minimum instance count is set to 0, there's no cost for the service. Just be aware that it takes a few seconds to start up the service, and the first user accessing the service needs to wait. Because the bot is not invoked by game-playing users, and this service is not always needed, it can be scaled down to 0. The game-apis service should always return answers fast; thus, the minimum scale should be set to 1. If there's no load on the service, there's an idle price. With this, the cost of the CPU is reduced to about 10% of the normal cost, but the memory (the application is still loaded in memory) has the normal price. To test the load with exactly one replica, set the game-apis service's minimum and maximum values to 1. Later, when scaling out, we'll increase the value of Max replicas again.

After changing the number of replicas in the YAML file, you can re-deploy the application using az up or just using az deploy.

We also need to make sure that the database allows the requests that are needed. With a load test, we can expect that we'll need more than the 400 RU/s. Before the first test runs, change the Azure Cosmos DB throughput to Autoscale with a maximum of 1,000 RU/s.

Now, we are ready to create a test.

Creating an Azure URL-based load test

To create a new load test, create the Azure Load Testing resource using the Azure portal. Specify a resource group name and the name of the resource.

Once the resource is available, open it in the portal and select **Tests | Tests** from the left bar. Then, click **Create** after choosing **Create a URL-based test**. Under **Basics**, specify **Test name** and **Test description** values and check the **Enable advanced settings** box, as shown in *Figure 12.4*:

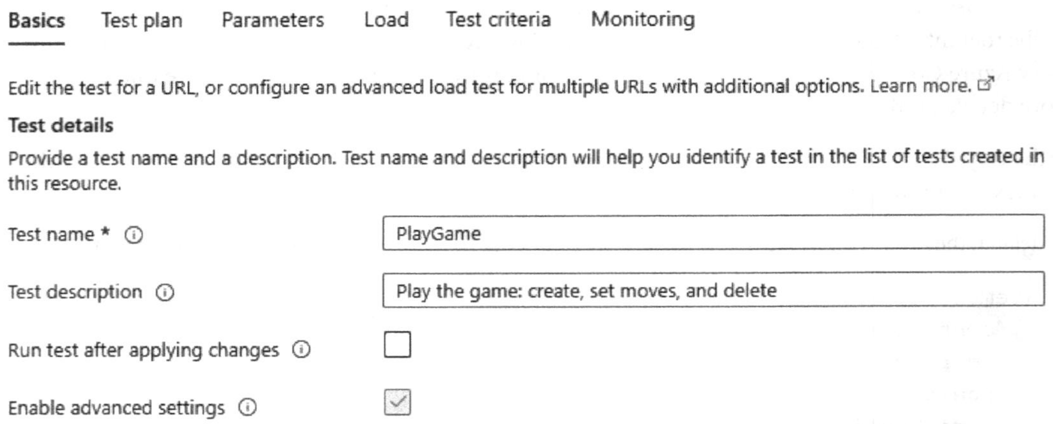

Basics	Test plan	Parameters	Load	Test criteria	Monitoring

Edit the test for a URL, or configure an advanced load test for multiple URLs with additional options. Learn more. ⧉

Test details

Provide a test name and a description. Test name and description will help you identify a test in the list of tests created in this resource.

Test name * ⓘ   `PlayGame`

Test description ⓘ   `Play the game: create, set moves, and delete`

Run test after applying changes ⓘ   ☐

Enable advanced settings ⓘ   ☑

Figure 12.4 – Load testing – basic settings

With Enable advanced settings selected, a test plan consisting of up to five HTTP requests can be created. So, in the Test plan section, add five requests, as shown in *Figure 12.5*:

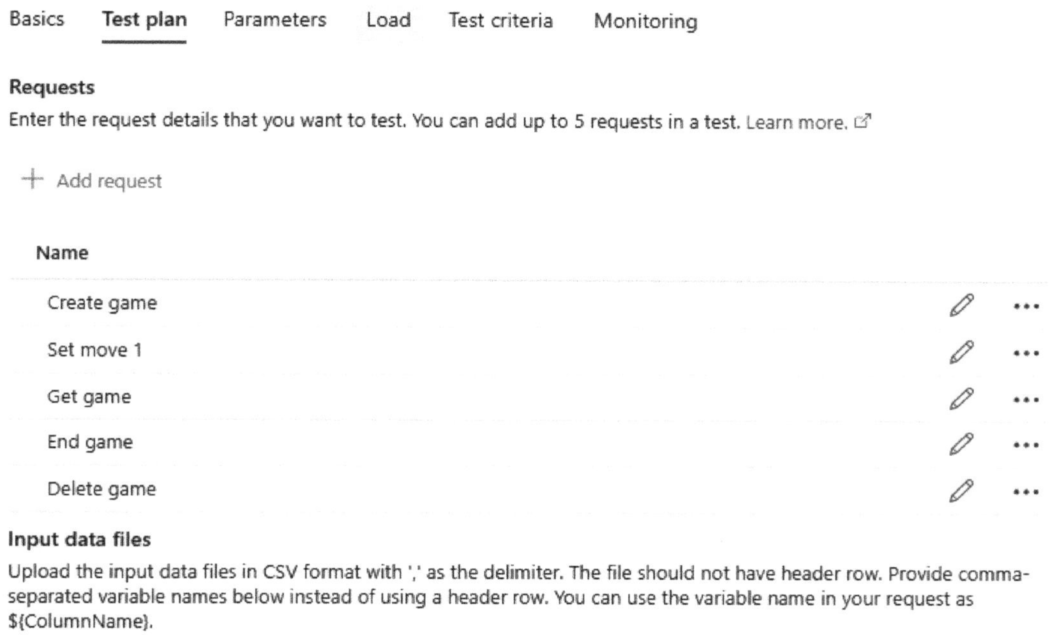

Basics   **Test plan**   Parameters   Load   Test criteria   Monitoring

Requests

Enter the request details that you want to test. You can add up to 5 requests in a test. Learn more. ☐

+ Add request

Name

Create game	🖉	•••
Set move 1	🖉	•••
Get game	🖉	•••
End game	🖉	•••
Delete game	🖉	•••

Input data files

Upload the input data files in CSV format with ',' as the delimiter. The file should not have header row. Provide comma-separated variable names below instead of using a header row. You can use the variable name in your request as ${ColumnName}.

Figure 12.5 – Load testing – test plans

The first request is a POST request to create the game. The second is a PATCH request to update the game with a move. This is followed by a GET request to get information about the game, a PATCH request to end the game, and a DELETE request to delete the game.

These requests can easily be configured with the UI, as shown in *Figure 12.6*:

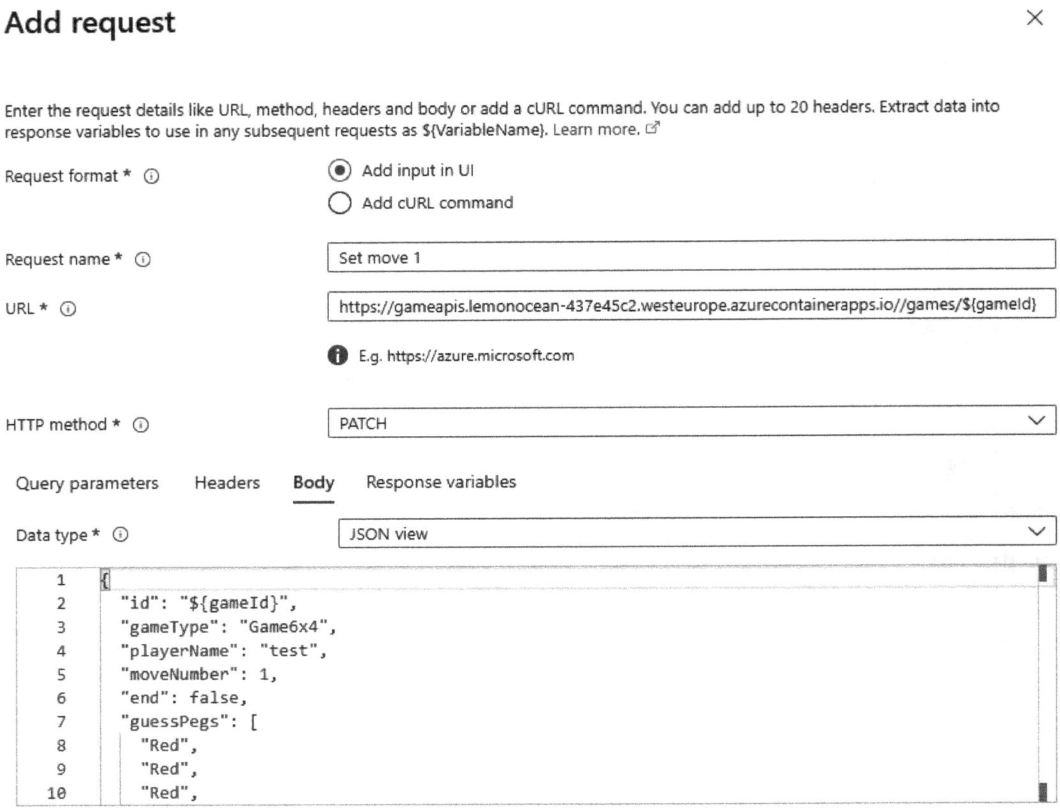

Add request   ✕

Enter the request details like URL, method, headers and body or add a cURL command. You can add up to 20 headers. Extract data into response variables to use in any subsequent requests as ${VariableName}. Learn more. ⬀

Request format * ⓘ	⦿ Add input in UI ◯ Add cURL command
Request name * ⓘ	Set move 1
URL * ⓘ	https://gameapis.lemonocean-437e45c2.westeurope.azurecontainerapps.io//games/${gameId}
	ⓘ E.g. https://azure.microsoft.com
HTTP method * ⓘ	PATCH ⌄

Query parameters   Headers   **Body**   Response variables

Data type * ⓘ   JSON view   ⌄

```
 1   {
 2     "id": "${gameId}",
 3     "gameType": "Game6x4",
 4     "playerName": "test",
 5     "moveNumber": 1,
 6     "end": false,
 7     "guessPegs": [
 8       "Red",
 9       "Red",
10       "Red",
```

Figure 12.6 – Load testing – adding requests

Instead of getting the request information from the OpenAPI description or the HTTP files, you can copy the requests from the README file of this chapter to the Body area. The requests, including their HTTP headers, are listed. Make sure you use the links to your Azure Container App when specifying the URL.

With the POST request, don't just specify the body – also define the use of the response. With the JSON result, id is returned; this can be accessed with the $.id expression. Set this to the gameId variable. Response variables can be used with later requests – and the game ID is needed with all the following requests. When setting the game move, use ${gameId} to pass the game ID to the URL string and the HTTP body. You can check the README file for this chapter for more details about the values you should specify with the different requests.

In the next dialogue, shown in *Figure 12.7*, the load can be specified:

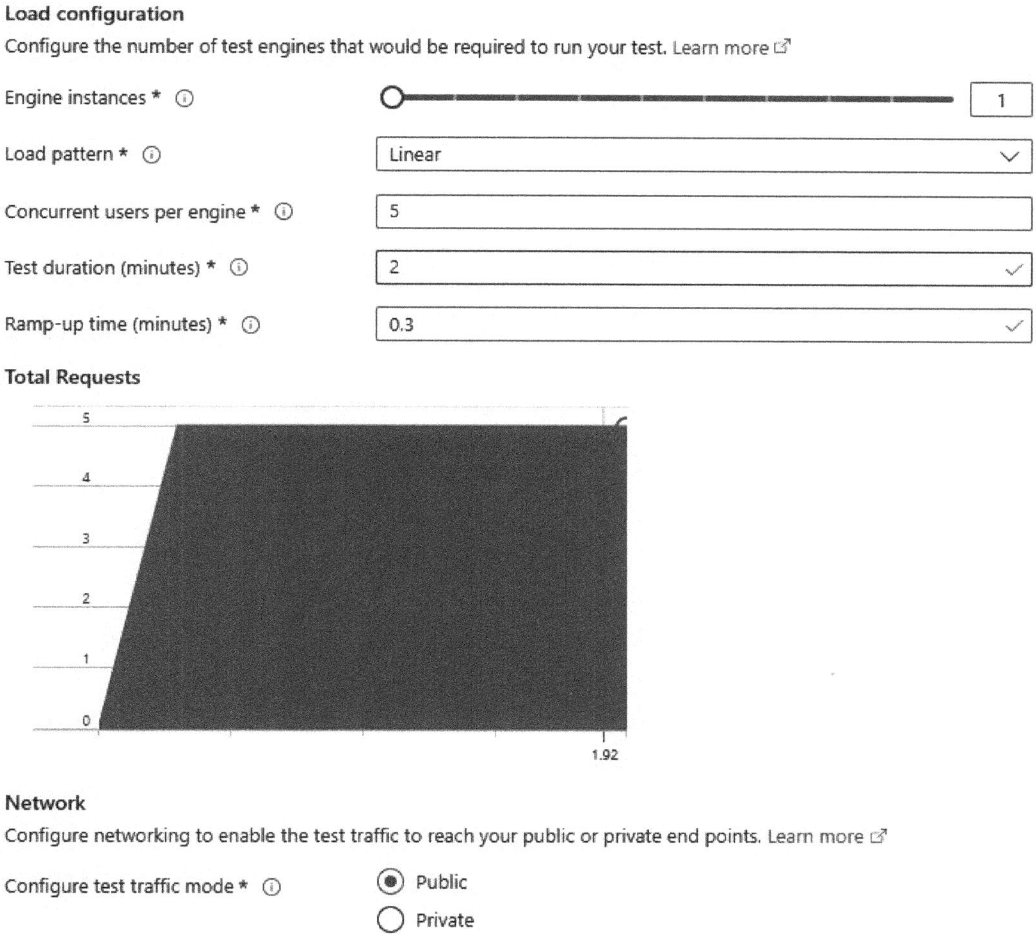

Load configuration

Configure the number of test engines that would be required to run your test. Learn more ⬈

Engine instances * ⓘ	⃝━━━━━━━━━━━━━━━━━━━━━━━━━ 1
Load pattern * ⓘ	Linear ⌄
Concurrent users per engine * ⓘ	5
Test duration (minutes) * ⓘ	2 ✓
Ramp-up time (minutes) * ⓘ	0.3 ✓

Total Requests

Network

Configure networking to enable the test traffic to reach your public or private end points. Learn more ⬈

Configure test traffic mode * ⓘ	◉ Public
	◯ Private

Figure 12.7 – Load testing – specifying the load

Here, we'll start small with just 5 concurrent virtual users and do other tests with more user loads and multiple engine instances, with a ramp-up time of 0.3 minutes. With one test engine instance, you can specify up to 250 virtual users and go up to 2,500 virtual users with 10 instances. The configuration also allows you to specify a Load pattern value, which increases the load over time. Having multiple test runs with different user numbers can give a good indication of what scaling rules should be used to increase the number of service instances.

Be aware of the cost you can incur when testing with 2,500 virtual users and 10 virtual machines behind the scenes. Contrary to the other resources we've used so far, with this, you can easily go over the subscription limits with the Visual Studio Enterprise Azure subscription or the free Azure subscription. Luckily, we only pay for the time the test runs and don't need to pay for physical machines that are only needed for a short time.

> **Note**
>
> Don't assume virtual users are the same as real users. Real users produce a lot less load than virtual users with Azure Load Testing. A real user needs to think between moves. Several seconds, if not minutes, are spent between each move. Virtual users just continuously invoke the APIs. With the JMeter tests that are used behind the scenes, the number of virtual users configures the number of threads to be used. How many real users you can calculate compared to virtual users depends on the type of application. You need to find out how long real users think on average with Codebreaker when monitoring the application in production. In *Chapter 10*, we created custom metric data to monitor the time spent between moves; this is a good value to use.

With the test criteria configuration (see *Figure 12.8*), you can specify when the test should fail – for example, when the response time takes too long. Before doing the first test run, you can leave the test criteria empty to see values that are reached with low load:

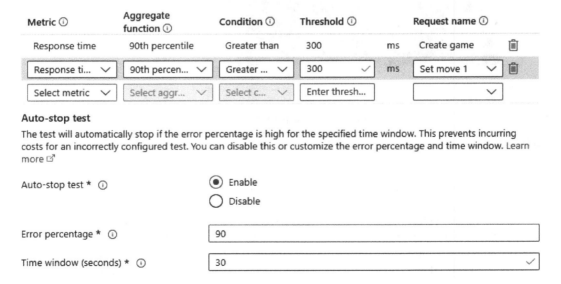

Figure 12.8 – Load testing – test criteria

For the last configuration, open the Monitoring settings, as shown in *Figure 12.9*:

Resources

Configure application components to monitor server-side metrics during the test run. Learn more ✎

+ Add/Modify

Resource Name	↑↓	Resource type	↑↓	Resource group	↑↓	
🌐 gameapis		Container App		rg-codebreaker-packt12		···
🌐 redis		Container App		rg-codebreaker-packt12		···
☁ codebreakercosmosy4qeblmjsvx	Azure Cosmos DB account		rg-codebreaker-packt12		···

Figure 12.9 – Specifying monitoring resources

Select the resources that are taking part in the test, such as the gameapis and redis Azure Container Apps, and the Azure Cosmos DB resource. You can easily filter the resources based on the resource group.

Now, we are ready to run the test.

Running a load with virtual users

When creating and changing a test, after clicking Save, you need to wait until the JMeter script is created; otherwise, the test will fail to start. To run the test, click the Run button and enter a test description – for example, 5 users 0.5 core.

After the test is completed, you will see client-side metrics from the test engine and server-side metrics from the selected Azure Container Apps service.

When I did my test run, 7,834 requests were sent (a lot more than five human users would do for 2 minutes), and up to 0.49 CPU cores and 354 MB of memory were used. The response time was below 116 milliseconds for 90% of the requests, and the throughput was 67.53 requests per second.

> **Note**
>
> Don't expect to get the same results with multiple runs. Many dependencies run these tests. What's the network performance and latency between the different Azure services being used? For my tests, I created the Azure Load Testing service in the same Azure region where the services are running. Even in the same Azure region, different resources could be running in the same or different data centers. These differences are not an issue. Users will be located outside an Azure data center. What we need to know is how many users can be served from one instance and what settings are best for the application, such as CPU and memory resources (scale up) or running multiple replicas (scale out). We also need to see what the real bottlenecks are, and what can be controlled.

Figure 12.10 shows the response time results with every API invocation:

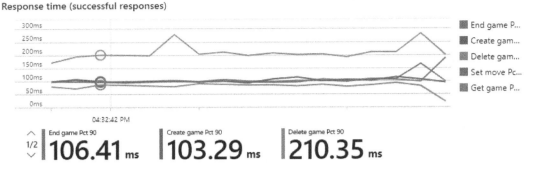

Figure 12.10 – Response time for five virtual users

With five virtual users, the response time is OK when considering all the requests. What might be interesting is that the delete request takes the most time to complete with Azure Cosmos DB.

Five virtual users is a good start, but let's add more load.

Reaching limits with a higher load

To change the load of a test, you can edit it. To do so, click **Load** and change **Concurrent users per engine value** to 25. Click **Apply** and wait for the JMeter script to be created with **Notifications** in the Azure portal. At this point, you can start the test again.

With my test run, increasing the number of virtual users to 25 resulted in just 11.701 total requests with 98.39 requests per second. The request to create a game needed 289 ms with a 90% percentile. *Figure 12.11* shows the number of requests per second for this test:

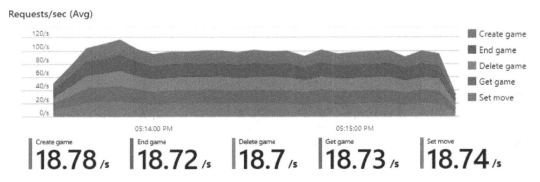

Figure 12.11 – Requests per second

Comparing the results to the five users' test runs, using 25 users resulted in just a small increase concerning the total requests and requests per second. As a result, the time for creating a game increased from 96 ms to 498 ms. This is not a good outcome. Why did this happen? The server-side metrics didn't reach Azure Container Apps limits with CPU cores and memory. The limit was not down to Azure Container Apps but the Azure Cosmos database, as shown in *Figure 12.12*:

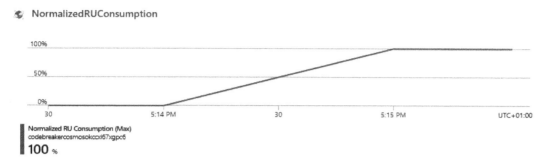

Figure 12.12 – Azure Cosmos DB RU consumption

When running this test with Azure Cosmos DB, the RU was configured with **autoscale** throughput, and a maximum 1,000 RU/s limit was reached. This can be seen in the preceding figure. You can also dig into **Application Map** on App Insights and check the different Azure Cosmos DB metrics, as shown in *Figure 12.13*:

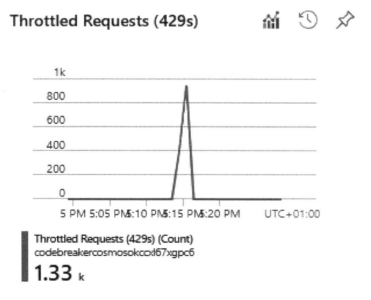

Figure 12.13 – Requests throttled with Azure Cosmos DB

As we can see, the requests have been throttled; an error code of 429 has been returned to the game-apis service. You can use **Kusto Query Language** (**KQL**) to query for these log messages (see *Chapter 11* for more information on KQL):

```
ContainerAppConsoleLogs_CL
| where  Log_s contains "Request rate is large"
```

The complete log message that's returned states *"Request rate is large. More Request Units may be needed, so no changes were made. Please retry this request later. Learn more: http://aka.ms/cosmosdb-error-429."*

Earlier in this chapter, we reduced the load on Azure Cosmos DB by removing requests from the database that were not needed by using a cache. Without this change, the limit would have been hit earlier.

While error code 429 has been returned to the game-apis service, the result of the invocation was still successful because of the built-in retry configuration with .NET Aspire. But of course, the time needed for the request increased.

When creating the test, we ensured we could see the metrics data for all the resources participating in the test. That's why we can see the Cosmos metrics with the test run and can easily fix it. Let's use Cosmos DB to increase the RU/s. With a maximum value of 1,000 RU/s, the minimum RU/s is 100. Increasing the maximum to 10,000 sets the minimum to 1,000. Make sure you only change the maximum to higher values for a short period while running the tests, and when needed. You can view the expected cost in the dialogue where you scale the RU/s. Make sure you reduce the scale limits as you don't need them anymore. It is possible to set the maximum to 1,000,000 RU/s, which sets the minimum to 100,000. You can view the price range when you change this throughput before clicking the **Save** button. Be aware that when changing the maximum value above 10,000 RU/s, it can take 4 to 6 hours for this compute power to become available.

With 25 virtual users, we reached the 1,000 RU/s limit. So, let's increase it to 10,000 RU/s. If not that many RU/s are required with a specific number of users, we'll see this in the test results and adjust the setting according to our needs after the test runs.

After increasing the RU/s limit and running the test again with 25 virtual users, Azure Cosmos DB is no longer the bottleneck. Just 12% of the RU/s are being used. So, let's increase the number of virtual users to 50.

Scaling up or scaling out services

Let's run the test with 50 virtual users and compare how the performance differs when increasing CPU and memory, as well as increasing the number of replicas.

Configuring scaling up

To scale up, we must increase the CPU and memory values.

> **Note**
>
> When using `azd up` to create a Container App, a *consumption-based* environment is created. There's also the option to create a *workload profile* with *dedicated hardware*. When using dedicated hardware, you can choose the type of virtual machine that will be used. At the time of writing this book, the virtual machines were in categories D (general purpose, 4 – 32 cores, 16 – 128 GiB memory) and E (memory optimized 4 – 32 cores, 32 – 256 GiB memory, and GPU enabled with up to 4 GPUs). The type of machine also defines the available network bandwidth. Depending on the workload you have, there are great options available.

To change CPU and memory in the Azure portal, within the Container App, select **Containers** from the left bar, click the **Edit and deploy** button, select the container image, and click **Edit**. This will open the container editor, where you can select the CPU cores and memory, as shown in *Figure 12.14*:

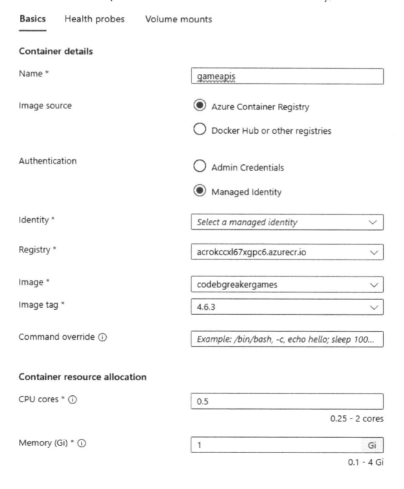

Figure 12.14 – Editing the container's settings

Here, we'll change the CPU and memory values. When using the consumption-based environment, be aware that configurations need to map – for example, 0.5 cores and 1.0 Gi memory, 1.0 cores and 2.0 Gi memory, up to 2.0 cores and 4.0 Gi memory. In the consumption workload profile, you can have up to 4.0 cores and 8.0 Gi memory with a container.

We can also configure this with the YAML template file:

```
template:
  containers:
    - image: {{ .Image }}
      name: gameapis
      resources:
        cpu: 1.0
        memory: 2Gi
# configuration removed for brevity
```

CPU and memory resources are specified within the `resources` category. After deciding on the best configuration, specifying this with the YAML file creates the right size.

Running the load test for 50 users for 2 minutes shows the following results based on the configuration:

	Total Requests	Throughput	Create Game Response	10,000 RU/s
0.5 Cores, 1 Gi	12,015	100.13/s	491 ms	16%
1 Cores, 2 Gi	20,621	171.84/s	383 ms	24%
2 Cores, 4 Gi	22,444	187.03/s	381 ms	26%

Table 12.1 – Scaling up load test results

With these configurations, we can see that increasing the compute resources to 1 core and 2 Gi of memory makes an improvement, whereas duplicating the compute resources again makes just a small improvement.

Now, let's change the replicas.

Configuring scaling out

You learned how to change the number of replicas earlier in this chapter. In this section, we'll change both the minimum and maximum count to the same values so that we can distribute the load across different instances.

We receive the following counts when the tests use 0.5 cores and 1 Gi of memory:

	Total Requests	Throughput	Create Game Response	10,000 RU/s
1 replica	12,015	100.13/s	491 ms	16%
2 replicas	16,291	135.76/s	490 ms	20%
4 replicas	27,704	230.87/s	299 ms	34%

Table 12.2 – Scaling out the load test results

Using two replicas with 0.5 cores and 1 Gi of memory uses the same CPU and memory resources as one replica with 1 core and 2 Gi of memory does. One instance with 1 core was the better performing option, with 20,621 requests compared to 16,291 requests. By adding more replicas, we can scale higher than what's possible by just adding CPU resources.

A big advantage of using multiple replicas is that we can dynamically scale based on the load. We'll create scale rules in the next section. Scale rules don't allow us to change CPU and memory resources.

One issue you need to be aware of when scaling multiple instances is whether the application is designed for this. When running the Codebreaker application while using the in-memory games store, the implementation was built with multi-threading in mind, but not with multiple machines. When one user starts a game and the next user sets a move, the first request might access the first machine where the game is stored in memory, and the second request might access the second machine where the game to set the move is not available. The Redis cache, which offers distributed memory, solves this issue. The sample application available in this chapter's GitHub repository includes the `DistributedMemoryGamesRepository` class, which can be configured with the `DataStore` configuration set to `DistributedMemory`. To test this on your local development environment, you can change the AppHost project:

Codebreaker.AppHost/Program.cs

```
var gameAPIs = builder.AddProject<Projects.Codebreaker_
GameAPIs>("gameapis")
  .WithReference(cosmos)
  .WithReference(redis)
  .WithReference(appInsights)
  .WithEnvironment("DataStore", dataStore)
  .WithEnvironment("StartupMode", startupMode)
  .WithReplicas(2);
```

When adding the `WithReplicas` method when configuring a project, the number of replicas can be specified. With a value of 2, when running the solution locally, the .NET Aspire dashboard (*Figure 12.15*) shows two instances of the `game-apis` service running. Each service has a port number that allows the specific instance to be accessed. The common port number, `9400`, is the port of the proxy client that references both `game-apis` service instances running with port numbers `49379` and `49376`. The port number that's used by the proxy is defined with the `launchsettings.json` file, while the port number for the instances randomly changes when a new application starts:

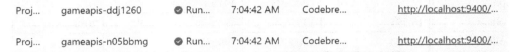

Figure 12.15 – Two replicas for game-apis

Now that we know about the improvements that we can make when running multiple replicas, let's scale dynamically.

Scaling dynamically with scale rules

With Azure Container Apps, scale rules can be defined based on concurrent HTTP requests, concurrent TCP requests, or custom rules. With custom rules, scaling can be based on CPU, memory, or many events based on different data sources.

A microservice isn't necessarily triggered based on HTTP requests. The service can also be triggered asynchronously, such as when a message arrives in a queue (for example, using Azure Storage Queue or Azure Service Bus) or when events occur (for example, using Azure Event Hub or Apache Kafka).

Azure Container Apps scale rules are based on **Kubernetes Event-driven Autoscaling (KEDA)**, which offers a large list of scalers. You can find the full list at `https://keda.sh`.

When using a KEDA scaler with the Azure Service Bus queue, you can specify how many messages should be in the queue when another replica should be started. What's common with all the KEDA scalers is the configuration of the polling interval – how often the values are checked (by default, this is 30 seconds), a scaling algorithm to calculate the number of replicas, and a cooldown period (300 seconds) – the time before replicas started can be stopped again.

In *Chapter 15*, we'll use communication with messages and events where autoscaling will be based on event-based KEDA scalers.

As we saw when we tested the load with a fixed number of instances, the best option to scale the services is using the number of HTTP requests. So, let's configure scaling with an HTTP rule. We can do this by using the Azure portal and the `azd infra synth` generated template YAML file. This is the output from the JSON content in the Azure portal:

```
"template": {
    ...
    "scale": {
        "minReplicas": 1,
        "maxReplicas": 8,
```

```
    "rules": [{
      "name": "http-rule",
      "http": {
        "metadata": {
          "concurrentRequests": "30"
        }
      }
    }]
  }
}
```

The default HTTP rule scales with 10 concurrent requests. Based on the tests, let's set this value to 30. The number of replicas is in the range of 1 to 8. What's important to know regarding HTTP scaling is that the number of requests is calculated every 15 seconds. The total requests from within the last 15 seconds are divided by 15 so that they can be compared to the `concurrentRequests` value. Based on this, the number of replicas is calculated. So, if there are 140 requests per second, the instance count will be set to 5.

When this scale rule is applied, and the instances are active, we can configure a load test with a dynamic pattern configuration, as shown in *Figure 12.16*:

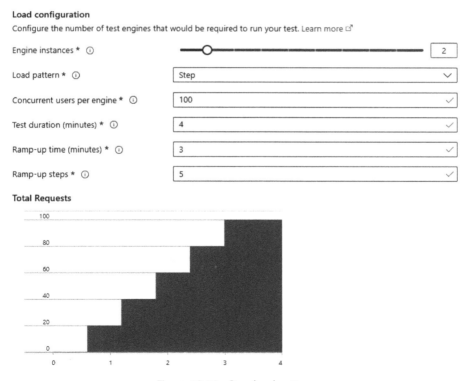

Figure 12.16 – Step load pattern

With this step load pattern, two engine instances are used to start 200 virtual users. The complete test duration is 4 minutes. The ramp-up time defines how long it takes until the 200 virtual users are reached – and 5 ramp-up steps are used to increment the users with 40 increments.

After the test runs are completed, you can see how the virtual users were added over time, as shown in *Figure 12.17*:

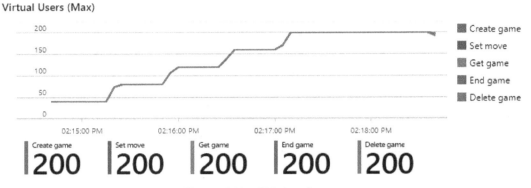

Figure 12.17 – 200 virtual users

The number of requests per second is shown in *Figure 12.18*:

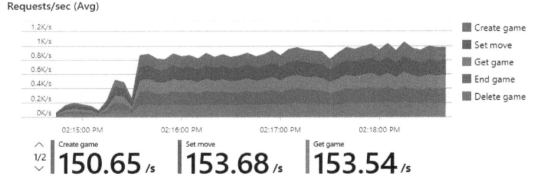

Figure 12.18 – Requests per second

Figure 12.19 shows the response time with longer times starting after 2:17 P.M. Do you have any idea what could have happened here?

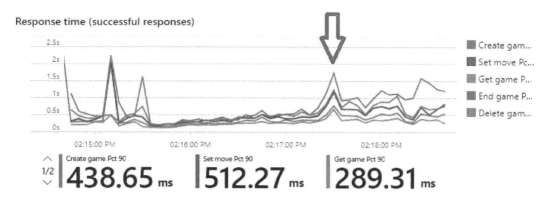

Figure 12.19 – Response time

The answer is that with this load, we reached the 10,000 RU/s limit using the Azure Cosmos database. This is shown in *Figure 12.20*:

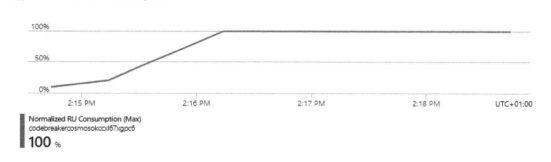

Figure 12.20 – Reaching 10,000 RU/s

The max RU/s were reached after 2:16 P.M. Responses with "too many requests" are not returned immediately when this limit is reached.

It's also interesting to see the replica count for the scale rule we've created. This is shown in *Figure 12.21*:

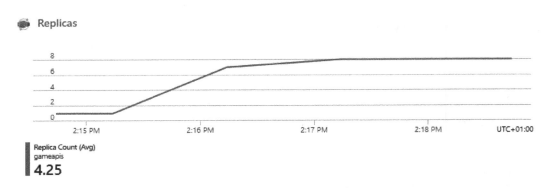

Figure 12.21 – Replica count

The replica count started at 1 and increased up to 8 – the maximum configured replica number. *Figure 12.19* showed another issue than the RU/s limit. Directly after the start, there are some peaks in the response time. This corresponds to the reduction in requests per second in *Figure 12.18*. I had to dig into this issue after this test run. Metrics data didn't reveal the reason, but checking the logs in the `ContainerAppSystemLogs_CL` table provided information about what the issue was. At that time, it was time to start a new replica. As indicated by the logs, a new replica was assigned, the image was pulled, and a container was created – but the *startup probe* failed, and the replica was unhealthy. Requests are not sent to such replicas. So, for the load we generated, we still only had one replica. Faulty replicas were started two times before the third one succeeded. This is why increasing the replica count took longer than it should have been, and that's the reason for the peak in the beginning. After this, every new replica that was created immediately succeeded. If you have some issues with a specific replica, you can use the replica's name to query the logs:

```
ContainerAppSystemLogs_CL
| where ReplicaName_s == "gameapis--tc47v0x-7cb88d64b8-np8cg"
| order by time_t asc
```

Next, we'll dive into health checks. This will help you understand startup probes.

Implementing health checks

The hosting platform needs to know if the service started successfully and is available to serve requests. While the service is running, the hosting platform continuously checks the service to see if it is running or broken and needs to be restarted. This is what health checks are for.

With Kubernetes, three probes can be configured:

- `Startup`: Is the container ready and did it start? When this probe succeeds, Kubernetes switches to the other probes.

- `Liveness`: Did the application crash or deadlock? If this fails, the pod is stopped, and a new container instance is created.

- `Readiness`: Is the application ready to receive requests? If this fails, no requests are sent to this service instance, but the pod keeps running.

Because Azure Container Apps is based on Kubernetes, these three probes can be configured with this Azure service as well.

Adding health checks to the DI container

Health checks can be configured with the DI container within the `AddDefaultHealthChecks` extension method:

Codebreaker.ServiceDefaults/Extensions.cs

```
public static IHostApplicationBuilder AddDefaultHealthChecks(this
IHostApplicationBuilder builder)
{
  builder.Services.AddHealthChecks()
    .AddCheck(
      "self",
      () => HealthCheckResult.Healthy(),
      tags: ["live"]);
  return builder;
}
```

The `AddHealthChecks` method registers the `HealthCheckService` class with the DI container that will be available for health checks. The `AddHealthChecks` method can be invoked multiple times to access `IHealthChecksBuilder`, which is used to register different implementations of checks. The fluently invoked `AddCheck` method uses a delegate to return the `HealthCheckResult`. `Healthy` result on invocation. The last parameter defines the `"live"` tag. Tags are used with middleware to specify with which route this health check should be used. As the name suggests, this tag is good for a `liveness` probe. When the service is accessible, a result is returned. If the service is not available, nothing is returned, and thus it will be restarted. The name `self` indicates that only the service itself is used with this health check, and external resources are only consulted with readiness health checks.

On startup of the `game-apis` service, using Azure Cosmos DB, the container is created, and with SQL Server, database migration can occur in case the database schema is updated. The application is not ready to receive requests before this is completed. With some applications, the cache needs to be filled with reference data before requests are accepted. To do this, a Boolean flag must be defined with the database update code that is set when the update is completed. Let's add a health check to the DI container configuration of `game-apis`:

Codebreaker.GameAPIs/Program.cs

```
builder.Services.AddHealthChecks().AddCheck("dbupdate", () =>
{
    return ApplicationServices.IsDatabaseUpdateComplete ?
        HealthCheckResult.Healthy("DB update done") :
        HealthCheckResult.Degraded("DB update not ready");
});
```

With this implementation, we return `Healthy` if the flag (the `IsDatabaseUpdateComplete` property) is true and `Degraded` otherwise.

When you try this out, the database migration might be too fast to see a degraded result – especially if the database has already been created Adding a delay to the migration of the database in the `Codebreaker.GameAPIs/ApplicationServices.cs` file helps here as you can view different health results.

> **Note**
> Health checks should be implemented quickly. These checks are invoked quite often and shouldn't result in a big overhead. Often, these checks involve opening a connection or checking a flag to be set.

Adding health checks with .NET Aspire components

All.NET Aspire components have health checks enabled – if the component supports health checks. Check out the documentation regarding these components to learn more.

Similar to metrics and telemetry configuration, health checks can be enabled and disabled with the configuration settings of the component.

This can be done programmatically:

Codebreaker.GameAPIs/ApplicationServices.cs

```
        builder.Services.AddDbContextPool<IGamesRepository,
GamesSqlServerContext>(options =>
```

```
{
  var connectionString = builder.Configuration.
GetConnectionString("CodebreakerSql") ??
    throw new InvalidOperationException("Could not read SQL Server
connection string");
  options.UseSqlServer(connectionString);
  options.UseQueryTrackingBehavior(QueryTrackingBehavior.NoTracking);
});
builder.EnrichSqlServerDbContext<GamesSqlServerContext>(
  static settings =>
  {
    settings.DisableTracing = false;
    settings.DisableRetry = false;
    settings.DisableHealthChecks = false;
  });
```

The parameter of the .NET Aspire SQL Server EF Core `EnrichSqlServerDbContext` component method allows us to override the default for the component settings, such as metrics, tracing, and health checks.

We can also specify this with the following .NET Aspire configuration:

```
{
  // code removed for brevity
  "Aspire": {
    "Microsoft": {
      "EntityFrameworkCore": {
        "SqlServer": {
          "DbContextPooling": true,
          "DisableTracing": false,
          "DisableHealthCheck": false,
          "DisableMetrics": false
        }
      }
    },
    "StackExchange": {
      "Redis": {
        "DisableTracing": false,
        "DisableHealthCheck": false
      }
    }
  }
}
```

This configuration shows the settings for the Redis and SQL Server EF Core components. Both components integrate with the ready probe. What's done with these health checks? There's no reading or writing to the database. Health checks should be fast and not put a high load on the system. The Redis component tries to open the connection. The SQL Server EF Core component invokes the EF Core `CanConnectAsync` method. What you need to be aware of is that when the idle pricing of an Azure Container App scales down to 1, with custom health checks, it might never be idle.

Using such a configuration ensures that this can be changed without the need to recompile the project.

Now that we've implemented and configured health checks, let's map these to URL requests.

Mapping health checks

Mapping health links to URLs allows us to access them. The shared `Codebreaker.ServiceDefaults` file contains the health endpoints that have been configured with the `MapDefaultEndpoints` method:

Codebreaker.ServiceDefaults/Extensions.cs

```
public static WebApplication MapDefaultEndpoints(this WebApplication
app)
{
  // code removed for brevity
  app.MapHealthChecks("/alive", new HealthCheckOptions
  {
    Predicate = r => r.Tags.Contains("live")
  });

  app.MapHealthChecks("/health");
  return app;
}
```

The `/alive` probe link has been configured to only use the health checks with the `live` tag, so the health checks are used to check if the service is alive. This link should be configured for live probing, and the service should be restarted if it does not return `Healthy`.

The `/health` probe link has been configured not to restrict the health checks based on a tag. Here, all health checks are invoked and need to be successful. This link should be used for a ready probe: is the service ready to receive requests? If this returns `Unhealthy` or `Degraded`, the service isn't stopped but doesn't receive requests.

> **Note**
>
> You might be wondering why .NET Aspire is not using the /healthz link for the ready probes, as it is typically used with Kubernetes. /healthz historically comes from Google's internal practices, z-pages, so that it doesn't get into conflicts. The .NET Aspire team had several iterations on deciding on the different links and included /liveness and /readiness, and finally ended up with /alive and /health.

Now that we've mapped the health checks to URIs, let's use these links.

Using the health checks with Azure Container Apps

You can automatically integrate the health probes after generating probe configuration with Azure Container Apps. You can also use these health probes with the Azure dashboard. However, this is not available with the first release of .NET Aspire and is planned for a later release. However, with a little customization, this can easily be done.

For this to work, you need to have initialized the solution with azd init, which you did previously, before publishing the solution to Azure. Now, from the folder that contains the solution, create the code that will publish these projects:

```
azd infra synth
```

With this, the AppHost project contains an infra folder with <app>.tmpl.yaml files. Within the gameapis.tmpl.yaml file, specify the probes section:

```
template:
  containers:
    - image: {{ .Image }}
      name: gameapis
      probes:
        - type: liveness
          httpGet:
            path: /alive
            port: 8080
          initialDelaySeconds: 3
          periodSeconds: 3
        - type: readiness
          httpGet:
            path: /health
            port: 8080
          initialDelaySeconds: 10
          periodSeconds: 5
        - type: startup
          tcpSocket:
```

```
        port: 8080
      initialDelaySeconds: 1
      periodSeconds: 1
      timeoutSeconds: 3
      failureThreshold: 30
    env:
    - name: AZURE_CLIENT_ID
# configuration removed for brevity
```

The `probes` section allows `liveness`, `readiness`, and `startup` probe types to be configured. The `liveness` probe is configured to invoke the `/alive` link and the `readiness` probe is configured to invoke the `/health` link with the port of the running Docker container. Azure Container Apps have default settings. However, as soon as you specify probing, you need to configure all probe types; otherwise, the probes that haven't been configured will be disabled. Thus, when specifying `liveness` and `readiness` probes, the `startup` probe should be configured as well. This probe uses a TCP connection to connect to the service to verify that the connection succeeds.

`initialDelaySeconds` specifies the seconds to wait until the first probe is done. If this fails, additional checks are done after the number of seconds specified by `periodSeconds`. A failure only counts until `failureTreshold` is reached.

The default startup probe uses a TCP probe to check the ingress target port with an initial delay of 1 second, a timeout of 3 seconds, and a period of 1 second. With the failure threshold, this multiplies, and the app can take some time until it's started successfully. When the `startup` probe has been successful once, only the `liveness` and `readiness` probes are used afterward.

After making this change, from the solution folder, run the following command:

```
azd deploy
```

This will deploy the service to Azure and configure the health checks.

Open the Azure portal, navigate to Azure Container Apps, and select **Containers** from the left pane. You'll see a tab called **Health probes**, as shown in *Figure 12.22*:

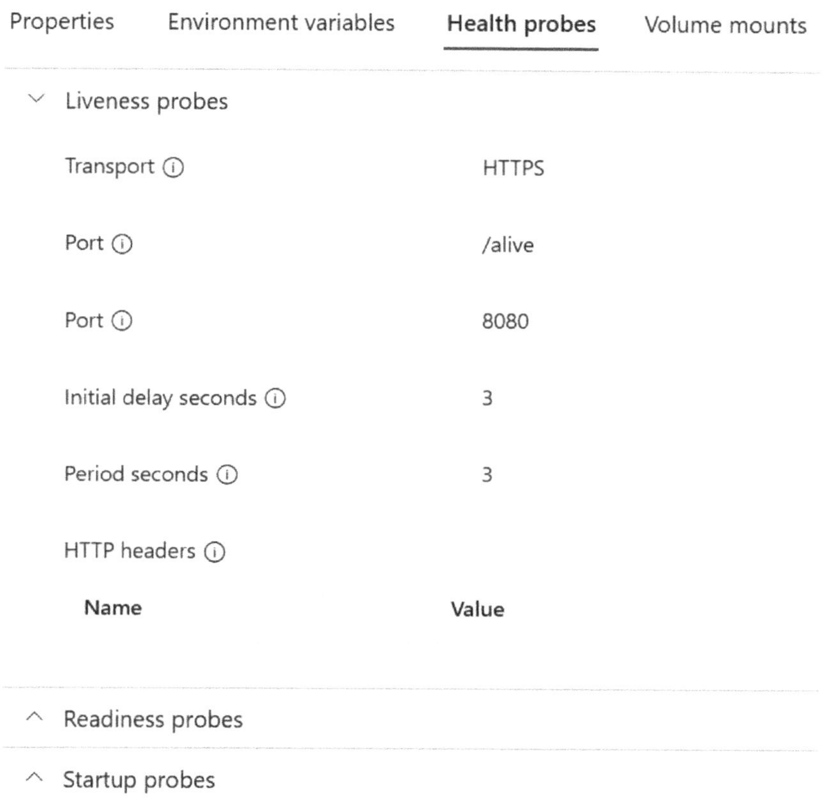

Properties	Environment variables	**Health probes**	Volume mounts

⌄ Liveness probes

Transport ⓘ   HTTPS

Port ⓘ   /alive

Port ⓘ   8080

Initial delay seconds ⓘ   3

Period seconds ⓘ   3

HTTP headers ⓘ

Name   **Value**

⌃ Readiness probes

⌃ Startup probes

Figure 12.22 – Health probes within Azure Container Apps

Here, we can see the configured settings for the liveness probe in the portal. You can also verify the readiness and startup probes. *Figure 12.23* shows the status of a container:

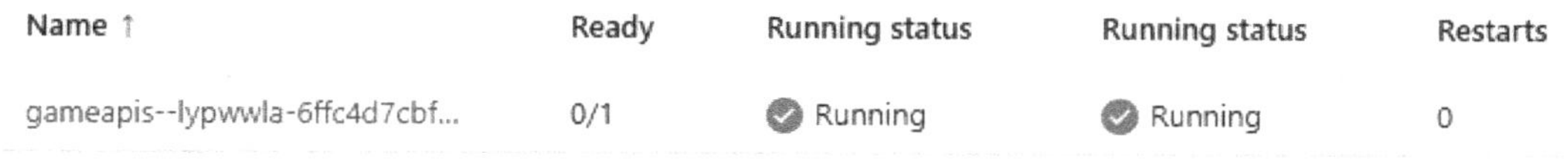

Name ↑	Ready	Running status	Running status	Restarts
gameapis--lypwwwla-6ffc4d7cbf...	0/1	✅ Running	✅ Running	0

Figure 12.23 – Container app running but not ready

Here, one replica is running, but this replica isn't ready. The readiness probe didn't return success at that time. If you configured the Redis component to offer health checks with .NET Aspire, you can stop this container using the Azure Container Apps environment. You'll see that the game-apis service isn't ready. Because the game-apis service has the component health check activated, an error is returned with the readiness check.

Summary

In this chapter, you learned how to use telemetry data and implemented a cache to reduce the number of database requests. You created health checks while differing between startup, liveness, and readiness checks. Liveness checks are used to restart services, while readiness checks are used to verify whether a service is ready to receive requests. Regarding readiness checks, you learned how to integrate .NET Aspire components. You also learned how to get information from load tests to find bottlenecks in the deployed application and to decide on the infrastructure you wish to use. By doing this, you learned how to configure the application so that it scales up when changes are made to CPU and memory, as well as how to scale out when running multiple replicas using scaling rules.

This chapter has uncovered an important reason for using microservices: with scaling, great flexibility can easily be achieved.

The next chapter will act as a starting point and implement different communication techniques with microservices. When adding more functionality to your application, you need to think about doing continuous load tests on the solution in a test environment and monitoring the changes.

Further reading

To learn more about the topics that were discussed in this chapter, please refer to the following links:

- *Distributed caching in ASP.NET Core*: `https://learn.microsoft.com/en-us/aspnet/core/performance/caching/distributed`
- *Database scalability: scaling out vs scaling up*: `https://azure.microsoft.com/en-au/resources/cloud-computing-dictionary/scaling-out-vs-scaling-up/`
- *Azure Virtual Machine Sizes*: `https://learn.microsoft.com/en-us/azure/virtual-machines/sizes`
- *Workload Profiles*: `https://learn.microsoft.com/en-us/azure/container-apps/workload-profiles-overview`
- *Container configuration*: `https://learn.microsoft.com/en-us/azure/container-apps/containers#configuration`
- *Azure Container Apps YAML specification*: `https://learn.microsoft.com/en-us/azure/container-apps/azure-resource-manager-api-spec`
- *Health checks in ASP.NET Core*: `https://learn.microsoft.com/en-us/aspnet/core/host-and-deploy/health-checks/`
- *Convention for /healthz*: `https://stackoverflow.com/questions/43380939/where-does-the-convention-of-using-healthz-for-application-health-checks-come-f`

Part 4: More communication options

In this part, the focus shifts towards leveraging various communication technologies and incorporating additional Azure services to enhance the application. Real-time messaging capabilities are implemented using SignalR to deliver instantaneous updates from the application to clients. gRPC is utilized for efficient binary communication between services, enabling seamless message exchange through queues and event publication. Azure services such as Azure SignalR Services, Event Hub, Azure Queue Storage, and Apache Kafka are integrated into the application ecosystem. Additionally, a detailed examination of considerations for production environments is provided, culminating in the deployment of the application to a Kubernetes cluster, specifically Azure Kubernetes Services, utilizing **Aspir8**.

This part has the following chapters:

- *Chapter 13, Real-time Messaging with SignalR*
- *Chapter 14, gRPC for Binary Communication*
- *Chapter 15, Asynchronous Communication with Messages and Events*
- *Chapter 16, Running Applications On-Premises and in the Cloud*

13

Real-Time Messaging with SignalR

Now that we've created various tests, added telemetry data, monitored the solution, and scaled our services, starting with this chapter, we'll continue creating services and using different communication technologies.

In this chapter, we'll use ASP.NET Core SignalR. SignalR is a technology that allows us to send real-time information from services to clients. The client initiates a connection to the service, which is then kept alive to send messages to the clients as information becomes available.

In this chapter, you'll learn how to do the following:

- Create a SignalR service
- Send real-time information to clients
- Create a SignalR client
- Use Azure SignalR Service

Technical requirements

In this chapter, like the previous chapters, you'll need an Azure subscription and Docker Desktop.

The code for this chapter can be found in this book's GitHub repository: `https://github.com/PacktPublishing/Pragmatic-Microservices-with-CSharp-and-Azure/`.

The `ch13` folder contains this chapter's projects, as well as their results. To add the functionality from this chapter, you can start with the source code from the previous chapter.

The projects we'll be considering are as follows:

- `Codebreaker.AppHost`: The .NET Aspire host project. The app model has been enhanced by an additional project running a SignalR hub and using Azure SignalR Service.

- `Codebreaker.Live`: This is a new project that hosts minimal APIs invoked by the `game-apis` service and the SignalR hub.

- `Codebreaker.GameAPIs`: This project has been enhanced and can forward completed games to `live-service`.

- `LiveTestClient`: This is a new console application that registers with the SignalR hub to receive completed games.

Creating a SignalR service

In this chapter, we'll create a new service that offers real-time information to return games that are played to every connected client. *Figure 13.1* shows how the services of the solution collaborate:

Codebreaker with SignalR

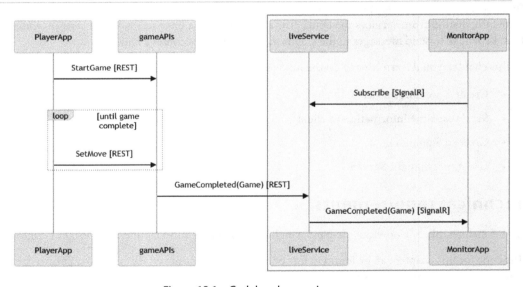

Figure 13.1 – Codebreaker services

The `game-apis` service has existed since *Chapter 2*. The client of `game-apis` invokes the REST API to start games and set moves. What's new is Codebreaker's `live-service`. This service offers a simple REST API that's invoked by the `game-apis` service every time a game completes. The main functionality of this service makes use of ASP.NET Core SignalR to offer real-time information to all connected clients. Clients need to subscribe before they can receive game completion messages.

What's offered by SignalR? As connections can never be started from a server to a client, and this is also true of SignalR, a client needs to connect to the SignalR service and subscribe to receive real-time information. The connection stays open, which allows the service to send information to the clients.

ASP.NET Core SignalR makes use of **WebSockets** if this technology is available – but the programming model with SignalR is a lot easier than using WebSockets directly. Contrary to HTTP, with WebSockets, the connection that's initiated by the client stays open, which allows the service to send messages when these are available.

WebSockets isn't always available and can be disabled by a proxy or a firewall and are not available everywhere. For example, **Azure Front Door** doesn't support WebSockets at this time.

If WebSockets isn't available, SignalR switches to other communication techniques, such as **server-sent events (SSE)** and **long polling**. With SSE, the client starts an HTTP request where it accepts `text/event-stream` data and asks the server to keep the connection alive (via the `Connection: keep-alive` HTTP header). All modern browsers support SSE nowadays. With polling, the client repeatedly asks the server if new data is available, opening new connections again and again. Long polling is a technique that reduces the number of requests from the client by the server not returning information immediately, stating that nothing new is available. Instead, the server waits for the timeout to nearly end before returning. If some new information is available while waiting, this new information is returned.

Using SignalR, there's no need to change programming constructs to decide between WebSockets, SSE, and long polling. This is done automatically by SignalR. However, in all these scenarios, compared to simple HTTP requests, the server has more overhead: connections with the clients need to stay alive and thus be kept in memory. To remove this overhead from our services, we'll use Azure SignalR Service later in this chapter.

To create a SignalR service, we just need to create an empty ASP.NET Core Web project. Because this service also offers a REST API, let's create a new Web API project:

```
dotnet new webapi --use-minimal-apis -o Codebreaker.Live
```

This creates a minimal API project, similar to what we have for the `game-apis` project. This project needs to be configured as a .NET Aspire project. With Visual Studio, use **Add | .NET Aspire Orchestrator Support**. Alternatively, add a project reference to the AppHost project so that it references the `Codebreaker.Live` project. Then, add a project reference to the `ServiceDefaults` project to the live project.

The `Codebreaker.Live` service project needs to be configured with the .NET Aspire app model:

Codebreaker.AppHost/Program.cs

```
// else path using Azure services
var live = builder.AddProject<Projects.Codebreaker_Live>("live")
```

```
  .WithExternalHttpEndpoints()
  WithReference(appInsights)
  .WithEnvironment("StartupMode", startupMode);


var gameAPIs = builder.AddProject<Projects.Codebreaker_
GameAPIs>("gameapis")
  .WithExternalHttpEndpoints()
  .WithReference(cosmos)
  .WithReference(redis)
  .WithReference(appInsights)
  .WithReference(live)
  .WithEnvironment("DataStore", dataStore)
  .WithEnvironment("StartupMode", startupMode);
  // code removed for brevity
```

By doing this, the project is added to the app model and referenced from the game-apis service as it needs the link from live-service to invoke the REST API on game completions.

Next, we'll add a SignalR hub that can be used by clients to connect.

Creating a SignalR hub

SignalR is already part of ASP.NET Core, so we don't need to add another NuGet package. Let's add the SignalR hub class – that is, LiveHub:

Codebreaker.Live/Endpoints/LiveHub.cs

```
public class LiveHub(ILogger<LiveHub> logger) : Hub
{
  public async Task SubscribeToGameCompletions(string gameType)
  {
    logger.ClientSubscribed(Context.ConnectionId, gameType);
    await Groups.AddToGroupAsync(Context.ConnectionId, gameType);
  }

  public async Task UnsubscribeFromGameCompletions(string gameType)
  {
    logger.ClientUnsubscribed(Context.ConnectionId, gameType);
    await Groups.RemoveFromGroupAsync(Context.ConnectionId, gameType);
  }
}
```

A SignalR hub derives from the `Hub` base class (the `Microsoft.AspNetCore.SignalR` namespace). This class defines the `OnConnectedAsync` and `OnDisconnectedAsync` methods, both of which can be overridden to react on client connects and disconnects. Here, we define the `RegisterGameCompletions` method with `gameType` as its parameter. This method is invoked by SignalR clients.

Using SignalR, a hub can send real-time information to all clients, one client, or a group of clients. With this implementation, we allow clients to register with a group. Game types are used to differentiate different groups. The `Hub` class defines a `Groups` property to subscribe and unsubscribe from a group. The `AddToGroupAsync` method adds the client and `RemoveFromGroupAsync` removes the client from the group. A connected client can be identified using `ConnectionId`, which can be accessed using the `Context` property.

To send information to connected clients, the `Hub` class offers the `Clients` property, which allows you to send to all clients (`Clients.All.SendAsync`), or a group (`Clients.Group("group-name").SendAsync`). However, in this case, we need to send the information outside of the `LiveHub` class (after receiving a REST invocation) from the `game-apis` service. We'll do this by implementing the `LiveGamesEndpoints` class.

Returning live information to the clients

The `LiveGamesEndpoints` class uses minimal APIs to implement a REST endpoint. What's special is that it can send information to connected clients:

Codebreaker.Live/Endpoints/LiveGamesEndpoints.cs

```
public static class LiveGamesEndpoints
{
  public static void MapLiveGamesEndpoints(this IEndpointRouteBuilder
routes, ILogger logger)
  {
    var group = routes.MapGroup("/live")
      .WithTags("Game Events API");

    group.MapPost("/game", async (GameSummary gameSummary,
      IHubContext<LiveHub> hubContext) =>
    {
      logger.LogInformation("Received game ended {type} {gameid}",
        gameSummary.GameType, gameSummary.Id);
      await hubContext.Clients.Group(gameSummary.GameType).
        SendAsync("GameCompleted", gameSummary);
      return TypedResults.Ok();
    })
    .WithName("ReportGameEnded")
```

```
        .WithSummary("Report game ended to notify connected clients")
        .WithOpenApi();
    }
}
```

The `GameSummary` class that's received by the `MapPost` method is implemented in the `CNinnovation.Codebreaker.BackendModels` NuGet package. This class contains summary information of a completed game. Along with this HTTP POST body parameter, the `MapPost` method receives an `IHubContext<LiveHub>` instance from the DI container. This interface is registered when the DI container is configured for SignalR to retrieve a context to the registered hub to send information to clients. Using the `Clients.Group` method, when passing the name of the group, an `IClientProxy` object is returned. This proxy is then used to send the `GameCompleted` method with the game summary.

Now, all we need to do is register SignalR and the hub with the DI container and the middleware.

Registering SignalR services and the hub

To use SignalR, and to make the hub available as an endpoint, we must implement the `ApplicationServices` class:

Codebreaker.Live/ApplicationServices.cs

```
public static class ApplicationServices
{
  public static void AddApplicationServices(this
IHostApplicationBuilder builder)
  {
    builder.Services.AddSignalR();
    // code removed for brevity
  }


  public static WebApplication MapApplicationEndpoints(this
WebApplication app, ILogger logger)
  {
    app.MapLiveGamesEndpoints(logger);
    app.MapHub<LiveHub>("/livesubscribe");
    return app;
  }
}
```

The AddApplicationServices method extends IHostApplicationBuilder to register the service classes that are needed for SignalR by invoking the AddSignalR method. The MapApplicationEndpoints method registers the SignalR hub and minimal API endpoint. The SignalR hub is registered with the endpoints by passing the Hub class with the generic parameter of the MapHub method. /livesubscribe is the link that's used by clients to connect to this service.

The methods of the ApplicationServices class are invoked from the top-level statements of the SignalR project:

Codebreaker.Live/Program.cs

```
var builder = WebApplication.CreateBuilder(args);

builder.AddServiceDefaults();
builder.AddApplicationServices();

builder.Services.AddEndpointsApiExplorer();
builder.Services.AddSwaggerGen();

var app = builder.Build();

app.MapDefaultEndpoints();
app.MapApplicationEndpoints(app.Logger);

app.Run();
```

The DI container is configured by invoking the AddServiceDefaults method, which is defined by the common ServiceDefaults project. This adds the DI container registration that's needed by all projects of the solution. The AddApplicationServices method adds the services that are needed from live-service, such as SignalR. Using builder.Build concludes the information that's needed via the DI container. The app instance starts configuring the middleware, which is where MapDefaultEndpoints and MapApplicationEndpoints are invoked. MapDefaultEndpoints registers links such as common health checks (covered in *Chapter 12*). MapApplicationEndpoints registers the endpoints offered by this service project.

Now that live-service is ready to go, let's call the API from the game-apis service.

Forwarding requests from the game-apis service

Recall the sequence diagram from earlier. We've updated this diagram to show how game-apis and live-service communicate:

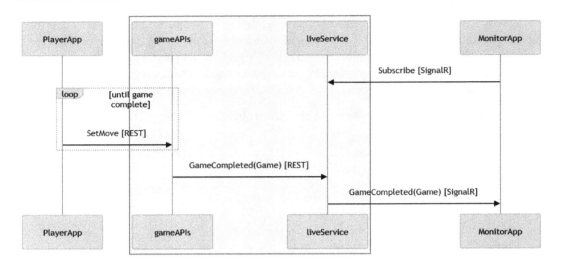

Figure 13.2 – REST calls to live-service

After a game is completed, the game-apis service invokes live-service with game completion information. To invoke live-service, create the LiveReportClient class, which injects HttpClient:

Codebreaker.GameAPIs/Services/LiveReportClient.cs

```
public class LiveReportClient(HttpClient httpClient,
ILogger<LiveReportClient> logger) : ILiveReportClient
{
  private readonly static JsonSerializerOptions s_jsonOptions = new()
  {
    PropertyNameCaseInsensitive = true
  };

  public async Task ReportGameEndedAsync(GameSummary gameSummary,
    CancellationToken cancellationToken = default)
  {
    try
    {
      await httpClient.PostAsJsonAsync("/live/game", gameSummary,
        options: s_jsonOptions, cancellationToken: cancellationToken);
    }
    catch (Exception ex) when (ex is HttpRequestException or
      TaskCanceledException or JsonException)
    {
      logger.ErrorWritingGameCompletedEvent(gameSummary.Id, ex);
```

```
      }
   }
}
```

The `ReportGameEndedAsync` method uses the `HttpClient` class to make an HTTP POST request to `/live/game` and send `GameSummary` information.

Let's configure the `HttpClient` class by updating the `ApplicationServices` class:

Codebreaker.GameAPIs/ApplicationServices.cs

```
public static void AddApplicationServices(this IHostApplicationBuilder
builder)
{
  // code removed for brevity
  builder.Services.AddScoped<IGamesService, GamesService>();
  builder.Services.AddHttpClient<ILiveReportClient,
    LiveReportClient>(client =>
  {
    client.BaseAddress = new Uri("https+http://live");
  });

  builder.AddRedisDistributedCache("redis");
}
```

Using .NET Aspire orchestration, the URL of the live client is retrieved by using service discovery with the `https+http://live` expression. This expression prefers the `https` schema, but if it isn't available, it uses the `http` schema. The name is resolved with service discovery via the app model, as explained in *Chapter 1*.

With that, `game-apis` has been configured to send game summary information. Now, all we need to do is create a client that receives real-time information from the SignalR service.

Creating a SignalR client

As a simple client receiving real-time information, all we need is a console application that connects to `live-service`. By doing this, it's simple to implement this functionality with any other client.

Start by creating a console project:

```
dotnet new console -o LiveTestClient
```

The `Microsoft.AspNetCore.SignalR.Client` NuGet package needs to be added to call a SignalR service. We must also add `Microsoft.Extensions.Hosting` for the DI container and `CNinnovation.Codebreaker.BackendModels` with the `GameSummary` type.

Create the `LiveClient` class, which will communicate with the SignalR service:

LiveTestClient/LiveClient.cs

```
internal class LiveClient(IOptions<LiveClientOptions> options) :
IAsyncDisposable
{
  // code removed for brevity
}

public class LiveClientOptions
{
  public string? LiveUrl { get; set; }
}
```

The `LiveClient` class specifies a constructor with `IOptions<LiveClientOptions>`. This will be configured with the DI container so that it can pass the URL string from `live-service`.

Add `appsettings.json` to configure the URL:

LiveTestClient/appsettings.json

```
{
  "Codebreaker.Live": {
    "LiveUrl": "http://localhost:5130/livesubscribe"
  }
}
```

For local testing, the port number needs to match the port number that's specified with `launchSettings.json`. Don't forget to configure so that `appsettings.json` is copied to the output directory.

The connection to the service is initiated by the `StartMonitorAsync` method:

LiveTestClient/LiveClient.cs

```
internal class LiveClient(IOptions<LiveClientOptions> options) :
IAsyncDisposable
{
  private HubConnection? _hubConnection;

  public async Task StartMonitorAsync(CancellationToken
    cancellationToken = default)
```

```
{
  string liveUrl = options.Value.LiveUrl ??
    throw new InvalidOperationException("LiveUrl not configured");
  _hubConnection = new HubConnectionBuilder()
    .WithUrl(liveUrl)
    .Build();

  _hubConnection.On("GameCompleted", (GameSummary summary) =>
  {
    string status = summary.IsVictory ? "won" : "lost";
    Console.WriteLine($"Game {summary.Id} {status} by {summary.
      PlayerName} after " +
      "{summary.Duration:g}  with {summary.NumberMoves} moves");
  });
  await _hubConnection.StartAsync(cancellationToken);
}
// code removed for brevity
public async ValueTask DisposeAsync()
{
  if (_hubConnection is not null)
  {
    await _hubConnection.DisposeAsync();
  }
}
}
```

To connect to the SignalR hub, `HubConnectionBuilder` is used to set up a connection. With this builder, the connection can be configured – for example, logging, server timeout, and reconnect behaviors can be set up. The connection is then initiated by invoking the `StartAsync` method.

The `On` method of `HubConnection` configures the receiving side: when a `GameCompleted` message is received, the `GameSummary` parameter specifies the data that's been received and writes a message about the game to the console. The name `GameCompleted` needs to match the name that has been passed with the `SendAsync` method of the service.

To subscribe to the messages from the service, implement the `SubscribeToGame` method:

LiveTestClient/LiveClient.cs

```
public async Task SubscribeToGame(string gameType, CancellationToken
cancellationToken = default)
{
  if (_hubConnection is null) throw new
```

```
InvalidOperationException("Start a connection first");

    await _hubConnection.InvokeAsync("SubscribeToGameCompletions",
      gameType, cancellationToken);
}
```

With this implementation, the `InvokeAsync` method of `HubConnection` is used. `SubscribeToGameCompletions` matches the name of the hub method, which uses the `game-type` parameter.

The top-level statements of the client application make use of the `LiveClient` class:

LiveTestClient/Program.cs

```
Console.WriteLine("Test client - wait for service, then press return
to continue");
Console.ReadLine();

var builder = Host.CreateApplicationBuilder(args);
builder.Services.AddSingleton<LiveClient>();
builder.Services.Configure<LiveClientOptions>(builder.Configuration.
GetSection("Codebreaker.Live"));
using var host = builder.Build();

var client = host.Services.GetRequiredService<LiveClient>();
await client.StartMonitorAsync();
await client.SubscribeToGame("Game6x4");

await host.RunAsync();

Console.WriteLine("Bye...");
```

After configuring the `LiveClient` class in the DI container, the `StartMonitorAsync` and `SubscribeToGame` methods are invoked.

With this in place, you can start the AppHost so that it runs all the services and multiple instances of the client application. Use `bot-service` to play multiple games. You'll see success messages from the bot, as shown in *Figure 13.3*.

```
Test client - wait for service, then press return to continue

info: Microsoft.Hosting.Lifetime[0]
      Application started. Press Ctrl+C to shut down.
info: Microsoft.Hosting.Lifetime[0]
      Hosting environment: Production
info: Microsoft.Hosting.Lifetime[0]
      Content root path: D:\codebreaker\ch13b\Pragmatic-Microservices-with-CSharp-and-Azure\
stClient\bin\Debug\net8.0
Game 0b1ca8ad-4258-4b42-8e78-a46be9b0ed51 won by Bot after 0:00:02,5387204 with 7 moves
Game b0302a54-da58-48e1-be1a-3e3e547ff19b won by Bot after 0:00:00,6862556 with 6 moves
Game b7b03a61-8a59-4967-b01d-36e90b754c2a won by Bot after 0:00:00,7572998 with 6 moves
Game 8d292eda-a23d-43d2-b32d-efa57a4bfcc9 won by Bot after 0:00:00,8845647 with 7 moves
Game 46977572-eaef-40f4-9947-7a4b71df9caa won by Bot after 0:00:04,7915443 with 5 moves
Game aa5613d0-4ff2-4784-8139-c1c7cb32cddf won by Bot after 0:00:00,2726738 with 2 moves
Game f14ecd52-43e2-4503-b258-5a56bdc91c68 won by Bot after 0:00:00,6572931 with 5 moves
Game d3e9d1f3-e397-4725-8f33-c2da4557daca won by Bot after 0:00:04,7725218 with 5 moves
Game abbbabe7-5d93-42ab-be9d-5af10f2b0a22 won by Bot after 0:00:00,6004578 with 5 moves
```

Figure 13.3 – Live client receiving game summaries

Using the bot, multiple games have been started with a think time of 0 and 1 seconds between game moves. These results show game wins that occur between 0.27 and 4.79 seconds.

Changing the serialization protocol

By default, SignalR serializes messages with JSON. Using the `Microsoft.AspNetCore.SignalR.Protocols.MessagePack` NuGet package, a binary serialization format can be used instead. This is an optimization that reduces the data that's sent.

To support this, all we need to do is update the DI configuration in the service, after adding the NuGet package:

Codebreaker.Live/ApplicationServices.cs

```
public static void AddApplicationServices(this IHostApplicationBuilder
builder)
{
  builder.Services.AddSignalR()
    .AddMessagePackProtocol();
}
// code removed for brevity
```

The `AddMessagePackProtocol` method adds `MessagePack` as another option for serialization. JSON is still available.

Regarding the client, the same NuGet package is required, but this time with the following configuration:

LiveTestClient/LiveClient.cs

```
string liveUrl = options.Value.LiveUrl ?? throw new
InvalidOperationException("LiveUrl not configured");
_hubConnection = new HubConnectionBuilder()
  .WithUrl(liveUrl)
  .ConfigureLogging(logging =>
  {
    logging.AddConsole();
    logging.SetMinimumLevel(LogLevel.Debug);
  })
  .AddMessagePackProtocol()
  .Build();
```

Like the server, the same NuGet package for the protocol is needed for the client, as well as the `AddMessagePackProtocol` API. With the client, logging is now turned on as well. Logging providers for SignalR can be configured using the `ConfigureLogging` method. Here, the console provider is added, and the minimum logging level is set to `LogLevel.Debug`. With this, we can see all communication between the client and the server, including the message protocols that are used and the ping messages that are sent.

When using MessagePack, you need to be aware of an important restriction: `DateTime.Kind` is not serialized. Thus, this type should be converted into UTC before being sent.

With this in place, you can start the solution again, start the bot to play some games, and start the SignalR client. When you look at the logging information, you'll see WebSockets and MessagePack in action:

Logging from the client application

```
dbug: Microsoft.AspNetCore.SignalR.Client.HubConnection[40]
  Registering handler for client method 'GameCompleted'.
// some log outputs removed for clarity
dbug: Microsoft.AspNetCore.Http.Connections.Client.HttpConnection[8]
  Establishing connection with server at 'http://localhost:5130/
livesubscribe'.
dbug: Microsoft.AspNetCore.Http.Connections.Client.HttpConnection[9]
  Established connection '1YXBdJ3Yi7A_86ZqoMKgiA' with the server.
info: Microsoft.AspNetCore.Http.Connections.Client.Internal.
WebSocketsTransport[1]
  Starting transport. Transfer mode: Binary. Url: 'ws://
localhost:5130/livesubscribe?id=CHpPUMdrJoxV0zLHsskN1Q'.
dbug: Microsoft.AspNetCore.Http.Connections.Client.HttpConnection[18]
```

```
        Transport 'WebSockets' started.
info: Microsoft.AspNetCore.SignalR.Client.HubConnection[24]
        Using HubProtocol 'messagepack v1'.
dbug: Microsoft.AspNetCore.Http.Connections.Client.Internal.
WebSocketsTransport[13]
        Received message from application. Payload size: 39.
```

Now that we've switched the serialization format, let's reduce the load from the service when a larger number of clients is connected by using Azure SignalR Service.

Using Azure SignalR Service

There's some overhead associated with SignalR since the server has open connections with all SignalR clients.

To remove this overhead from our services, we can use **Azure SignalR Service**. This service acts as an intermediary between clients and the SignalR service, as depicted in *Figure 11.4*:

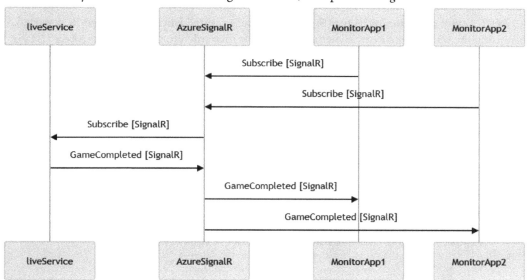

Figure 13.4 – Using Azure SignalR Service

The preceding figure shows multiple monitor clients that each have a connection open to Azure SignalR Service. `live-service` just needs to deal with a single connection. Azure SignalR subscribes to the events and forwards them to a single client, a group of clients, or all clients, as specified by `live-service`.

The load for every client connection is handled by Azure SignalR Service, while this service just acts as a single client to the SignalR service.

A free version of this service without SLA that's limited to 20 connections and 20,000 messages per day can be used for development purposes. Standard and premium SKUs can scale up to 1,000 connections per unit, 100 units, and unlimited messages.

To activate Azure SignalR Service within the app model, we need to update app-model within the AppHost project:

Codebreaker.AppHost/Program.cs

```
var builder = DistributedApplication.CreateBuilder(args);

var signalR = builder.AddAzureSignalR("signalr");
// code removed for brevity
var live = builder.AddProject<Projects.Codebreaker_Live>("live")
  .WithExternalHttpEndpoints()
  .WithReference(appInsights)
  .WithReference(signalR);
```

With .NET Aspire provisioning, Azure SignalR Service is created when the application is started. When using the WithReference method, the URI is forwarded to the Codebreaker.Live service. Here, the Microsoft.Azure.SignalR NuGet package is required to connect this service:

Codebreaker.Live/ApplicationServices.cs

```
public static void AddApplicationServices(this IHostApplicationBuilder
builder)
{
  var signalRBuilder = builder.Services.AddSignalR()
    .AddMessagePackProtocol();
  if (Environment.GetEnvironmentVariable("StartupMode") !=
"OnPremises")
  {
      signalRBuilder.AddNamedAzureSignalR("signalr");
  }
}
```

Using AddNamedAzureSignalR, the connection string is retrieved via service discovery, and the SignalR hub is connected to this Azure service.

Now, start the application again and check the Azure portal to see that the service has been created. Use the Aspire dashboard to see the environment variable that's been assigned to the Codebreaker.Live service, and check the logs to see the connections that have been made to Azure SignalR Service. Run the bot so that it plays several games and then start several SignalR client (LiveTestClient) processes.

When you open the Azure portal, open the resource group of the developer environment (rg-aspire-<yourhost>-codebreaker.apphost) and select Azure SignalR Service. In the **Monitoring** category in the left bar, select **Live trace settings**. Click the **Enable Live Trace** checkbox and choose to collect information for **ConnectivityLogs**, **MessagingLogs**, and **HttpRequestLogs**. Then, click the **Open Live Trace Tool** button. You'll receive the following output:

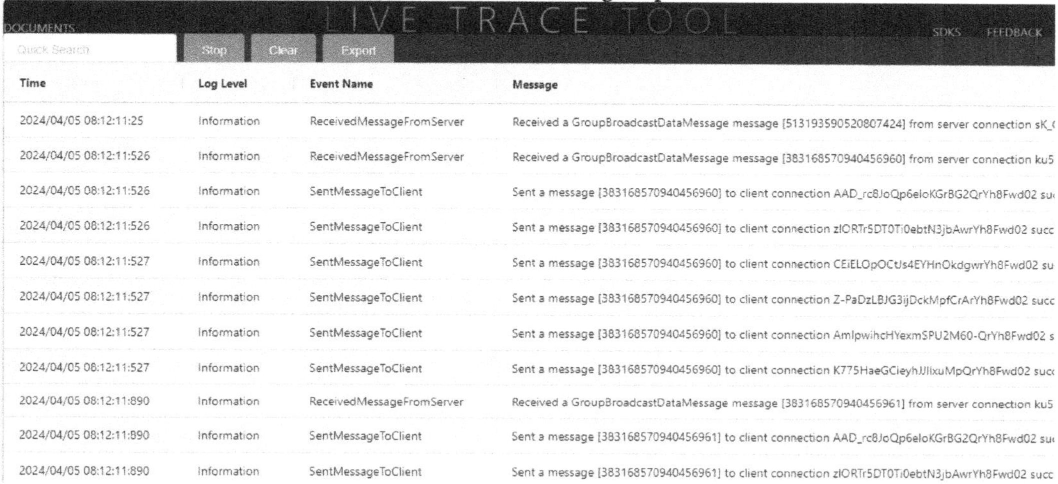

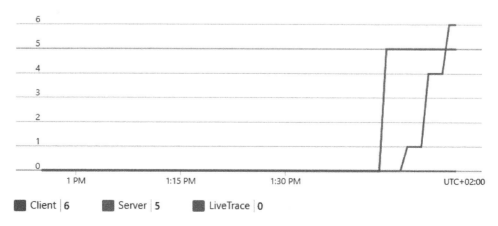

Figure 13.5 – Azure SignalR Service live trace. This screenshot only intends
to show the output result page; text readability is not essential

With the Azure SignalR Service live trace, you can see all the messages that have been sent from the Codebreaker live service, as well as the messages that have been sent to the subscribing clients.

To see metrics data, go back (or open a new browser window) to the **Overview** page of Azure SignalR Service. There, you can see the number of connections that have been opened, as shown in *Figure 13.6*:

Figure 13.6 – Azure SignalR Service connection metrics

Figure 13.7 shows the number of messages that have been sent:

Message Count (total)

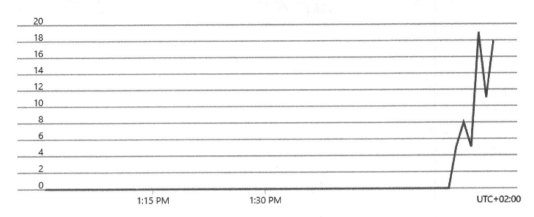

Figure 13.7 – Azure SignalR Sevice message metrics

Now that everything is up and running, you've earned a well-deserved break so that you can play some games (which you can monitor as well).

Summary

In this chapter, you learned how to offer real-time data using SignalR. You created a live service containing a SignalR hub that offers real-time information about completed games. Clients can register to a subset – a group – of the information offered. You also created a simple console application that acts as a client. The same functionality can be implemented in other clients. You can check this out in the Blazor client application provided in this book's GitHub repository, which contains the SignalR client functionality.

Then, you learned how to use Azure SignalR Service, which reduces the load on the service hosting the SignalR hub as the clients directly interact with Azure SignalR Service while this service acts as one client to SignalR.

With the implementation of this chapter, we created a REST API that's invoked by the `game-apis` service to send completed games. REST is great for easy communication with all clients, but it doesn't offer the best performance. The only client for the `Codebreaker.Live` service is the `game-apis` service.

Regarding service-to-service communication, instead of using REST APIs, less overhead is used when binary serialization is used alongside protocols such as gRPC. This will be covered in the next chapter.

Further reading

To learn more about the topics that were discussed in this chapter, please refer to the following links:

- *ASP.NET Core SignalR*: `https://learn.microsoft.com/en-us/aspnet/core/signalr/introduction`

- *Message Pack protocol*: `https://learn.microsoft.com/en-us/aspnet/core/signalr/messagepackhubprotocol`

- *Azure SignalR Service*: `https://learn.microsoft.com/en-us/azure/azure-signalr/signalr-overview`

- *.NET Aspire support for Azure SignalR Service*: `https://learn.microsoft.com/en-us/dotnet/aspire/real-time/azure-signalr-scenario`

gRPC for Binary Communication

Service-to-service communication does not need to be via REST passing JSON data. Performance and cost are important factors to consider when it comes to using **gRPC**, a binary and platform-independent communication technology. Reducing the data that's transferred can increase performance and reduce costs.

In this chapter, we'll change some services in the Codebreaker solution so that they offer gRPC instead or in addition to REST services. You'll learn how gRPC differs from REST, as well as how to create services and clients using this binary communication technology.

In this chapter, you'll learn how to do the following:

- Configure a service project to use gRPC
- Create a platform-independent communication contract with Protobuf
- Create gRPC services
- Create clients that call gRPC services

Technical requirements

In this chapter, like the previous chapters, you'll need an Azure subscription and Docker Desktop.

The code for this chapter can be found in this book's GitHub repository: `https://github.com/PacktPublishing/Pragmatic-Microservices-with-CSharp-and-Azure/`.

The `ch14` folder contains the projects we'll be looking at in this chapter, as well as their results. To add the functionality from this chapter, you can start with the source code from the previous chapter.

The projects we'll be considering are as follows:

- `Codebreaker.AppHost`: The .NET Aspire host project. The app model has been updated to use HTTPS with the `game-apis` service and `live-service` so that it supports gRPC.

- `Codebreaker.Live`: The project we created in the previous chapter has been changed to offer a gRPC service instead of a REST service.

- `Codebreaker.GameAPIs`: This project has been updated so that it includes a gRPC client to invoke `live-service`. In addition to the REST service used by many different clients, a gRPC service has been added as an alternative. This is invoked by the bot service.

- `Codebreaker.Bot`: The bot service has been updated to use a gRPC client instead of REST to invoke the `game-apis` service.

- `LiveTestClient`: You will need to use the live test client from the previous chapter to verify the SignalR service.

Before digging into gRPC, let's compare it with **Representational State Transfer (REST)**.

Comparing REST with gRPC

The most important difference between REST and gRPC is that REST is a *guideline* that builds upon HTTP, whereas gRPC is a *protocol*. Both are used for communication between clients and services. Let's take a closer look.

Communication style

REST is a guideline that defines services to be stateless, makes use of HTTP verbs (GET, POST, PUT, DELETE) to manipulate resources, typically uses a human-readable format such as JSON or XML, and is commonly used with web APIs.

gRPC uses a *strongly defined contract* to specify the operations that are available with the services. While other specifications can be used as well, most services make use of **Protocol Buffers (Protobuf)** to specify the contract. With this, gRPC has a compact payload size and efficient serialization.

Performance

When it comes to textual representation, REST has a higher overhead and a higher latency. With gRPC, because of its efficient serialization, latency is lower, and the binary serialization reduces the payload's size. With gRPC, multiplexing allows for concurrent requests across a single connection.

Flexibility

REST adds flexibility by using URIs for resources and is not strictly based on HTTP and HTTPS. Other protocols can be used as well that fulfill REST principles.

With gRPC, contracts strictly specify the communication. gRPC is based on HTTP/2, which offers some advantages compared to HTTP/1, such as multiplexing concurrent calls over a single connection. gRPC-Web is an alternative that allows a subset of gRPC to be used with HTTP/1.

Language support

REST just needs HTTP and works with any language that supports HTTP. With gRPC, based on the protobuf contract, code is created – and this requires supported languages to be used. Check out the list of supported languages at `https://grpc.io/docs/languages/`. It includes C#, C++, Dart, Java, Go, Python, PHP, and others.

Security

REST relies on transport security (HTTPS). Authentication and authorization are done at the application level. With gRPC, transport security (TLS/SSL) is supported, and authentication via OAuth and JWT is built-in. gRPC supports per-message encryption.

Use cases

REST allows for easy interoperability. Only HTTP calls need to be made. It's used with web APIs and simple services, as well as for interoperability with existing services.

With cloud services, we pay for compute and memory resources. With a lot of communication going on, the number of instances needed can be reduced by using memory and CPU-efficient technologies. Communication between services can be done using gRPC.

Let's start by updating the solution so that it makes use of gRPC (with the logging collectors, gRPC is already in use).

Updating a service project so that it uses gRPC

When using .NET templates to create new projects, a gRPC service can be created by running the following command:

```
dotnet new grpc
```

When using such a project, you can check the project file to see the NuGet packages and other configurations that are needed.

Because we already have existing projects, we'll update these to offer a gRPC service. But first, let's have a look at the communication between the Codebreaker services shown in *Figure 14.1*:

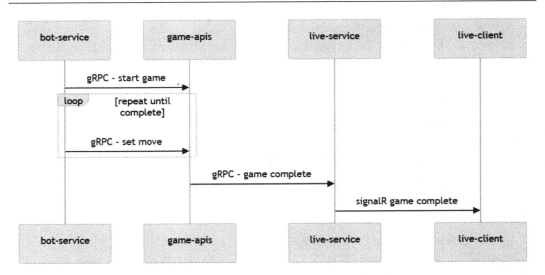

Figure 14.1 – Codebreaker communication technologies

Let's start with the right-hand side. In *Chapter 13*, we implemented the Codebreaker live service using a SignalR hub, which notifies SignalR clients. For the SignalR client, we created a console client application. Also in *Chapter 13*, we used minimal ASP.NET Core APIs which allowed us to call the game-apis service to send completed games. This is a form of service-to-service communication that can be replaced by gRPC. The game-apis service itself is called by clients and the bot service. With this service, a REST API must be invoked by any client technology. The communication between the bot and the game-apis service can be done using gRPC as well. Thus, when it comes to the game-apis service, we'll offer an alternative communication technology so that communication between the game-apis service and live-service will be replaced. Thus, in this chapter, we'll do the following:

- Replace the minimal API implementation of live-service with gRPC
- Add an alternative option for the game-apis service so that we offer gRPC as well
- Implement a gRPC client with the game-apis service to invoke live-service
- Implement a gRPC client with clients of the game-apis service

Let's start by creating a gRPC service contract for live-service that's invoked by the game-apis service.

Creating service contracts

First, we need to add the Grpc.AspNetCore NuGet package to the Codebreaker.Live project. Then, we must add a Protobuf file as a contract to the service. The contract file is language and platform-independent, so a .NET service can communicate with a Dart application using the same Protobuf file.

Using Visual Studio, add a `Protos` folder, and use the Visual Studio template to create a **Protobuf file**. When using the command line, with the current directory set to the project folder of `Codebreaker.Live`, you can run the following command to create the `LiveGame.proto` file inside the `Protos` folder:

```
dotnet new proto -o Protos -n LiveGame
```

Now, let's create contracts for both `live-service` and `game-apis`.

Creating a gRPC service contract for live-service

The simpler contract is the one for `live-service`:

Codebreaker.Live/Protos/LiveGame.proto

```
syntax = "proto3";
option csharp_namespace = "Codebreaker.Live.Grpc";
package ReportGame;
import "google/protobuf/empty.proto";
import "google/protobuf/duration.proto";
import "google/protobuf/timestamp.proto";

service ReportGameService {
  rpc ReportGameCompleted (ReportGameCompletedRequest)
    returns (google.protobuf.Empty);
}

message ReportGameCompletedRequest {
  string id = 1;
  string gameType = 2;
  string playerName = 3;
  bool isCompleted = 4;
  bool isVictory = 5;
  int32 numberMoves = 6;
  google.protobuf.Timestamp startTime = 7;
  google.protobuf.Duration duration = 8;
}
```

The `syntax` keyword specifies the Protobus version that should be used. Version 3 added features such as maps and `oneof` fields. The `option` keyword allows us to add language-specific features. `option csharp_namespace` sets the C# namespace that should be used by the generated classes, postfixed with the name set by the `package` keyword. The `service` keyword describes the list of operations offered by the gRPC service. Every operation within the service uses the `rpc` keyword (remote procedure call). In our code, the operation is called `ReportGameCompleted`, uses

ReportGameCompletedRequest as a parameter, and returns google.protobuf.Empty. google.protobuf.Empty is one of the well-known Protobuf types that's available. When used, this type must be imported using the import keyword. The google.protobuf.Duration and google.protobuf.Timestamp types are imported as well. ReportGameCompletedRequest is a message that's specified using the message keyword. Every field within a message needs a unique identifier. The serializer and deserializers use this number to get a match. Thus, if you specify a contract once, don't change the number in future versions, as this breaks existing clients or services. The types used need to be platform-independent as the same Protobuf file can be used by all platforms that support gRPC. string, bool, and int32 are types that are defined by the Protobuf specification. With .NET, these types map to string, bool, and int, respectively. Id is a GUID, but there's no representation for GUIDs with Protobuf. string can be used for the identifier. With the .NET GameSummary class, the StartTime property is of the DateTime type and the Duration property is of the TimeSpan type. To map these types with Protobuf, you can use Timestamp and Duration. These types are defined in the Google.Protobuf.WellKnownTypes .NET namespace. Timestamp and Duration offer conversion methods to convert to and from DateTime and TimeSpan.

To create .NET classes from the Protobuf file, a Protobuf entry needs to be added to the project file, like so:

Codebreaker.Live/Codebreaker.Live.csproj

```
<ItemGroup>
  <Protobuf Include="Protos\LiveGame.proto"
    GrpcServices="Server" />
</ItemGroup>
```

This Protobuf entry references the Protobuf file with the Include attribute and specifies the code to be generated for the server by setting GrpcServices to Server. With the server, classes for all the defined messages and base classes for every service specified are generated. Later in this chapter, we'll use the Protobuf element to generate classes for the client.

In the previous chapter, we created the GameSummary class to report game completions. To convert this class into the gRPC-generated ReportGameCompletedRequest class, we must define a conversion method:

Codebreaker.Live/Extensions/ReportGameCompletedRequestExtensions.cs

```
public static class ReportGameCompletedRequestExtensions
{
    public static GameSummary ToGameSummary(
        this ReportGameCompletedRequest request)
```

```
{
    Guid id = Guid.Parse(request.Id);
    DateTime startTime = request.StartTime.ToDateTime();
    TimeSpan duration = request.Duration.ToTimeSpan();
    return new GameSummary(
        id,
        request.GameType,
        request.PlayerName,
        request.IsComleted,
        request.IsVictory,
        request.NumberMoves,
        startTime,
        duration);
}
}
```

In this implementation, we create a new GameSummary instance and use conversion methods – for example, we parse a string to a Guid value and invoke the ToDateTime and ToTimeSpan methods to convert the Timestamp and Duration values.

> **Note**
>
> To see the generated code with Visual Studio, click on the ReportGameCompletedRequest class, use the context menu, and select **Go To Implementation**. This directly opens the generated code.

With ReportGameService, we just have a very simple contract. To allow game-apis to be called and moves to be made by the bot service, we need a more complex contract.

Creating a gRPC service contract for the game-apis service

The gRPC contracts of the game-apis service are lengthier than for live-service. Here, we'll focus on some specific parts of the contract. Check out this book's GitHub repository for the complete definition.

The service contract specifies operations to play the game:

Codebreaker.GameAPIs/Protos/GameService.proto

```
syntax = "proto3";
option csharp_namespace = "Codebreaker.Grpc";
package GamesAPI;

import "google/protobuf/duration.proto";
```

```
import "google/protobuf/timestamp.proto";

service GrpcGame {
  rpc CreateGame(CreateGameRequest)
    returns (CreateGameResponse);
  rpc SetMove(SetMoveRequest) returns (SetMoveResponse);
  rpc GetGame(GetGameRequest) returns (GetGameResponse);
}
// code removed for brevity
```

The alternative option to the REST interface, `GrpcGameService`, defines operations to create a game (`CreateGame`), set a move (`SetMove`), and get information about a game (`GetGame`).

Most messages that are used to send a request to the service just contain scalar values. The `SetMoveRequest` message is different. This message contains a list of guess pegs:

Codebreaker.GameAPIs/Protos/GameService.proto

```
message SetMoveRequest {
  string id = 1;
  string gameType = 2;
  int32 moveNumber = 3;
  bool end = 4;
  repeated string guessPegs = 5;
}
```

Lists are specified using the `repeated` keyword.

Not only pre-defined types can be repeated – it's also possible to repeat an inner message type:

Codebreaker.GameAPIs/Protos/GameService.proto

```
message GetGameResponse {
  string id = 1;
  string gameType = 2;
  string playerName = 3;
  // code removed for brevity
  repeated Move moves = 14;
}

message Move {
  string id = 1;
  int32 moveNumber = 2;
  repeated string guessPegs = 3;
```

```
    repeated string keyPegs = 4;
}
```

The `GetGameResponse` message type contains a repeated list of `Move` messages. The `Move` message type contains a list of strings for the guess pegs and the key pegs.

Protobuf also defines a list of keys and values via the `map` type:

Codebreaker.GameAPIs/Protos/GameService.proto

```
message FieldNames {
  repeated string values = 1;
}

message CreateGameResponse {
  string id = 1;
  string gameType = 2;
  string playerName = 3;
  int32 numberCodes = 4;
  int32 maxMoves = 5;
  map<string, FieldNames> fieldValues = 6;
}
```

With a map, the key and value types are specified. With the `fieldValues` field, the key is a string. The corresponding REST API specifies a string array for the value. Using `repeated` with the value type is not possible with Protobuf. Instead, `FieldMessage` is defined to contain a `repeated string`, and this is used with the `map` value.

The message contracts create .NET classes that are specific to gRPC. When it comes to the local service classes, it's better if they're independent of communication technologies. So, we need to create conversion methods.

Creating conversion methods

The gRPC service of `live-service` receives `ReportGameCompletedRequest`. This is forwarded to the SignalR service we created in the previous chapter as a `GameSummary` method. So, we need to convert `ReportGameCompletedRequest` into a `GameSummary` method:

Codebreaker.Live/Extensions/GrpcExtensions.cs

```
internal static class GrpcExtensions
{
  public static GameSummary ToGameSummary(
    this ReportGameCompletedRequest request)
```

```
{
    Guid id = Guid.Parse(request.Id);
    DateTime startTime = request.StartTime.ToDateTime();
    TimeSpan duration = request.Duration.ToTimeSpan();
    return new GameSummary(
        id,
        request.GameType,
        request.PlayerName,
        request.IsCompleted,
        request.IsVictory,
        request.NumberMoves,
        startTime,
        duration);
    }
}
```

This is done in the form of an extension method where we extend the `ReportGameCompletedRequest` type. With the implementation of the `ToGameSummary` method, simple scalar types can be passed on, creating a `GameSummary` object. Google's `Timestamp` and `Duration` types offer the `ToDateTime` and `ToTimeSpan` methods to convert `DateTime` and `TimeSpan`.

> **Note**
>
> Libraries such as `AutoMapper`, `Mapster`, and others can be used to automatically implement such conversions. While this works out of the box with simple properties without the need to add custom code, some customization is needed when converting different types. What you need to be aware of is that mapper libraries that use .NET reflection instead of source generators increase memory and CPU usage and cannot be used with native AOT. Depending on the types you need to map, you might prefer a custom extension method.

Check out this book's GitHub repository for additional conversion methods that can be used with the `game-apis` service.

With the conversion methods in place, let's create the gRPC service implementations.

Creating gRPC services

To implement a gRPC service for the `Codebreaker.Live` project, create the `GRPCLiveGameService` class:

Codebreaker.Live/Endpoints/GRPCLiveGameService.cs

```
using Codebreaker.Grpc;
using Google.Protobuf.WellKnownTypes;
```

```
using Grpc.Core;

namespace Codebreaker.Live.Endpoints;

public class GRPCLiveGameService(
  IHubContext<LiveHub> hubContext,
  ILogger<LiveGameService> logger) :
    ReportGame.ReportGameBase
{
  async public override Task<Empty> ReportGameCompleted(
    ReportGameCompletedRequest request,
    ServerCallContext context)
  {
    logger.LogInformation("Received game ended {type} " +
      "{gameid}", request.GameType, request.Id);
    await hubContext.Clients.Group(request.GameType)
      .SendAsync("GameCompleted", request.ToGameSummary());
    return new Empty();
  }
}
```

With the GRPCLiveGameService class, using constructor injection, the hub context we created in the previous chapter is injected to send a GameSummary method to all connected clients participating with the group that is named with game-type. The GRPCLiveGameService class needs to derive from the base class – that is, ReportGame.ReportGameBase. ReportGameBase is implemented as an inner class of ReportGame based on the Protobuf contract.

Next, use the GRPCLiveGameService class to map it as a gRPC endpoint:

Codebreaker.Live/ApplicationServices.cs

```
public static WebApplication MapApplicationEndpoints(this
WebApplication app)
{
  app.MapGrpcService<GRPCLiveGameService>();
  app.MapHub<LiveHub>("/livesubscribe");
  return app;
}
```

You can remove the mapping of the minimal API endpoint and just configure the gRPC endpoint with the WebApplication class known as MapGrpcService, passing the service class as a generic parameter.

gRPC requires HTTP/2. So, we need to configure the `Kestrel` server:

Codebreaker.Live/appsettings.json

```json
{
  "Kestrel": {
    "EndpointDefaults": {
      "Protocols": "Http1And2"
    }
  }
}
```

Configuring `Protocols` to `Http1And2` starts the `Kestrel` server and ensures it supports both HTTP/1 and HTTP/2. The gRPC service needs HTTP/2. When it comes to `live-service`, SignalR is offered from the same server. To allow the SignalR service to connect via HTTP/1 or HTTP/2, the server must be configured to offer both versions.

The `game-apis` service's implementation has similarities. In the project file, we need to add the `Grpc.AspNetCore` package, add a `Protobuf` element to the project file, and specify that we wish to create classes for the server:

Implementing this aspect of the gRPC service is simple: we can inject the `IGameService` interface. This uses the same classes we already used to implement the minimal API service:

Codebreaker.GameAPIs/GrpcGameEndpoints.cs

```csharp
public class GrpcGameEndpoints(
  IGamesService gamesService,
  ILogger<GrpcGameEndpoints> logger) :
  Grpc.GrpcGame.GrpcGameBase
{
  public override async
    Task<Grpc.CreateGameResponse> CreateGame(
    Grpc.CreateGameRequest request,
    ServerCallContext context)
  {
    logger.GameStart(request.GameType);
    Game game = await gamesService.StartGameAsync(
      request.GameType, request.PlayerName);
    return game.ToGrpcCreateGameResponse();
  }
```

```
public override async Task<SetMoveResponse> SetMove(
  SetMoveRequest request, ServerCallContext context)
{
  Guid id = Guid.Parse(request.Id);
  string[] guesses = request.GuessPegs.ToArray();
  (Game game, Models.Move move) =
    await gamesService.SetMoveAsync(
      id, request.GameType, guesses, request.MoveNumber);
  return game.ToGrpcSetMoveResponse(move);
}

public override async Task<GetGameResponse> GetGame(
  GetGameRequest request, ServerCallContext context)
{
  Guid id = Guid.Parse(request.Id);
  Game? game = await gamesService.GetGameAsync(id);
  if (game is null)
    return new GetGameResponse()
    {
      Id = Guid.Empty.ToString()
    };
  return game.ToGrpcGetGameResponse();
}
}
```

In gRPC, we can derive from the generated base class, `GrpcGame.GrpcGameBase`, override the base class methods that have been specified with the service contracts, and use the conversion methods to convert the input and output types into their corresponding gRPC representations.

Similar to `live-service`, the `game-apis` service needs gRPC to be added to the DI container and mapped to the endpoint, and `Kestrel` needs to be configured so that it supports both HTTP/1 and HTTP/2.

With the services implemented, let's consider the gRPC clients.

Creating gRPC clients

If you're using Visual Studio 2022, you can take advantage of its built-in support to add a gRPC client. From Solution Explorer, select the project, open the context menu, and select **Add | Connected Service**. This opens the dialogue shown in *Figure 14.2*:

Add service reference

Select a service reference to add to your application

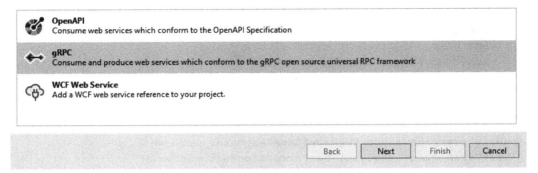

Figure 14.2 – Add service reference

Select **gRPC** and click **Next**. This opens the dialogue shown in *Figure 14.3*:

Add new gRPC service reference

Select a file or URL

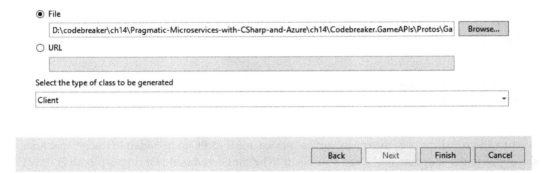

Figure 14.3 – Add new gRPC service reference

Select the Protobuf file, then select **Client** from the **Select the type of class to be generated** dropdown to create the classes for messages and the code for the client.

If you're not using Visual Studio, you can use a .NET command-line tool called dotnet. To install this tool, run the following command:

```
dotnet tool install -g dotnet-grpc
```

With this tool globally installed, you can use the `dotnet-grpc` command to create the proxy classes for the `game-apis` client:

```
cd Codebreaker.GameAPIs
dotnet-grpc add-file ..\Codebreaker.Live\Protos\LiveGame.proto
```

You can run the following for the bot service client:

```
cd ..
cd Codebreaker.Bot
dotnet-grpc add-file ..\Codebreaker.GameAPIs\Protos\GameService.proto
```

What happened with these commands or with the Visual Studio integration?

- The `Grpc.AspNetCore` NuGet package was added to the project
- A `Protobuf` element was added to the project file

Using the command-line tool, code for both the client and the server is created when the `Protobuf` entry is created. To only create code for the client, the `GrpcServices="Client"` attribute needs to be added:

Codebreaker.GameAPis/Codebreaker.GameAPIs.csproj

```
<ItemGroup>
  <Protobuf
    Include="..\Codebreaker.Live\Protos\LiveGame.proto"
    GrpcServices="Client" />
  <Protobuf Include=".\Protos\GameService.proto" GrpcServices="Server"
/>
</ItemGroup>
```

Regarding the `game-apis` service, the project file now includes two `Protobuf` entries. One is used to create the service part (which we did in the previous section), while the new entry is for the client code.

The bot invokes the `game-apis` service. So, the proto file of the `game-apis` service needs to be referenced in the bot project file:

Codebreaker.Bot/Codebreaker.Bot.csproj

```
<ItemGroup>
  <Protobuf Include=
    "..\Codebreaker.GameApis\Protos\GameService.proto"
    GrpcServices="Client" />
</ItemGroup>
```

Again, for the client, `GrpcServices` is set to `Client`.

Upon using the proto files, client proxy classes are created, offering methods to invoke the service with the names of the operations. These proxy classes can be injected. In the following code snippet, the generated `ReportGameClient` class is being injected into the `GrpcLiveReportClient` class:

Codebreaker.GameApis/GrpcLiveReportClient.cs

```
public class GrpcLiveReportClient(
  ReportGame.ReportGameClient client,
  ILogger<LiveReportClient> logger) : ILiveReportClient
{
  public async Task ReportGameEndedAsync(GameSummary gameSummary,
    CancellationToken cancellationToken = default)
  {
    try
    {
      ReportGameCompletedRequest request = gameSummary.
        ToReportGameCompletedRequest();
      await client.ReportGameCompletedAsync(request);
    }
    catch (Exception ex) when (ex is RpcException or
      SocketException)
    {
      logger.ErrorWritingGameCompletedEvent(
        gameSummary.Id, ex);
    }
  }
}
```

`ReportGameClient` implements the `ILiveReportClient` interface – the same interface we defined and implemented in the previous chapter to invoke the SignalR service on completion of a game. When implementing the same interface, we just need to change the configuration of the DI container so that it invokes the service via gRPC instead of using the REST interface. With the implementation of the `ReportGameEndedAsync` method, we can invoke the generated method from the proxy and need to convert the parameter with the help of an extension method:

Codebreaker.GameApis/ApplicationServices.cs

```
builder.Services.AddSingleton<ILiveReportClient,
  GrpcLiveReportClient>()
  .AddGrpcClient<ReportGame.ReportGameClient>(
    grpcClient =>
    {
```

```
        grpcClient.Address = new Uri("https://live");
    });
// code removed for brevity
```

Similarly, for the bot service, `GrpcGamesClient` implements the `IGamesClient` interface and injects `GrpcGame.GrpcGameClient`:

Codebreaker.Bot/GrpcGamesClient.cs

```
public class GrpcGamesClient(
  GrpcGame.GrpcGameClient client,
  ILogger<GrpcGamesClient> logger) : IGamesClient
{
  // code removed for brevity
  public async Task<(string[] Results, bool Ended,
    bool IsVictory)> SetMoveAsync(
    Guid id, string playerName, GameType gameType,
    int moveNumber, string[] guessPegs,
    CancellationToken cancellationToken = default)
  {
    SetMoveRequest request = new()
    {
      Id = id.ToString(),
      GameType = gameType.ToString(),
      MoveNumber = moveNumber,
      End = false
    };
    request.GuessPegs.AddRange(guessPegs);
    var response = await client.SetMoveAsync(request,
      cancellationToken: cancellationToken);
    return (response.Results.ToArray(), response.Ended,
      response.IsVictory);
  }
}
```

The `IGamesClient` interface was the same one that's implemented by `GameClient`, which calls the REST API. This interface is injected into `CodebreakerGameRunner`, so no changes are required when switching to gRPC.

The `SetMoveAsync` method makes a request to the gRPC service by invoking the `SetMoveAsync` method of `GrpcGame.GrpcGameClient`. Similar to before, it aims to convert the parameters into the ones needed by gRPC.

For help with implementing the other methods of the interface, check out this book's GitHub repository. Note that it's similar to what we did previously.

To glue `IGamesClient` to the new implementation, and to configure the gRPC client, we need to update the DI container's configuration:

Codebreaker.Bot/ApplicationServices.cs

```
builder.Services.AddSingleton<IGamesClient,
    GrpcGamesClient>()
    .AddGrpcClient<GrpcGame.GrpcGameClient>(client) =>
    {
        client.Address = new Uri("https://gameapis");
    });
```

The `AddGrpcClient` method configures the generated class with the address of the game-apis service.

Now, run the application and start one or multiple instances of the live test client to see if completed games show up. Start the bot service and send requests to the bot service to play several games. At the same time, use another client to play a game. How many games does the bot run until you complete one? Of course, this depends on the think time you configure with the bot. Do the results show up in the console of the live test client? Check the logs and the environment variables of the different services in the .NET Aspire dashboard.

Summary

In this chapter, you learned the differences between REST APIs and gRPC, as well as the advantages when using gRPC with service-to-service communication.

You created a service contract using Protobuf syntax to define services and messages. Contrary to REST, gRPC is strict when it comes to messages and service operations. You created servers and clients using the classes that were generated with the proto files.

To reduce cost in the cloud, it can be cost-effective to use protocols with lower overhead for service-to-service communication. However, other options are available as well. Upon completion of the game, it's not required to inform the listener immediately. There's also a price aspect to this: with the current implementation, the live service is running when accessed from the game APIs service. If nobody is listening, this is not required. By using asynchronous communication, the `live-service` can register to receive information when it's started – this is when a listener is active. Asynchronous communication will be covered in the next chapter.

Further reading

To learn more about the topics that were discussed in this chapter, please refer to the following links:

- *Protobuf Language Guide*: `https://protobuf.dev/programming-guides/proto3/`
- *gRPC on .NET-supported platforms*: `https://learn.microsoft.com/en-us/aspnet/core/grpc/supported-platforms`
- *grpc-dotnet GitHub repository*: `https://github.com/grpc/grpc-dotnet`
- *Call gRPC services with the .NET client*: `https://learn.microsoft.com/en-us/aspnet/core/grpc/client`

15
Asynchronous Communication with Messages and Events

In the previous chapter, we updated our services using binary communication. However, some services don't need connected services: the client and the server do not need to be connected at the same time, which means communication can be done asynchronously. This communication can be done by sending messages to a queue or publishing events.

In this chapter, we'll use Azure services for asynchronous communication – that is, Azure Queue Storage and Azure Event Hubs. We'll also use Apache Kafka as an alternative option.

You'll learn how to do the following:

- Differentiate message queues and events
- Send and receive messages using a queue
- Publish and subscribe events with Azure Event Hubs
- Use Apache Kafka for event processing

Technical requirements

In this chapter, like the previous chapters, you'll need an Azure subscription and Docker Desktop.

The code for this chapter can be found in this book's GitHub repository: `https://github.com/PacktPublishing/Pragmatic-Microservices-with-CSharp-and-Azure/`.

The `ch15` folder contains the projects for this chapter, along with their outputs. To add the functionality from this chapter, you can start with the source code from the previous chapter.

We'll be considering the following projects in this chapter:

- `Codebreaker.AppHost`: The .NET Aspire host project. The app model has been enhanced by adding Azure Storage, Azure Event Hubs, and Apache Kafka services.

- `Codebreaker.BotQ`: This is a new project that contains nearly the same code as `Codebreaker.Bot`. However, instead of using a REST API to trigger gameplay, a message queue is used.

- `Codebreaker.GameAPIs`: This project has been updated so that it doesn't forward completed games to `live-service` directly. Instead, it publishes events to Azure Event Hubs or Apache Kafka depending on the launch profile startup.

- `Codebreaker.Live`: This project has been changed so that it subscribes to events from Azure Event Hubs using async streams. The SignalR implementation has also been changed so that it uses async streams.

- `Codebreaker.Ranking`: This is a new project that receives events from Azure Event Hubs or Kafka, writes this information to an Azure Cosmos DB database, and offers a REST service to retrieve the rank of the day. With Event Hub, we have a different way to receive events than we do when using `live-service`.

Comparing messages and events

In the previous chapter, we used connected network communication with all the services. First, we look into the communication between the bot and the game APIs as shown in *Figure 15.1*:

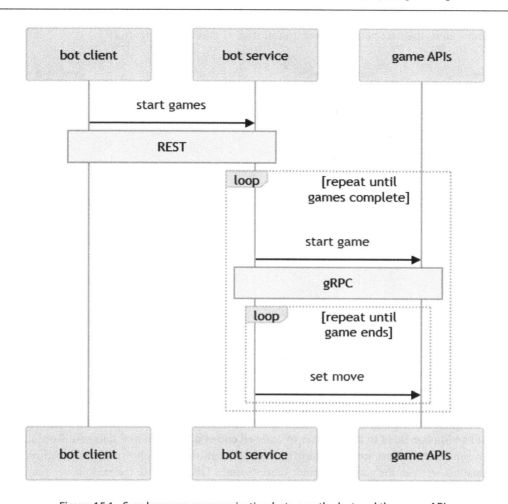

Figure 15.1 - Synchronous communication between the bot and the game APIs

The bot service can be accessed via REST. The bot service itself invokes the game APIs service via gRPC (all other clients use REST with the game APIs service). The bot service then continues communication with the game APIs service, sends moves until the game is complete, and continues with the next game until a specified number of games is played. The bot client invokes the bot service via REST which is (like gRPC) synchronous communication, with request/reply. The bot service here doesn't have a synchronous implementation, as the bot client doesn't need to wait until all the games are played – the HTTP protocol would timeout during this time. Instead, the bot service returns an HTTP ACCEPTED answer (status code 202) with a unique identifier which can be used by the client to check for a status. The protocol itself is synchronous, as the client waits for answer 202.

When a game ends, the next part of the communication is shown with *Figure 15.2*.

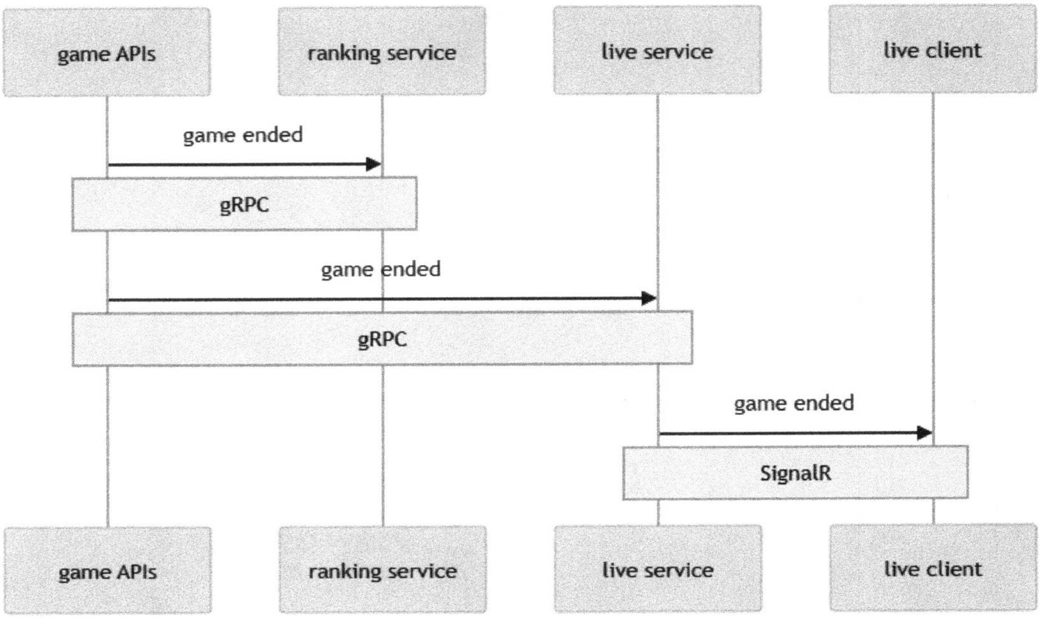

Figure 15.2 - Synchronous communication initiated from the game APIs

The game APIs service informs the ranking service and the live service using gRPC. The live service continues communication via SignalR to inform all connected clients about the game end. The ranking service will be implemented in this chapter to write all ended games to a new database. To simplify this image, some services that are used within the communication are not shown. There's synchronous communication between the game APIs and Azure Cosmos DB, similar to the ranking service.

With synchronous communication, if there's a delay in one of the services, the delay goes back to the original caller. If there's an error in one of the services, the client does not receive a successful response.

Microsoft Azure offers several services that can be used to create asynchronous communication: Azure Queue Storage, the Azure Service Bus, Azure Event Gird, and Azure Event Hub. Let's have a look at a new version of the new sequence with the communication with the bot client and the bot service in *Figure 15.3*.

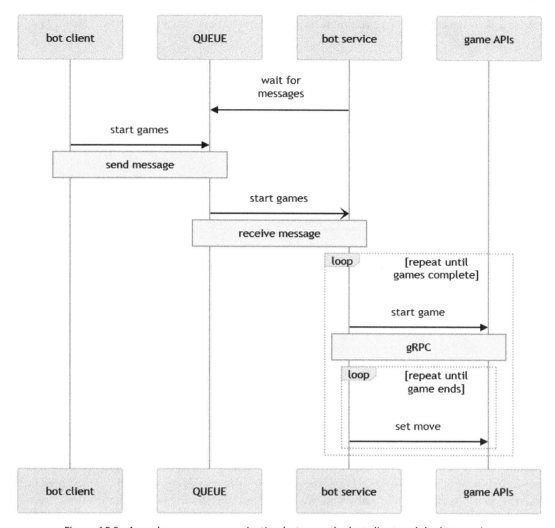

Figure 15.3 - Asynchronous communication between the bot client and the bot service

With the new implementation the Azure Queue Storage comes into play. The bot service registers with the queue to receive messages. The bot client, instead of using HTTP to communicate with the bot service, sends a message to the queue. The bot client does not need to wait if anybody started to work on this message. For the client, the job is done. The bot service, because it registered to receive messages, receives the message from the queue and plays the games in the same way as before, this doesn't change.

Next, we look at the asynchronous communication initiated by the game APIs service in *Figure 15.4*.

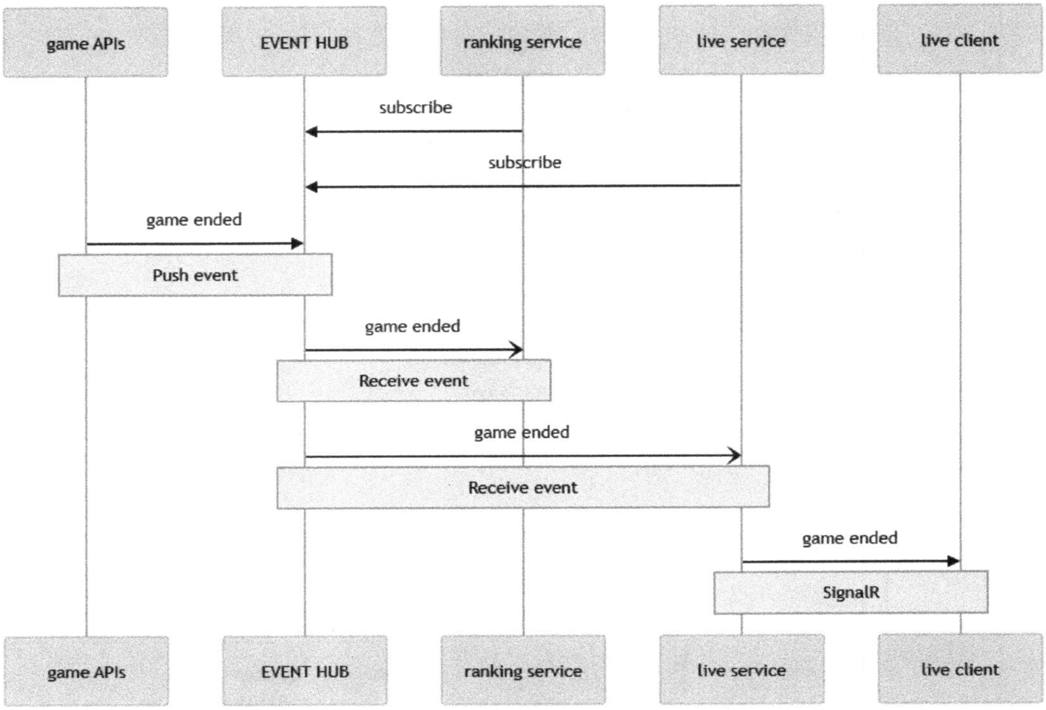

Figure 15.4 – Asynchronous communication initiated from the game APIs

Here, the Azure Event Hub comes into play. The game APIs service, instead of doing synchronous communication both with the ranking and the live services, just communicates with the event hub. A game-ended event is pushed to this service. The game APIs doesn't need to know who is interested in this event, who is receiving this event. Here, two subscribers, the ranking service and the live service are registered, and receive this event. From here on, communication is the same as before. The ranking service writes the received information to a database (which is not shown), and the live service forwards this information to clients that are subscribed to the live service – if they are subscribed to the same game type that is stored with the event.

The important difference between using message queues and events can be seen in this scenario. When sending a message to a queue, just one receiver deals with the message. Multiple readers can be connected to the same queue (for performance reasons), but only one reader processes the message. If the message is processed successfully, it's removed from the queue. With events, multiple subscribers receive and process the same event.

Let's look at the different options that are available with Microsoft Azure for messages and events.

Message queues

Microsoft Azure offers Azure Queue Storage (part of the Azure Storage Account) and Service Bus queues that can be used for queuing messages. Azure Queue Storage is the simpler and cheaper option to use, but Azure Service Bus offers a lot more features, such as ordering guarantees, atomic operations, sending messages in a batch, duplicate detection, and more. See `https://learn.microsoft.com/en-us/azure/service-bus-messaging/service-bus-azure-and-service-bus-queues-compared-contrasted` for details.

Events

To publish and subscribe to events, you can use Azure Event Grid and Azure Event Hubs. **Azure Event Grid** is easy to use to subscribe to events with most Azure resources. In the Azure portal, with most resources, you can see the **Events** menu in the left bar. For example, when opening an Azure Storage Account, after clicking on **Events**, click **Event Subscription**. With a storage account, you'll see event types such as **Blob Created**, **Blob Deleted**, **Blob Renamed**, **Directory Created**, **Directory Deleted**, **Directory Renamed**, **Blob Tier Changed**, and others, as shown in *Figure 15.5*:

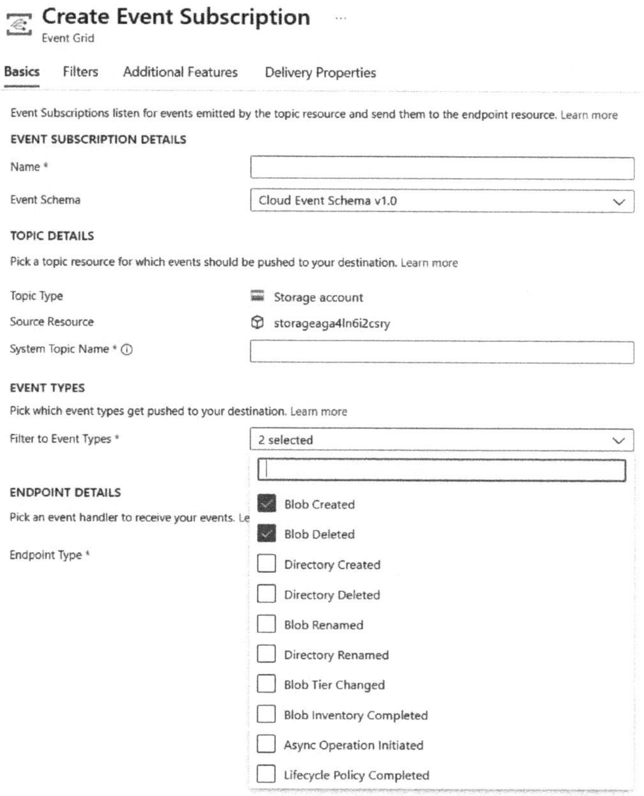

Figure 15.5 – Create Event Subscription

Event types are predefined by Azure resources. With event subscriptions, you can specify where the event should be fired. You can choose this with the endpoint, which can be an Azure function, a storage queue, a hybrid connection, a Webhook, and so on.

Event Grid also allows you to specify custom topics to be defined, high throughput with up to 10,000,000 events per second, and 100,000 operations a month for free. This service is available as a PaaS offering running on Azure with the name **Event Grid on Kubernetes with Azure Arc** on Kubernetes.

To support even more events, massive scaling with partitions, and a big data streaming platform with low latency, **Azure Event Hubs** can be used. This service offers reliable event delivery where events are stored for up to 7 days in case they have not been delivered. Event Hubs also has great integration with streaming analytics.

Let's update the Codebreaker solution with Azure Queues and Azure Event Hubs.

> **Note**
> At the time of writing, .NET Aspire has planned support with Azure Event Grid. Event Hubs, Queue Storage, and Azure Service Bus are already available.

Let's start reading from Azure Queue Storage with the Codebreaker bot service.

Creating a service that reads from Azure Queue Storage

The `Codebreaker.Bot` project we used previously offers a minimal API service. With the updates, a REST API isn't needed – a simple console application will do. Just create a new console application (`dotnet new console -o Codebreaker.BotQ`) and copy the source code from `Codebreaker.Bot`. The new bot will also use gRPC for communication with the game APIs service. Because this isn't an ASP.NET Core application, these NuGet packages are needed for gRPC:

- `Google.Protobuf`
- `Grpc.Net.ClientFactory`
- `Grpc.Tools`

For the DI container, we also need `Microsoft.Extensions.Hosting`, and for the .NET Aspire Storage Queues component, we need `Aspire.Azure.Storage.Queues`.

Next, we'll update the app model.

Defining app-model for Azure Storage

With the AppHost project, reference the newly created project, `Codebreaker.BotQ`, and add the `Aspire.Hosting.Azure.Storage` NuGet package so that you can use the Azure Storage resource.

Invoke the `AddAzureStorage` method to specify Azure Storage with the app model:

Codebreaker.AppHost/Program.cs

```
// code removed for brevity
if (startupMode == "OnPremises")
{
}
else
{
  var storage = builder.AddAzureStorage("storage");
  var botQueue = storage.AddQueues("botqueue");
}
```

The Azure Storage resource supports queues, tables, and blobs. This time, we'll use queues, hence why we're invoking the `AddQueues` extension method.

The project configuration references the queue:

Codebreaker.AppHost/Program.cs

```
string botLoop =
  builder.Configuration.GetSection("Bot")["Loop"] ??
    "false";
string botDelay =
  builder.Configuration.GetSection("Bot")["Delay"] ??
    "1000";
// code removed for brevity
builder.AddProject<Projects.Codebreaker_BotQ>("bot")
  .WithReference(insights)
  .WithReference(botQueue)
  .WithReference(gameAPIs)
  .WithEnvironment("Bot__Loop", botLoop)
  .WithEnvironment("Bot__Delay", botDelay);
```

Pay attention to using the new bot project instead of the old one. The new project references the queue to pass the connection string. In addition, we specify the `Loop` and `Delay` parameters, which are read from the configuration and set as environment variables on starting `bot-service`.

These values are specified within the AppHost development configuration:

Codebreaker.AppHost/appsettings.Development.json

```
{
  // configuration removed for brevity
  "Bot": {
    "Loop": true,
    "Delay": 2000
  }
}
```

The new bot-service can read values from the storage queue in a loop – which is configured here. When published with Azure, the loop isn't needed. This will be covered later in the *Deploying the solution to Microsoft Azure* section.

Now that the app model has been specified, let's continue with the new bot project.

Using the storage queue component

With the previous bot project, we received values so that we could start playing a sequence of games. The same information is needed with the new bot:

Codebreaker.BotQ/Endpoints/BotQueueClient.cs

```
public record class BotMessage(
  int Count, int Delay, int ThinkTime);
```

The Count property is for the number of games to play, the Delay property is for the delay between games, and the ThinkTime property is for the value of the think time between game moves.

The BotQueueClientOptions class is used to receive the configuration values that are passed from the AppHost:

Codebreaker.BotQ/Endpoints/BotQueueClient.cs

```
// code removed for brevity
public class BotQueueClientOptions
{
  public bool Loop { get; set; } = false;
  public int Delay { get; set; } = 1000;
}
```

Within the constructor of the `BotQueueClient` class, `options`, `logger`, the previously used `CodebreakerTimer`, and `QueueServiceClient` are injected:

Codebreaker.BotQ/Endpoints/BotQueueClient.cs

```
public class BotQueueClient(
  QueueServiceClient client,
  CodebreakerTimer timer,
  ILogger<BotQueueClient> logger,
  IOptions<BotQueueClientOptions> options)
{
// code removed for brevity
```

The `QueueService` client class is from the `Azure.Storage.Queues` namespace and communicates with the Azure Storage queue resources to get information about queues, as well as to create queues. With the implementation of `CodebreakerTimer`, a timer is used to play game after game. It uses the values we receive in the message from the queue.

The `RunAsync` method kicks off the work:

Codebreaker.BotQ/Endpoints/BotQueueClient.cs

```
public async Task RunAsync()
{
  var queueClient = client.GetQueueClient(«botqueue»);
  await queueClient.CreateIfNotExistsAsync();
  var deadLetterClient = client.GetQueueClient(
    «dead-letter»);
  await deadLetterClient.CreateIfNotExistsAsync();

  bool repeat = options.Value.Loop;
  do
  {
    await ProcessMessagesAsync(
    queueClient, deadLetterClient);
    await Task.Delay(options.Value.Delay);
  } while (repeat);
}
// code removed for brevity
```

To read messages from the queue, we use the `QueueClient` class. The `QueueServiceClient` method, `GetQueueClient`, returns `QueueClient` to communicate with the queue named `botqueue`. With the app model we specified earlier, only the storage account is created, not the queue itself. We create the queue if it doesn't already exist. Then – in a loop – we invoke

`ProcessMessagesAsync`. If the loop isn't set, messages are retrieved only once. This can be used when publishing Azure Container Apps jobs, as will be discussed in the *Deploying the solution to Microsoft Azure* section.

> **Note**
>
> A dead letter queue can be checked to find out if there have been issues with messages. When a message cannot be successfully processed a few times, for example, when the receiver throws, the message is written to the dead letter queue.

Next, `ProcessMessageAsync` reads a message from the queue:

Codebreaker.BotQ/Endpoints/BotQueueClient.cs

```
private async Task ProcessMessagesAsync(
  QueueClient queueClient,
  QueueClient deadLetterClient)
{
  QueueProperties properties =
    await queueClient.GetPropertiesAsync();
  if (properties.ApproximateMessagesCount > 0)
  {
    QueueMessage[] messages =
      await queueClient.ReceiveMessagesAsync();
    foreach (var encodedMessage in messages)
    {
      if (encodedMessage.DequeueCount > 3)
      {
        await deadLetterClient.SendMessageAsync(
          encodedMessage.MessageText);
        await queueClient.DeleteMessageAsync(
          encodedMessage.MessageId,
          encodedMessage.PopReceipt);
        continue;
      }
      byte[] bytes = Convert.FromBase64String(
        encodedMessage.MessageText);
      string message = Encoding.UTF8.GetString(bytes);
      var botMessage =
```

```
            JsonSerializer.Deserialize<BotMessage>(message);
          timer.Start(
            botMessage.Delay,
            botMessage.Count,
            botMessage.ThinkTime);
          await queueClient.DeleteMessageAsync(
            encMessage.MessageId, encMessage.PopReceipt);
        }
      }
    }
// code removed for brevity
```

First, attributes of the queue are checked to see if there's a message available using the `ApproximateMessagesCount` property. If this is the case, messages are retrieved using `ReceiveMessagesAsync`. This method reads the messages from the queue, at which point the messages can no longer be seen by others. The time the message is not visible can be set with the `visibilityTimeout` parameter. The default is 30 seconds. When successfully deserializing the message, it is deleted using `DeleteMessageAsync`. The `timer.Start` method starts a task to play the games asynchronously. So, if a game takes longer to play (with many games or with higher think times), this does not influence deleting the message. If the message returns to the queue, it can be processed again. The implementation checks for the dequeue count of the message that's retrieved. If it's read three times, the message goes to a dead-letter queue and can be manually checked for issues.

Next, let's configure the DI container:

Codebreaker.BotQ/ApplicationServices.cs

```
public static void AddApplicationServices(this IHostApplicationBuilder
builder)
{
  builder.AddAzureQueueClient("botqueue");
  builder.Services.AddScoped<BotQueueClient>();
  var botConfig = builder.Configuration.GetSection("Bot");
  builder.Services.Configure<BotQueueClientOptions>(
    section);
  builder.Services.AddScoped<CodebreakerTimer>();
  builder.Services.AddScoped<CodebreakerGameRunner>();
  builder.Services.AddSingleton<IGamesClient,
    GrpcGamesClient>()
    .AddGrpcClient<GrpcGame.GrpcGameClient>(
```

```
      client =>
      {
        client.Address = new Uri("https://gameapis");
      });
  }
```

The `AddAzureQueueClient` method is defined with the `Aspire.Azure.Storage.Queues` NuGet package. This method configures the Aspire component and registers the `QueueService` client with the DI container. With the environment variables passed by the AppHost, these values are retrieved using `builder.Configuration.GetSection` and configured with the `BotQueueClientOptions` class, which defines the loop's behavior. Other than this, the timer, game runner, and gRPC are configured in the same way as the previous bot service implementation.

Now, we are ready to run the application and test queues.

Running the application

You can set breakpoints within the new bot service project to verify the functionality of the queue. When you start the application, an Azure Storage Account is created. With the initialization of `BotQueueClient`, message queues are created. This can verified in the Azure portal, as shown in *Figure 15.6*:

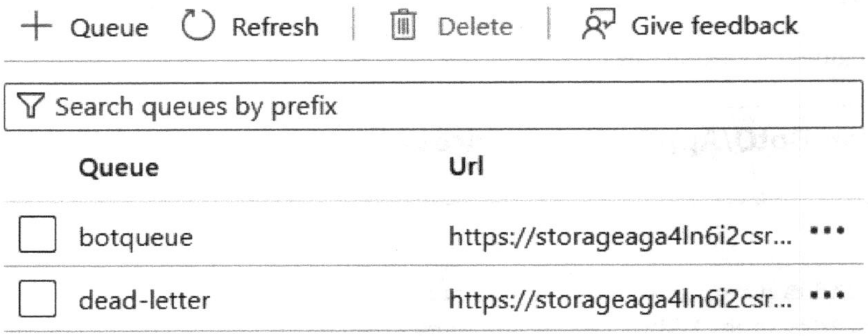

Figure 15.6 – Storage queues created

The `botqueue` and `dead-letter` storage queues have been created. Now, open `botqueue` to pass a valid JSON message, as shown in *Figure 15.7*:

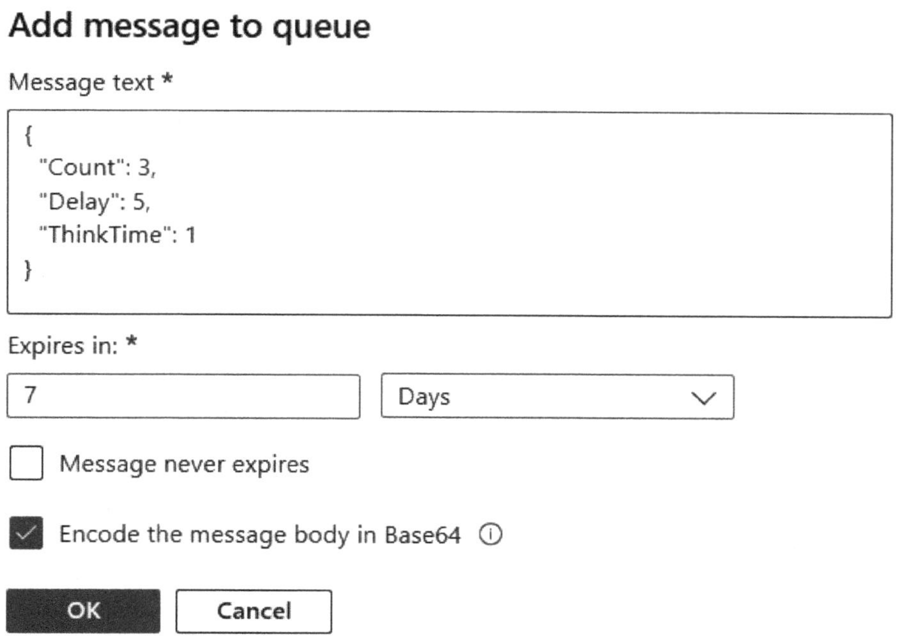

Add message to queue

Message text *

```
{
  "Count": 3,
  "Delay": 5,
  "ThinkTime": 1
}
```

Expires in: *

| 7 | | Days | ⌄ |

☐ Message never expires

☑ Encode the message body in Base64 ⓘ

[OK] [Cancel]

Figure 15.7 – Add message to queue

Encode the message body in Base64 is selected by default – we used this encoding to read the message. The message needs to be in a valid JSON format so that it can be deserialized by the BotMessage class:

```
{
  "Count": 3,
  "Delay": 5,
  "ThinkTime": 1
}
```

With a valid JSON message, you will see that the message is processed. After sending a message that's not in JSON format, you'll see the message in the dead-letter queue – after some retries.

As the bot now starts playing games when we send messages, let's get into the next enhancement – using Azure Event Hubs.

Publishing messages to Azure Event Hubs

To use Azure Event Hubs, we'll implement the game APIs service so that we can publish events.

Defining app-model for Event Hubs

To use Azure Event Hubs with the AppHost project, the `Aspire.Hosting.Azure.EventHubs` NuGet package is required.

Here, Event Hubs needs to be added to the app model:

Codebreaker.AppHost/Program.cs

```
var eventHub =
  builder.AddAzureEventHubs("codebreakerevents")
    .AddEventHub("games");
// code removed for brevity
```

The `AddAzureEventHubs` method, adds an Azure Event Hubs namespace, `AddEventHub`, as an event hub. A namespace is a management container with network endpoints and access control. The default Event Hub namespace that's created is in the Standard tier. For development, you can change this to the Basic tier. Event hubs are created within namespaces. For scalability, event hubs use one or more partitions. By default, the event hub is created with four partitions. A partition contains an ordered stream of events. The number of partitions doesn't change the price, but the number of throughput units does. With the number of throughput units, you specify a number of events per second. The number of partitions should be equal to or higher than the number of throughput units. Throughput units can be changed as needed; the number of partitions can only be changed in premium and dedicated tiers.

With the event hub specified, we can reference it:

Codebreaker.AppHost/Program.cs

```
var gameAPIs =
  builder.AddProject<Projects.Codebreaker_GameAPIs>(
    "gameapis")
    .WithExternalHttpEndpoints()
    .WithReference(cosmos)
    .WithReference(redis)
    .WithReference(insights)
    .WithReference(eventHub)
    .WithEnvironment("DataStore", dataStore);
```

With the game APIs service, we replace the referenced live service with the event hub. The reference to live service is no longer needed.

With this, we can look at the game APIs service.

Using the .NET Aspire Event Hubs component to produce events

To use the .NET Aspire Event Hubs component, we must add the `Aspire.Azure.Messaging.EventHub` NuGet package.

Using this package, we can configure the DI container:

Codebreaker.GameAPIs/ApplicationServices.cs

```
builder.AddAzureEventHubProducerClient(
  "codebreakerevents",settings =>
  {
    settings.EventHubName = "games";
  });
```

From the game APIs service, to send information about completed games, in previous chapters, we created the `LiveReportClient` class to call a REST service and the `GrpcLiveReportClient` class to invoke a gRPC service. Now, we can implement the same interface we used earlier – `IGameReport` with the `EventHubReportProducer` class.

Sending events can easily be done, as shown here:

Codebreaker.GameAPIs/Services/EventHubReportProducer.cs

```
public class EventHubReportProducer(
  EventHubProducerClient producerClient,
  ILogger<EventHubLiveReportClient> logger) :
  IGameReport
{
  public async Task ReportGameEndedAsync(
    GameSummary game,
    CancellationToken cancellationToken = default)
  {
    var data = BinaryData.FromObjectAsJson(game);
    await producerClient.SendAsync(
      [ new EventData(data) ],
      cancellationToken);
    // code removed for brevity
  }
}
```

The `EventHubReportProducer` class injects the `EventHubProducerClient` class to send events to the event hub. `GameSummary` is converted into `BinaryData` with `BinaryData.FromObjectAsJson`. The `EventData` class from the `Azure.Messaging.EventHubs` namespace allows us to pass a string, `BinaryData`, and `ReadOnlyMemory<byte>`. Then, the event is published by invoking the `SendAsync` method.

Now that we've published some events, let's subscribe to them.

Subscribing to Azure Event Hubs events

The `Codebreaker.Live` project previously offered a gRPC service that was invoked by the game APIs service to publish completed games via SignalR. Instead of offering a gRPC service, we can subscribe to events.

Create a new `Codebreaker.Ranking` project so that you can offer minimal APIs. This project will receive the same events as `Codebreaker.Live` but write them to a database to offer ranks for games based on days, weeks, and months.

To create the `Codebreaker.Ranking` project, use the following command:

```
dotnet new webapi -minimal -o Codebreaker.Ranking
```

Add the newly created project as a reference to `Codebreaker.AppHost`, and reference `Codebreaker.ServiceDefaults` to configure the service defaults. Now, we can update app model.

Defining app-model for Event Hubs subscribers

With the AppHost project, the live and ranks projects reference the event hub, similar to the events publishing project:

Codebreaker.AppHost/Program.cs

```
var storage = builder.AddAzureStorage("storage");
var blob = storage.AddBlobs("checkpoints");

var live =
  builder.AddProject<Projects.Codebreaker_Live>("live")
  .WithExternalHttpEndpoints()
  .WithReference(insights)
  .WithReference(eventHub)
  .WithReference(signalR);

builder.AddProject<Projects.Codebreaker_Ranking>("ranking")
  .WithExternalHttpEndpoints()
```

```
    .WithReference(cosmos)
    .WithReference(insights)
    .WithReference(eventHub)
    .WithReference(blob);
  // code removed for brevity
```

Subscribing to events can be done in two ways, either with the event hub consumer client or the event processor client. The event hub consumer client is simpler to use and supports async streams. The event processor client is more powerful and supports receiving from multiple partitions in parallel. The second option needs to save checkpoints in an Azure blob storage account. We use the same account that we already use for queues.

We will implement both versions. The Codebreaker live service uses async streams, and the event hub consumer client with the class `EventHubConsumerClient` fits its need. The Codebreaker ranking service makes use of the event processor client, using `EventProcessorClient`.

Using the Event Hubs component with async streaming

When using the `Codebreaker.Live` project, references to the `Aspire.Azure.Messaging.EventHubs` NuGet package are required:

> **Note**
> The `Codebreaker.Live` project was created in *Chapter 13*. Here, we'll create a new SignalR hub to offer streaming.

Codebreaker.Live/ApplicationServices.cs

```
public static void AddApplicationServices(this IHostApplicationBuilder
builder)
{
  builder.Services.AddSignalR()
    .AddMessagePackProtocol()
    .AddNamedAzureSignalR("signalr");
  builder.AddAzureEventHubConsumerClient("codebreakerevents",
  settings =>
  {
    settings.EventHubName = "games";
  });
}
```

The `AddAzureEventHubConsumerClient` method configures the `EventHubConsumerClient` class as a singleton within the DI container.

Now, we must create a new SignalR hub to inject `EventHubConsumerClient`:

Codebreaker.Live/Endpoints/StreamingLiveHub.cs

```
public class StreamingLiveHub(
  EventHubConsumerClient consumerClient,
  ILogger<StreamingLiveHub> logger) : Hub
{
  // code removed for brevity
```

By using the primary constructor, `EventHubConsumerClient` is injected to retrieve the events.

Now, create the `SubscribeToGameCompletions` method:

Codebreaker.Live/Endpoints/StreamingLiveHub.cs

```
public async IAsyncEnumerable<GameSummary>
  SubscribeToGameCompletions(
    string gameType,
    [EnumeratorCancellation] CancellationToken
      cancellationToken)
{
  await foreach (PartitionEvent ev in
    consumerClient.ReadEventsAsync(cancellationToken))
  {
    GameSummary gameSummary;
    try
    {
      logger.ProcessingGameCompletionEvent();
      gameSummary = ev.Data.EventBody
        .ToObjectFromJson<GameSummary>();
    }
    catch (Exception ex)
    {
      logger.ErrorProcessingGameCompletionEvent(
        ex, ex.Message);
      continue;
    }
    if (gameSummary.GameType == gameType)
    {
      yield return gameSummary;
    }
    else
    {
```

```
        continue;
      }
    }
  }
```

SignalR supports async streaming with methods returning `IAsyncEnumerable`. The `SubscribeToGameCompletions` method receives a game type parameter that only returns game completions of this game type. `EventHubConsumerClient` supports async streaming by invoking the `ReadEventsAsync` method. If the received game summary is of the requested game type, it's returned to the client via the async stream.

At this point, the middleware needs to be configured so that it references the new hub:

Codebreaker.Live/ApplicationServices.cs

```
app.MapHub<LiveHub>("/livesubscribe");
app.MapHub<StreamingLiveHub>("/streaminglivesubscribe");
```

We also need to update the client by using async streaming and the new link:

LiveTestClient/StreamingLiveClient.cs

```
public async Task SubscribeToGame(string gameType, CancellationToken
cancellationToken = default)
{
  if (_hubConnection is null) throw new
InvalidOperationException("Start a connection first!");

  try
  {
    await foreach (GameSummary summary in
      _hubConnection.StreamAsync<GameSummary>(
        "SubscribeToGameCompletions",
        gameType,
        cancellationToken))
    {
      string status = summary.IsVictory ? "won" : "lost";
      Console.WriteLine($"Game {summary.Id} {status} " +
        $"by {summary.PlayerName} after " +
        $"{summary.Duration:g} with " +
        $"{summary.NumberMoves} moves");
    }
  }
  catch (HubException ex)
```

```
    {
        logger.LogError(ex, ex.Message);
        throw;
    }
    catch (OperationCanceledException ex)
    {
        logger.LogWarning(ex.Message);
    }
}
```

With the same SignalR initialization configuration we created in *Chapter 13*, the client now uses the `StreamAsync` method from the SignalR `HubConnection` class to async stream the results that are returned from the service.

With these changes, you can already test and run the solution, starting from the message queue up to the SignalR streaming client, to receive completed games. However, let's add another Event Hubs client to process messages, this time with the Event Hubs processor.

Using the .NET Aspire Event Hubs component to process messages

The `Codebreaker.Ranking` project receives events, writes those events to an Azure Cosmos database, and offers minimal APIs to get ranking information from the players. This project references the .NET Aspire `Aspire.Azure.Messaging.EventHubs` and `Aspire.Azure.Storage.Blobs` components:

Codebreaker.Ranking/ApplicationServices.cs

```
public static void AddApplicationServices(this IHostApplicationBuilder builder)
{
    // code removed for brevity
    builder.AddKeyedAzureBlobClient("checkpoints");
    builder.AddAzureEventProcessorClient("codebreakerevents",
        settings =>
        {
            settings.EventHubName = "games";
            settings.BlobClientServiceKey = "checkpoints";
        });
    builder.Services.AddDbContextFactory<RankingsContext>(
        options =>
        {
            string connectionString =
                builder.Configuration.GetConnectionString(
```

```
        "codebreakercosmos") ??
        throw new InvalidOperationException(
          "Could not read the Cosmos connection-string");
      options.UseCosmos(connectionString, "codebreaker");
    });
  builder.EnrichCosmosDbContext<RankingsContext>();
  builder.Services
    .AddSingleton<IGameSummaryEventProcessor,
      GameSummaryEventProcessor>();}
```

AddAzureEventProcessorClient registers a singleton instance of the EventProcessorClient class. We connect to the same namespace and event hub, so this configuration is the same. What's different is that AddKeyedAzureBlobClient is a method from the .NET Aspire Blob Storage component. This method registers a singleton instance with the DI container to read and write blobs. The storage is connected to the event hub by setting BlobClientServiceKey to write checkpoints.

You can also simplify the configuration by not registering a keyed configuration. The one default storage component that's registered is automatically used from the event hub component.

Other than the event hub configuration, an EF Core context must be configured to write the received game summary information to an Azure Cosmos DB database. Check out *Chapter 3* for more details. Contrary to *Chapter 3*, we register an EF Core context factory with the DI container, which allows us to inject this into a singleton object and create the context objects with a shorter lifetime.

The registered GameSummaryEventProcessor is our implementation for dealing with events:

Codebreaker.Ranking/Services/GameSummaryEventProcessor.cs

```
public class GameSummaryEventProcessor(
  EventProcessorClient client,
  IDbContextFactory<RankingsContext> factory,
  ILogger<GameSummaryEventProcessor> logger)
{
  public async Task StartProcessingAsync(
    CancellationToken = default)
  {
    // code removed for brevity
  }
  public Task StopProcessingAsync(
    CancellationToken cancellationToken = default)
  {
  }
}
```

The class injects `EventProcessorClient` and the EF Core context factory. This class implements methods to start and stop the processing of events.

The `StartProcessingAsync` method is shown in the following code snippet:

Codebreaker.Ranking/Services/GameSummaryEventProcessor.cs

```csharp
public async Task StartProcessingAsync(CancellationToken
cancellationToken = default)
{
  // code removed for brevity
  client.ProcessEventAsync += async (args) =>
  {
    GameSummary summary = args.Data.EventBody
      .ToObjectFromJson<GameSummary1>();

    using var context = await factory.CreateDbContextAsync(
      cancellationToken);

    await context.AddGameSummaryAsync(summary,
      cancellationToken);
    await args.UpdateCheckpointAsync(cancellationToken);
  };

  client.ProcessErrorAsync += (args) =>
  {
    logger.LogError(args.Exception,
      "Error processing event, {error}",
      args.Exception.Message);
    return Task.CompletedTask;
  };
  await client.StartProcessingAsync(cancellationToken);
}
```

Once you start processing events by invoking the `StartProcessingAsync` method, the `EventProcessorClient` class fires .NET events that are invoked when messages are received, and on errors: `ProcessEventAsync` and `ProcessErrorAsync`. A received message is converted from binary into a `GameSummary` object and written to the database. In addition to that, the checkpoint in the storage account is written so that we know which event message was processed last.

When this is in place, start the application, open the Azure portal to send messages to the bot queue to let the bot play some games, and monitor how events are sent. *Figure 15.8* shows the Azure portal showing Event Hub metrics:

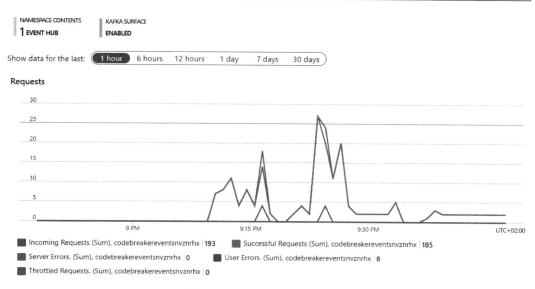

Figure 15.8 – Event Hub metrics

Apart from checking the Event Hub metrics, verify the data that's been written to the rankings database. Also, start the SignalR client application so that you can monitor event data with async streaming.

Open the Event Hub instance in the Azure portal and select **Configuration** within the **Settings** category (see *Figure 15.9*):

Figure 15.9 – Event Hub configuration

Here, you can see the configured partition count and can disable the hub. Regarding **Retention time (hrs)**, you can configure how long the messages should be kept for. The maximum time depends on the configured SKU of the Event Hub namespace. With the standard SKU, the maximum retention is 7 days. If `ranking-service` is not running on one day, there's still enough time to process the games afterward.

You can also configure how data is captured (**Features | Capture**) with an Azure Storage Account (**Avro** serialization format) or Azure Data Lake (**Parquet** or **Delta Lake** serialization format). Make sure you check out the pricing options before configuring capturing or other SKUs.

Mentioning prices, what do you need to be aware of when you're deploying the solution to Microsoft Azure?

Deploying the solution to Microsoft Azure

When using a low load, the complete solution is not expensive when it's running on Microsoft Azure. CPU power typically results in higher cost. How many containers are running with Azure Container Apps? `bot-service`, the `game-apis` service, `live-service`, `ranking-service`, and the Redis container. The `game-apis` service should scale with the minimum value set to 1, which provides a fast response for the first user so that they get a fast first answer. If you scale to 1 when the service is idle, there's an idle price that reduces the cost of the CPU considerably. `bot-service`, `live-service`, and `ranking-service` can scale down to 0, which means there's no cost in terms of CPU and memory. However, be aware of custom health checks (covered in *Chapter 12*), which can play against scaling to 0. With `live-service`, if there's no listener to query for running games, you can't subscribe to events. Thus, cost only applies if clients are connected.

The bot contains a loop that keeps running and checks the queue repeatedly. This is not necessary with the Azure Container Apps environment. Here, we can create an **Azure Container Apps job** resource. This resource is only started based on a trigger – for example, a cron time or an event such as a message available in a storage queue.

Creating Azure Container Apps jobs is not yet supported with .NET Aspire out of the box. However, this is possible with some customization. Here's what you need to do:

1. Initialize the project with `azd init`.
2. Use `azd infra synth` to create Bicep files and manifest files.
3. Create Azure resources using `azd provision`.
4. Change the manifest file of the project that should be deployed as a Container App Job instead of a Container App.
5. Deploy the projects using `azd deploy` (you can deploy project by project using `azd deploy <service>` or deploy all).

Because of the quick updates that are made to .NET Aspire, check out the README file in this chapter's GitHub repository for the latest updates.

Next, let's look at an alternative option to using Azure services.

Using Apache Kafka for event processing

Apache Kafka can be an alternative to using Azure Queue Storage and Azure Event Hubs – especially when it comes to on-premises solutions. This technology is used by many companies in their on-premises environments for high performance application-to-application messaging, supports a scalable multiple producers and consumers environment (like Event Hub), and supports a read-only-once scenario like message queues.

Starting the AppHost with the `OnPremises` launch profile will now use the previously created `Codebreaker.Bot`. This uses a REST API instead of message queues, replaces the event publishing mechanism of the `game-apis` service, and makes the event subscription from `ranking-service` use Apache Kafka.

First, we'll change `app-model`.

Configuring Apache Kafka with app-model

To use the Apache Kafka resource with `app-model`, we must add the `Aspire.Hosting.Kafka` NuGet package:

Codebreaker.AppHost/Program.cs

```
var kafka = builder.AddKafka("kafkamessaging");
// code removed for brevity
var gameAPIs = builder.AddProject<Projects.Codebreaker_
GameAPIs>("gameapis")
  .WithExternalHttpEndpoints()
  .WithReference(sqlServer)
  .WithReference(redis)
  .WithReference(kafka)
  .WithEnvironment("DataStore", dataStore)
  .WithEnvironment("StartupMode", startupMode);
builder.AddProject<Projects.Codebreaker_Ranking>("ranking")
  .WithExternalHttpEndpoints()
  .WithReference(cosmos)
  .WithReference(kafka)
  .WithEnvironment("StartupMode", startupMode);
```

The `AddKafka` method adds a Docker container for local development. This resource is referenced from the `game-apis` service and `ranking-service` to forward the connection. `StartupMode` is configured with the launch profile and forwarded as an environment variable to both of these projects so that they can choose between Azure Event Hubs and Apache Kafka.

Next, we'll use an Aspire component to publish events.

Publishing Apache Kafka events

When it comes to publishers and subscribers, the `Aspire.Confluent.Kafka` NuGet package is used. Within this package, in the `Confluent.Kafka` namespace, the `IProducer` interface is defined. An object of this type is injected with the `KafkaGameReportProducer` class to publish completed games:

Codebreaker.GameAPIs/Services/KafkaGameReportProducer.cs

```
public class KafkaGameReportProducer(
  IProducer<string, string> producerClient,
  ILogger<KafkaLiveReportProducer> logger)
  : IGameReport
{
  public async Task ReportGameEndedAsync(
    GameSummary game,
    CancellationToken cancellationToken = default)
  {
    Message<string, string> message = new()
    {
      Key = game.Id.ToString(),
      Value = JsonSerializer.Serialize(game)
    };
    string[] topics = ["ranking", "live"];
    foreach (var topic in topics)
    {
      _ = producerClient.ProduceAsync(topic, message,
        cancellationToken);
    }

    producerClient.Flush(TimeSpan.FromSeconds(5));
    logger.GameCompletionSent(game.Id, "Kafka");
    return Task.CompletedTask;
  }
}
```

`KafkaGameReportProducer` implements the same interface that was used before – that is, `IGameReport`. The generic parameters of the `IProducer` interface define types for the key and the value. With Kafka, serializers can be specified, which allows for custom serialization. We can use simple strings that easily work across different platforms. With .NET, we can use the `System.Text.Json` serializer to serialize `GameSummary` objects to strings.

The IProducer interface defines the ProduceAsync method to publish messages. The first parameter names a topic. Upon invoking the ProduceAsync method, the message is sent to a Kafka broker service. The message is kept there until it is read by a subscriber for the topic – up to a retention period. The default retention period is 1 week. To send messages to multiple subscribers (live-service and ranking-service), a list of topics is used.

The ProduceAsync method returns DeliveryResult when the message is delivered to the broker. We don't wait for the message to be delivered; instead, we use a loop to send the same message with multiple topics. Task.WhenAll could be used to wait for all deliveries or the Flush method to wait until a timeout is reached. The Flush method returns the number of items in the queue. Before the producer gets disposed of, you need to make sure that all messages are delivered to the broker. Because the producer is configured as a singleton with the DI container, we can keep the flush for later.

Now, we must configure the DI container:

Codebreaker.GameAPIs/ApplicationServices.cs

```
// code removed for brevity
string? mode = builder.Configuration["StartupMode"];
if (mode == "OnPremises")
{
  builder.AddKafkaProducer<string, string>(
    "kafkamessaging", settings =>
  {
    settings.Config.AllowAutoCreateTopics = true;
  });
  builder.Services.AddSingleton<IGameReport,
    KafkaGameReportProducer>();
}
```

The AddKafkaProducer method registers the IProducer interface as a singleton. kafkamessaging is the string that's used with app-model to get the connection string to the Kafka server. With the KafkaProducerSettings parameter, telemetry configuration and producer settings can be configured. Here, the AllowAutoCreateTopics setting is set to true – which is the default with producers. With consumers, this value is false by default. The previously created KafkaGameReportProducer class is registered as a singleton as well. The IGameReport interface is already used by the GameService class to report completed games, regardless of how this reporting is implemented.

Now, let's subscribe to these events with the ranking service.

Subscribing to Apache Kafka events

Subscriber applications such as `Codebreaker.Ranking` need the same .NET Aspire component package to be referenced. When it comes to consumer classes, the `IConsumer` interface must be injected:

Codebreaker.Ranking/GameSummaryKafkaConsumer.cs

```
public class GameSummaryKafkaConsumer(
  IConsumer<string, string> kafkaClient,
  IDbContextFactory<RankingsContext> factory,
  ILogger<GameSummaryEventProcessor> logger)
  : IGameSummaryProcessor
{
  public async Task StartProcessingAsync(
    CancellationToken cancellationToken = default)
  {
    kafkaClient.Subscribe("ranking");
    try
    {
      while (!cancellationToken.IsCancellationRequested)
      {
        try
        {
          var result = kafkaClient.Consume(
            cancellationToken);
          var value = result.Message.Value;
          var summary =
            JsonSerializer.Deserialize<GameSummary>(value);
          // code removed for brevity
          using var context = await
            factory.CreateDbContextAsync(
              cancellationToken);
          await context.AddGameSummaryAsync(
            summary, cancellationToken);
        }
        catch (ConsumeException ex) when
          (ex.HResult == -2146233088)
        {
```

```
        logger.LogWarning("Consume exception {Message}",
          ex.Message);
        await Task.Delay(TimeSpan.FromSeconds(10));
      }
    }
  }
}
```

The ranking service subscribes to messages with the `ranking` topic using the `Subscribe` method. The topic was used when publishing messages. If the topic doesn't exist because a message hasn't been written yet, the `Consume` method throws a `ConsumeException` error. This exception is caught, and the `Consume` method is repeated after a delay. A game might not be completed when the ranking service starts up and the Docker container of the Kafka service hasn't been configured to keep state.

When a message is received, it is written to the database, as we saw earlier with Azure Event Hubs.

Now, we just need to configure the DI container:

Codebreaker.Ranking/ApplicationServices.cs

```
string? mode = builder.Configuration["StartupMode"];
if (mode == "OnPremises")
{
  builder.AddKafkaConsumer<string, string>(
    "kafkamessaging", settings =>
  {
    settings.Config.GroupId = "Ranking";
  };
  builder.Services.AddSingleton<IGameSummaryProcessor,
    GameSummaryKafkaConsumer>();
}
```

To register the `IConsumer` interface, we use the `AddKafkaConsumer` method. `GroupId` needs to be configured with a Kafka consumer client. Groups are used for scalability. Similar to Azure Event Hubs, Kafka makes use of partitions. Multiple subscribers using the same group ID receive messages from different partitions. This allows for high scalability.

Now, start the solution with the `OnPremises` launch profile. Start the Open API page of the bot to let it play some games and debug and monitor the services. *Figure 15.10* shows metrics counts for the `game-apis` service with the bytes transmitted:

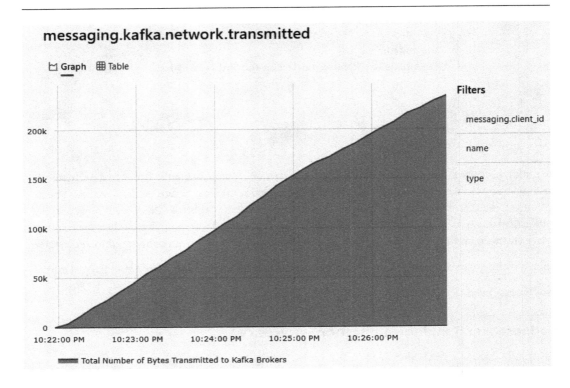

Figure 15.10 – Kafka metrics

Check the bytes, the messages that have been published and subscribed to, and the queue sizes for the publisher and the subscriber. Now is also a good time to take a break and play a few rounds of Codebreaker.

Summary

In this chapter, you learned how to decouple services by using asynchronous communication technologies, messages, and events. With Microsoft Azure, we used queues from an Azure Storage Account and events from Azure Event Hubs. Instead of using these PaaS services, you can also run Kafka within a Docker container in an Azure Container Apps environment, but you need to configure this with the app model.

You also learned the differences between using message queues and a publish/subscribe event model with multiple subscribers.

Be sure to check out the .NET Aspire component for Azure Service Bus in the *Further reading* section. This service offers more features with message queues, and you'll learn about some concepts you already know about from Apache Kafka.

After all the different services, in the next chapter we'll look into what more we should think about when deploying the application to the production environment, and we'll deploy the solution to a Kubernetes cluster.

Further reading

To learn more about the topics that were discussed in this chapter, please refer to the following links:

- *.NET Aspire Azure Blob Storage component*: `https://learn.microsoft.com/en-us/dotnet/aspire/storage/azure-storage-blobs-component`

- *.NET Aspire Azure Queue Storage component*: `https://learn.microsoft.com/en-us/dotnet/aspire/storage/azure-storage-queues-component`

- *Azure Event Hubs documentation*: `https://learn.microsoft.com/en-us/azure/event-hubs/`

- *.NET Aspire Azure Event Hubs component*: `https://github.com/dotnet/aspire/tree/main/src/Aspire.Hosting.Azure.EventHubs`

- *.NET Aspire Azure Service Bus component*: `https://learn.microsoft.com/en-us/dotnet/aspire/messaging/azure-service-bus-component`

- *.NET Aspire RabbitMQ component*: `https://learn.microsoft.com/en-us/dotnet/aspire/messaging/rabbitmq-client-component`

- *.NET Aspire Apache Kafka component*: `https://learn.microsoft.com/en-us/dotnet/aspire/messaging/kafka-component`

- *Apache Kafka .NET Client*: `https://docs.confluent.io/kafka-clients/dotnet/current/overview.html`

16

Running Applications On-Premises and in the Cloud

Up to the last chapter, we added additional functionality to the Codebreaker application; in *Chapter 15*, we added services communicating with asynchronous communication. We used Azure Storage queues and Azure Event Hubs with the Azure Codebreaker variant; with the on-premises version, we added a Kafka container.

In this chapter, we look at what needs to be known when deploying the solution to Microsoft Azure and to on-premises environments. Using Azure, we deployed the solution to an Azure Container Apps environment from *Chapter 5* onward. The Azure Container Apps environment uses Kubernetes behind the scenes. In this chapter, we directly deploy to a Kubernetes cluster, which can easily be used in on-premises environments and in any cloud.

In this chapter, you will learn about the following:

- Customizing deployment with C# and Aspire
- Creating a Kubernetes cluster with Azure
- Deploying the application to Kubernetes with Aspir8

Technical requirements

With this chapter, like the previous chapters, you need an Azure subscription, .NET 8 with .NET Aspire, and Docker Desktop. In this chapter, we'll use a new tool, Aspir8, to deploy the application to a Kubernetes cluster.

The code for this chapter can be found in this GitHub repository: `https://github.com/PacktPublishing/Pragmatic-Microservices-with-CSharp-and-Azure/`.

In the `ch16` folder, you'll see the projects that can be deployed. The most important project for this chapter is `Codebreaker.AppHost`, which defines the app model using Azure native cloud services, as well as a configuration that can be used with an on-premises environment. This configuration is also used to deploy the solution to a Kubernetes cluster.

Thinking about deployment in production

The Codebreaker solution uses several different native Azure cloud services. In *Chapter 8*, you saw how we can use **GitHub Actions** to deploy to different environments, such as development, testing, staging, and production environments using approvals. As more and more services have been added in the last chapters, the deployments need to be updated as well.

With many organizations, deployments to production environments are somewhat disconnected from the development environment. Often, a different team from the development organization manages these deployments using different tools.

Continuous Integration (**CI**) and **Continuous Development** (**CD**) are often used in repositories separated from the source code. Different products such as GitHub Actions, Azure DevOps pipelines, and many third-party offerings are used.

From the pipelines, it's possible to trigger the Azure Developer CLI (azd), use Bicep scripts, directly use the Azure CLI or PowerShell scripts, or use third-party offerings such as Terraform, Ansible, Chef, and Puppet.

When deciding between the different products, it's also necessary to think about the requirements for the production environment, and what's different from the development environment. With the production environment, different loads are expected. For a load test, it's useful to run the same infrastructure as used with the production environment. With this, it's important that the complete infrastructure for an environment needs to be easily creatable.

The infrastructure needs to map the needs of the business – what income is lost if things are not working as expected? We need to think about these topics:

- **Scalability**: Adapting to changing demands. Demand might increase slightly over time, or there might also be spikes in demand.

- **Reliability**: Making sure that the services work as expected.

- **Availability**: Making sure that the services are available from where the customers are. Availability metrics are **Mean Time Between Failures** (**MTBF**) – how long until a failure happens – and **Mean Time To Repair** (**MTTR**) – how long it takes until it's running again.

- **Recovery**: If there's an outage, recovery metrics that can be used are **Recovery Point Time** (**RTO**) – the acceptable time for apps to be unavailable – and **Recovery Point Objective** (**RPO**) – the maximum allowed time for a data loss.

These requirements need to be compared to the business needs. With **redundancy**, resources are replicated, and multiple services are running. There's not a single point of failure. Data can be replicated within one data center in an Azure region, between different data centers in an Azure region (Azure **availability zones**), and across different Azure regions (using a **multi-region** architecture).

Another requirement for the production environment is to enhance security. Data protection needs to make sure personal user data is safe. With **encryption at rest**, data is stored encrypted in the database. Instead of a service-managed key, customer-managed keys can be used. Using customer-managed keys is possible with many Azure services, but usually, different (more expensive) SKUs are required to enable customer-managed keys. Virtual networks are another option to enhance security. With subnets, it's possible to restrict access to the database server. **Private endpoints** can be used to restrict access only to a specific service and prevent data exfiltration. IP firewall rules can be configured.

We can't discuss all the different requirements here, but an important takeaway is that with production environments, we might need some additional Azure resources (such as virtual networks), different configurations, and other SKUs. See *Figure 16.1* for the Codebreaker application making use of multiple Azure regions.

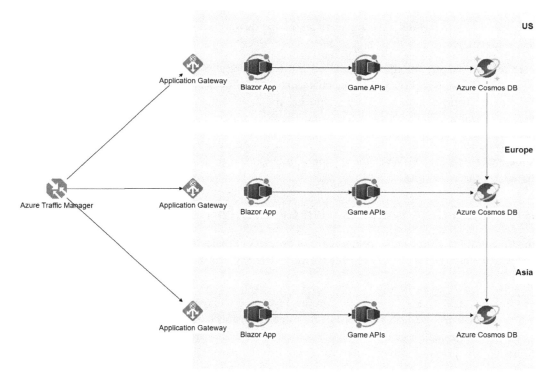

Figure 16.1 – Codebreaker with virtual networks

This figure shows Azure regions in the US, Europe, and Asia, and Azure Cosmos DB replicated across the regions. The database is accessed from Container Apps running in the same region of the database. The frontends (Blazor) and backends (Game APIs) in one region can run in one Azure Container Apps environment, with the Game APIs service only accessible internally. Azure Application Gateway with a firewall configured is used to access the Blazor web application. Azure Traffic Manager can route across different regions.

How does the Codebreaker solution fulfill all the requirements? With scalability, reliability, and security, all the resources used by the application need to be verified. In *Chapter 12*, we added a huge load on the Codebreaker services to test scaling up and scaling out. Due to the stateless nature of the developed services, the resources used also scale accordingly, and we anticipate no issues in meeting all the requirements. The Azure Cosmos DB database can replicate worldwide, even with multi-region writes to store games near the user for the best performance. We paid attention to the partition key, which doesn't block other gamers' writes to the database. Azure Event Hubs (added in *Chapter 15*) offers a lot more performance than needed. The Standard SKU supports 1,000 events per second with one throughput unit. Additional throughput units can be added, and a switch to the Premium tier, which offers even more, could be done. An important aspect is to see what's going on to react early, which was covered in *Chapter 11*.

While many organizations have separate teams for development and infrastructure, this has some disadvantages.

Using the .NET Aspire manifest created from the app model, we covered creating Bicep scripts in *Chapter 6*. These Bicep scripts can be customized to fulfill the requirements of the production environment. Using customized Bicep scripts has the disadvantage that changes on the app model don't automatically reflect with the Bicep script. The Bicep script needs to be manually updated again.

It would be great to use C# code to completely define the Azure infrastructure configuration with all the different aspects needed. When the app model is updated with this, the infrastructure configuration is changed at the same time.

Customizing deployments with C# and .NET Aspire

At the time of this writing, enhancements are in progress to make this happen. Currently, it's just in experimental mode, and the APIs available are likely to change, thus we will only look briefly into this.

To define the .NET Aspire app model, APIs have an overload with a delegate parameter. For example, the `AddAzureKeyVault` method we used so far is an extension method for the `IDistributedApplicationBuilder` interface and uses a `name` parameter. A second overload specifies an additional `Action` delegate parameter. This overload has the `Experimental` attribute applied to mark that the API may change. The parameters used with this delegate are `IResourceBuilder<AzureKeyVaultResource>`, `ResourceModuleConstruct`, and `KeyVault`. This allows us to configure a secret retrieved from a parameter when creating Azure Key Vault:

```
#pragma warning disable AZPROVISION001
var aSecret = builder.AddParameter("aSecret", secret: true);
```

```
var keyVault = builder.AddAzureKeyVault("keyvault",
  (_, construct, _) =>
  {
    var secret = new KeyVaultSecret(construct,
      name: "secret1");
    secret.AssignProperty(p => p.Properties.Value,
      aSecret);
  });
#pragma warning restore AZPROVISION001
```

With the method used here, the first and third parameters of the delegate are ignored. The second parameter of the `ResourceModuleConstruct` type specifies the scope of creating `KeyVaultSecret` – it's created for this Azure Key Vault.

Another sample shows configuring properties and invoking method of a builder with an Azure Storage Account:

```
var storage = builder.AddAzureStorage("storage",
  (builder, _, account) =>
  {
    builder.AddQueues("botqueue");
    builder.AddBlobs("checkpoints");

    account.AssignProperty(p => p.AccessTier, "Hot");
    account.AssignProperty(p => p.Sku.Name,
      "Standard_LRS");
  });
```

When creating the Azure Storage account, the `IResourceBuilder<AzureStorageAccount>` and `StorageAccount` parameters are used with the `Action` delegate, and the second parameter is ignored. `IResourceBuilder` is used to create a queue and a blob container with the storage account. We used these `AddQueues` and `AddBlobs` methods already without the experimental API invoking these methods with the return of `AddStorageAccount`. The `AddAzureStorage` method returns a builder. This is just for convenience defining this within this code block. The `StorageAccount` parameter is used to specify properties, setting the SKU to local redundancy, and the access tier to hot, which is cheaper for operations but more expensive for the storage.

Many organizations are in the process of changing the way to deploy and manage their infrastructure. Knowing about these developments can be useful to decide what direction should be taken, and what tools best fit the needs of the organization.

For now, it's very likely that the API will change – so use it with care. With the fast development pace of .NET Aspire, new features can improve fast, and this feature might be released not too far away (at the time of this writing). Check the README file of this chapter for updates.

Next, we'll look into easy deployment to Kubernetes.

Creating a Kubernetes cluster with Microsoft Azure

While the Azure Container Apps environment is based on Kubernetes, the Kubernetes tool (**kubectl**) cannot be used; the Kubernetes functionality is abstracted for simplification. Kubernetes is an open source system to scale and manage containerized applications and is used by many companies in their on-premises environment. With this, for many companies, it's important to have the possibility to run services on-premises and in any cloud environment. See the *Further reading* section for links to learn more about Kubernetes.

The Codebreaker application has been built with two launch profiles. We'll publish the OnPremises launch profile to a Kubernetes cluster. With this launch profile, for example, Kafka is used instead of Azure Event Hubs.

By having Docker Desktop installed, you can enable Kubernetes. This single-node cluster is just for a small test scenario. Instead, we'll use a managed version of Kubernetes: The **Azure Kubernetes Service (AKS)**. Compared to a self-installed cluster, installation and management are a lot easier.

Before creating the cluster, we need a new resource group, and an **Azure Container Registry (ACR)**.

Using the Azure CLI, create a new resource group:

```
az group create -l westeurope -n rg-codebreaker-kubernetes
```

Specify an Azure region of your choice and specify a resource group name. Then, create a new ACR using az acr create:

```
az acr create -g rg-codebreaker-kubernetes --sku Basic -l <yourregion>
-n <youracr>
```

Use the previously created resource group, specify an SKU (the cheapest version, Basic, fits the purpose), and use a unique name for the registry.

With this, create a new AKS in the Azure portal (https://portal.azure.com) – see *Figure 16.2*.

Project details

Select a subscription to manage deployed resources and costs. Use resource groups like folders to organize and manage all your resources.

Subscription * ⓘ

| Visual Studio - MVP | ⌄ |

Resource group * ⓘ

| rg-codebreaker-kubernetes | ⌄ |

Create new

Cluster details

Cluster preset configuration *

| 🧪 Dev/Test | ⌄ |

To quickly customize your Kubernetes cluster, choose one of the preset configurations above. You can modify these configurations at any time. Compare presets

Kubernetes cluster name * ⓘ

| akscodebreaker |

Region * ⓘ

| (Europe) West Europe | ⌄ |

Availability zones ⓘ

| None | ⌄ |

AKS pricing tier ⓘ

| Free | ⌄ |

Figure 16.2 – Basic AKS configuration

Select the resource group just created with the first dialog. With **Cluster details**, you can choose a preset configuration of either **Production Standard**, **Dev/Test**, **Production Economy**, and **Production Enterprise**. The virtual machine sizes are different based on the presets, and some features are differently configured. For example, **Production Enterprise** has a **private cluster** where the API server is only accessible from an internal network. Select the **Dev/Test** preset for our test environment. Enter a cluster name and select the region of the cluster. All the other **Basics** settings can stay as their defaults – including the AKS pricing tier, **Free**. With the **Free** offering, a cost only applies for the nodes where our built Docker images are running and other services configured, such as managed Prometheus and Grafana. Be aware that every node instance you configure is a virtual machine that needs to be paid for. The **Dev/Test** preset setting is best for experimenting and testing with fewer than 10 nodes. With the **Standard** pricing tier, you can run up to 5,000 nodes in a cluster.

After the configuration of the **Basics** settings, click **Next** to configure the node pools (*Figure 16.3*).

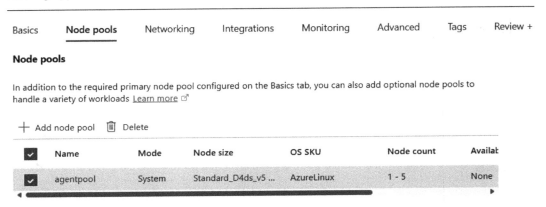

Figure 16.3 – AKS node pools

The default configuration for the node pool is **2 - 5** nodes. In this chapter we don't create load tests; therefore, to reduce the cost, you can reduce the minimum number of nodes to 1. System node pools need Linux for the operating system. These node pools run system pods. To run the applications, user node pools are preferred. For a cheaper test, we just use one node pool – a system node pool.

When selecting the configuration of the pool, you can select the OS, the VM size, auto or manual scaling, the minimum and maximum node count, and the maximum pods per node. The allowed range is from 30–250 pods per node. One pod can run one or more containers. In most Kubernetes configurations, a pod runs one container. If the pod or the node where the pod runs fails, Kubernetes creates a replica.

With the **Node pools** configuration, you can also enable virtual nodes. Virtual nodes make use of Azure Container Instances, which allow the fast startup of containers if more load is needed.

> **Note**
> Creating user node pools allows you to select Windows for a node pool. This allows running legacy applications on Kubernetes. This is a difference AKS has to offer that's not available with Azure Container Apps.

After the **Node pools** configuration, clicking **Next** leads to the **Networking** configuration. Leave this with the default settings. Clicking **Next** again opens the **Integrations** settings (see *Figure 16.4*).

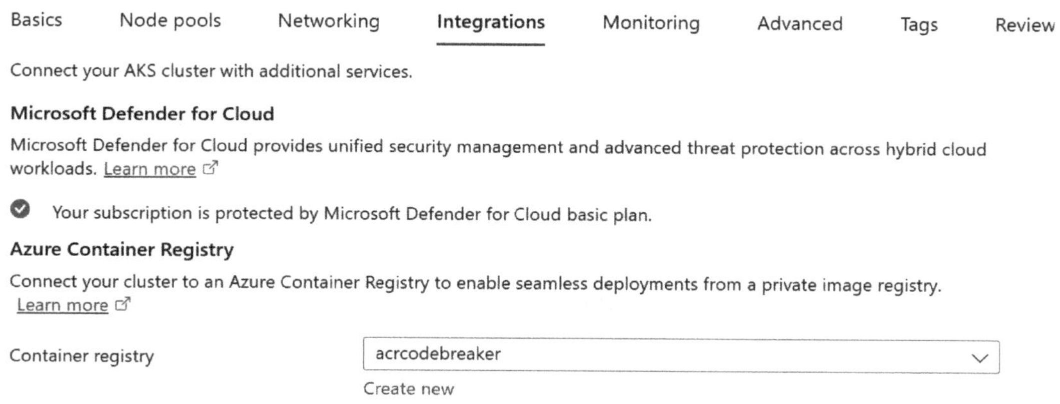

Basics   Node pools   Networking   **Integrations**   Monitoring   Advanced   Tags   Review

Connect your AKS cluster with additional services.

Microsoft Defender for Cloud

Microsoft Defender for Cloud provides unified security management and advanced threat protection across hybrid cloud workloads. Learn more ☑

✅   Your subscription is protected by Microsoft Defender for Cloud basic plan.

Azure Container Registry

Connect your cluster to an Azure Container Registry to enable seamless deployments from a private image registry. Learn more ☑

Container registry | acrcodebreaker   ∨ |
                                        Create new

Figure 16.4 – AKS Integrations settings

With the **Integrations** settings, select the previously created ACR. With AKS, a direct integration with the registry is offered.

Clicking **Next** opens the **Monitoring** settings. Enabling Container Insights creates a Log Analytics namespace that you know from *Chapter 11*. With the OnPremises launch profile, Docker containers for Grafana and Prometheus are configured. Alternatively, the Azure services Managed Prometheus and Managed Grafana could be used.

Leave the remaining settings as default. By clicking on **Review + create**, the final checks are done. If this succeeds, click the **Create** button. Creating an AKS takes several minutes – but it's a lot faster than creating a Kubernetes cluster manually.

After deployment to the Kubernetes cluster succeeds, connect the Kubernetes command-line client, kubectl, to AKS. With Docker Desktop, this tool is installed with it. To connect kubectl to this AKS installation, use the following:

```
az aks get-credentials --resource-group <your resource group> --name
<your aks name>
```

This adds the connection to AKS to the %HOMEPATH%/.kube/config configuration file. Now, you can use the kubectl tool:

```
kubectl get nodes
```

This returns the running nodes from the AKS service.

Next, let's publish our application.

Using Aspir8 to deploy to Kubernetes

With .NET Aspire, we created the app model to define all the dependencies between the different resources that are used. First, in *Chapter 1*, you saw the Aspire manifest that's created from an app model. This manifest file is independent of any technology where to deploy it. The Azure Developer CLI creates Bicep scripts for deploying the solution (see *Chapter 6* and *Chapter 8*). The open source tool **Aspirate** (**Aspir8**) (see `https://github.com/prom3theu5/aspirational-manifests`) converts the Aspire manifest file to Docker Compose or Kubernetes with **Helm** charts or **kustomize** manifests.

You can create an Aspire manifest for every launch profile, like so:

```
cd Codebreaker.AppHost
dotnet run --launch-profile OnPremises -- --publisher manifest
--output-path onpremises-manifest.json
```

Our app model is defined with two different versions. One version uses cloud-native Azure services, while the other option is independent of any cloud environment. The second one is configured by starting the application with the `OnPremises` launch profile.

With `dotnet run`, we pass the `--launch-profile OnPremises` option to start the application using the profile as specified with the `launchprofiles.json` file. The `--` option is a separator to specify arguments to the running application. The `--publisher manifest` option creates the Aspire manifest file.

> **Note**
>
> We have a strict separation with the Codebreaker app model definition. It's also possible in a somehow mixed mode. For example, you can use a solution running on-premises to use Azure Application Insights running within Azure to get the advantages of this cloud service offering. You can also use Azure Functions to run on an on-premises Kubernetes cluster. Many options are available to choose the service that best fits your needs.

Before using the `aspirate` tool, it needs to be installed:

```
dotnet tool install -g aspirate --prerelease
```

At the time of this writing, this tool is not released, thus it's necessary to set the `--prerelease` option. The `-g` option installs this tool as a global tool.

> **Note**
>
> At the time of this writing, the `aspirate` tool is in a prerelease state, and changes are expected. Check the README file of *Chapter 16* from the book's repository for the latest updates deploying the Codebreaker application to Kubernetes.

Optionally, you can specify an initial configuration with Aspir8:

```
cd Codebreaker.AppHost
aspirate init --launch-profile OnPremises
```

The `aspirate` tool allows specifying a launch profile similar to the .NET CLI to customize the configuration accordingly. By using `aspirate init`, you can specify a container builder and select between Docker Desktop and Podman. The default setting is Docker Desktop. With a fallback value for the container registry, enter the URL of the ACR you created. `aspirate init` creates the `aspirate-state.json` file with the configuration specified. You can rerun `aspirate init`, which overwrites this configuration file.

Creating Kubernetes manifests

Let's now use the app model with the launch profile to generate manifests for publishing to Kubernetes:

```
aspirate generate --launch-profile OnPremises --output-path ./
kustomize-output --skip-build --namespace codebreakerns
```

`aspirate generate` can create Kubernetes manifests for deployment, as well as build and publish Docker images. Here, we don't build Docker images by using the `--skip-build` option. With the `--launch-profile` option, the `AppHost` project with the app model is directly used. `aspirate generate` can also reference the previously generated .NET Aspire manifest with the `--aspirate-manifest` option instead. By setting `--output-path`, a different folder is specified to create the output result. The `--namespace` option is Kubernetes-related to define a namespace for the services deployed. This makes it easier to differentiate between the different services running on the cluster.

Note

`aspirate` supports generating manifests using Helm and `kustomize`. Helm is a package manager that uses a packaging format named **charts**. This is a collection of YAML files and templates. With Helm, installation, upgrades, and rollbacks can be done with simple commands. `kustomize` is a configuration manager natively built into `kubectl` with a template-free approach to patch and merge YAML files.

Check the result of the `kustomize-output` folder. For every project specified, a folder is created (e.g., `gameapis`, `bot`, and `redis`) that contains `deployment.yaml`, `service.yaml`, and `kustomization.yaml`.

A deployment defines a declarative configuration for a pod and a replica set. The "desired state" of a pod is described by the deployment. In this file, you can read and change the number of replicas used, and the containers running in a pod.

A service defines a network application. This specifies the ports used with the application. A service runs in one or more pods.

The `kustomization.yaml` file references both `deployment.yaml` and `service.yaml`, and specifies configuration values such as the environment variables you've seen with the .NET Aspire dashboard.

Having the manifest files ready, we can create Docker images and push them to the ACR.

Creating and pushing Docker images

Using `aspirate build`, we can build and publish Docker images to the registry. With the `aspirate` tool, it's possible to specify username and password values to push images to private registries. When using ACR, this is not necessary because Aspir8 makes use of `dotnet publish`. Just make sure to log in to the ACR using the following:

```
az acr login –name <yourregistry>
```

Then, you can use `aspirate build`:

```
aspirate build --launch-profile OnPremises --container-image-tag 3.8
--container-image-tag latest --container-registry <yourregistry>.
azurecr.io
```

Starting this command, specify the name of your registry. Specifying multiple tags will add them to the repository as shown in *Figure 16.5*.

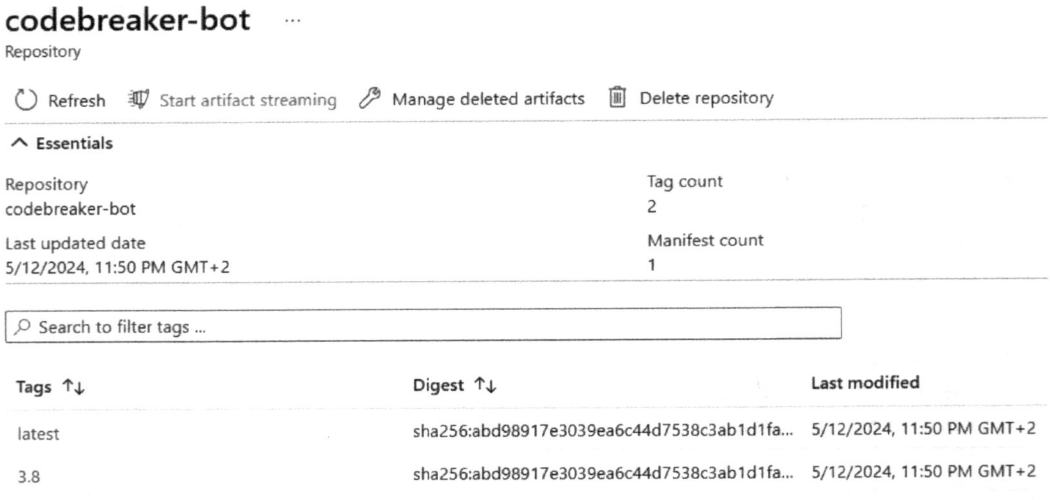

codebreaker-bot ...

Repository

◯ Refresh   ⚡ Start artifact streaming   ✏ Manage deleted artifacts   🗑 Delete repository

∧ Essentials

Repository	Tag count
codebreaker-bot	2
Last updated date	Manifest count
5/12/2024, 11:50 PM GMT+2	1

🔎 Search to filter tags ...

Tags ↑↓	Digest ↑↓	Last modified
latest	sha256:abd98917e3039ea6c44d7538c3ab1d1fa...	5/12/2024, 11:50 PM GMT+2
3.8	sha256:abd98917e3039ea6c44d7538c3ab1d1fa...	5/12/2024, 11:50 PM GMT+2

Figure 16.5 – AKS repository

The images are pushed to the ACR and show the `latest` and `3.8` tags, as specified with `aspirate build`. Next, deploy the images with the Kubernetes manifests to the cluster.

Deploying to Kubernetes

Now, we can apply the manifests to the Kubernetes cluster:

```
aspirate apply --input-path kustomize-output
```

The `aspirate apply` command uses the previously created manifest files to apply the services and deployments to the Kubernetes cluster by using the `kubectl apply` command. Just make sure to have AKS configured as the default Kubernetes environment (using the previously used command after creating AKS: `az aks get-credentials`).

Now, you can use the following command:

```
kubectl get deployments --namespace codebreakerns
```

This command shows the deployments from the `codebreakerns` namespace. You can see the deployments that are available and ready.

Similarly, use this command to see the services:

```
kubectl get services --namespace codebreakerns
```

Here, you see the services with the IP addresses running, and the ports registered.

Now, you can configure an **ingress** controller to access a service and test it running with Kubernetes. Configuring ingress controllers is planned with `aspirate`. For now, check the *Further reading* section to see how this can be done.

> **Note**
>
> Aspir8 also supports Docker Compose besides Kubernetes with `kustomize` and Helm. By using `aspirate generate`, you can supply `compose` with the `--output-format` option. This creates a simple Docker Compose file that you can start with the Docker CLI.

Summary

In this chapter, you learned about some final considerations for deploying applications using a microservices architecture in a production environment. You now have awareness of running the solution in multiple regions and using availability zones and can discuss the impact in your organization.

You learned about AKS as a managed option to host a Kubernetes cluster and deploy the solution by using the .NET Aspire manifest to create deployments with Aspir8.

By reaching *Chapter 16* of this book, you've done an impressive tour, starting with minimal APIs, and adding more services from chapter to chapter using different technologies.

With the book's repository, the solution is planned to be updated to newer .NET and .NET Aspire versions. As newer versions become available, the book version will stay available in the `dotnet8` branch.

To see more developments with Codebreaker, check the `https://github.com/codebreakerapp` organization. There, you can see further developments of the solution, as well as a list of client applications. Also, check `https://codebreaker.app` to play a few games – of course, now you can also use a version running in your (hosted) Kubernetes cluster.

Further reading

To learn more about the topics discussed in this chapter, you can refer to the following links:

- *Business continuity and disaster recovery*: `https://learn.microsoft.com/en-us/azure/cloud-adoption-framework/ready/landing-zone/design-area/management-business-continuity-disaster-recovery`

- *Azure load-balancing options*: `https://learn.microsoft.com/en-us/azure/architecture/guide/technology-choices/load-balancing-overview`

- Kubernetes: `https://kubernetes.io/`

- Learn Kubernetes basics: `https://kubernetes.io/docs/tutorials/kubernetes-basics/`

- Helm – the package manager for Kubernetes: `https://helm.sh`

- The `kubectl` command-line tool: `https://kubernetes.io/docs/reference/kubectl/`

- Aspirate GitHub repository: `https://github.com/prom3theu5/aspirational-manifests`

- *aspir8 from scratch*: `https://github.com/devkimchi/aspir8-from-scratch`

- *Configure ingress with Azure Kubernetes Services*: `https://learn.microsoft.com/en-us/azure/aks/app-routing`

- *Deploy a .NET microservice to Kubernetes manually*: `https://learn.microsoft.com/en-us/training/modules/dotnet-deploy-microservices-kubernetes/`

- `kubectl` – the definitive pronunciation guide: `https://www.youtube.com/watch?v=2wgAIvXpJqU`

Index

Packtpub.com

Subscribe to our online digital library for full access to over 7,000 books and videos, as well as industry leading tools to help you plan your personal development and advance your career. For more information, please visit our website.

Why subscribe?

- Spend less time learning and more time coding with practical eBooks and Videos from over 4,000 industry professionals

- Improve your learning with Skill Plans built especially for you

- Get a free eBook or video every month

- Fully searchable for easy access to vital information

- Copy and paste, print, and bookmark content

Did you know that Packt offers eBook versions of every book published, with PDF and ePub files available? You can upgrade to the eBook version at packtpub.com and as a print book customer, you are entitled to a discount on the eBook copy. Get in touch with us at customercare@packtpub.com for more details.

At www.packtpub.com, you can also read a collection of free technical articles, sign up for a range of free newsletters, and receive exclusive discounts and offers on Packt books and eBooks.

Other Books You May Enjoy

If you enjoyed this book, you may be interested in these other books by Packt:

Microservices Design Patterns in .NET

Trevoir Williams

ISBN: 978-1-80461-030-5

- Use Domain-Driven Design principles in your microservice design
- Leverage patterns like event sourcing, database-per-service, and asynchronous communication
- Build resilient web services and mitigate failures and outages
- Ensure data consistency in distributed systems
- Leverage industry standard technology to design a robust distributed application
- Find out how to secure a microservices-designed application
- Use containers to handle lightweight microservice application deployment

Refactoring with C#

Matt Eland

ISBN: 978-1-83508-998-9

- Understand technical debt, its causes and effects, and ways to prevent it
- Explore different ways of refactoring classes, methods, and lines of code
- Discover how to write effective unit tests supported by libraries such as Moq
- Understand SOLID principles and factors that lead to maintainable code
- Use AI to analyze, improve, and test code with the GitHub Copilot Chat
- Apply code analysis and custom Roslyn analyzers to ensure that code stays clean
- Communicate tech debt and code standards successfully in agile teams

Packt is searching for authors like you

If you're interested in becoming an author for Packt, please visit authors.packtpub.com and apply today. We have worked with thousands of developers and tech professionals, just like you, to help them share their insight with the global tech community. You can make a general application, apply for a specific hot topic that we are recruiting an author for, or submit your own idea.

Share Your Thoughts

Now you've finished *Pragmatic Microservices with C# and Azure*, we'd love to hear your thoughts! Scan the QR code below to go straight to the Amazon review page for this book and share your feedback or leave a review on the site that you purchased it from.

https://packt.link/r/1835088295

Your review is important to us and the tech community and will help us make sure we're delivering excellent quality content.

Download a free PDF copy of this book

Thanks for purchasing this book!

Do you like to read on the go but are unable to carry your print books everywhere?

Is your eBook purchase not compatible with the device of your choice?

Don't worry, now with every Packt book you get a DRM-free PDF version of that book at no cost.

Read anywhere, any place, on any device. Search, copy, and paste code from your favorite technical books directly into your application.

The perks don't stop there, you can get exclusive access to discounts, newsletters, and great free content in your inbox daily

Follow these simple steps to get the benefits:

1.   Scan the QR code or visit the link below

https://packt.link/free-ebook/9781835088296

2.   Submit your proof of purchase

3.   That's it! We'll send your free PDF and other benefits to your email directly

Download a free PDF copy of this book

Thanks for purchasing this book!

Do you like to read on the go but are unable to carry your print books everywhere?

Is your eBook purchase not compatible with the device of your choice?

Don't worry, now with every Packt book you get a DRM-free PDF version of that book at no cost.

Read anywhere, any place, on any device. Search, copy, and paste code from your favorite technical books directly into your application.

The perks don't stop there, you can get exclusive access to discounts, newsletters, and great free content in your inbox daily

Follow these simple steps to get the benefits:

1. Scan the QR code or visit the link below

https://packt.link/free-ebook/9781835088296

2. Submit your proof of purchase
3. That's it! We'll send your free PDF and other benefits to your email directly